Mechanisms Linking Aging, Diseases and Biological Age Estimation

Mechanisms Linking Aging, Diseases and Biological Age Estimation

Editor

Sara C. Zapico

Smithsonian Institution, National Museum of Natural History
Anthropology Department
Washington, DC
USA

CRC Press
Taylor & Francis Group
Boca Raton London New York

CRC Press is an imprint of the
Taylor & Francis Group, an **informa** business

A SCIENCE PUBLISHERS BOOK

Cover illustrations reproduced by kind courtesy of Dr. Joe Adserias Garriga and Dr. Isaac Antolín González.

CRC Press
Taylor & Francis Group
6000 Broken Sound Parkway NW, Suite 300
Boca Raton, FL 33487-2742

First issued in paperback 2021

ISBN-13: 978-0-367-78242-9 (pbk)
ISBN-13: 978-1-4987-0969-9 (hbk)

Library of Congress Cataloging-in-Publication Data

Names: Zapico, Sara C., editor.
Title: Mechanisms linking aging, diseases, and biological age estimation / editor, Sara C. Zapico, Smithsonian Institution, National Museum of Natural History, Anthropology Department, Washington, DC, USA.
Description: Boca Raton, FL : CRC Press, 2016. | "A science publishers book." | Includes bibliographical references and index.
Identifiers: LCCN 2016036650| ISBN 9781498709699 (hardback : alk. paper) | ISBN 9781498709705 (e-book : alk. paper)
Subjects: LCSH: Aging. | Age factors in disease. | Human beings--Age determination.
Classification: LCC QH529 .M43 2016 | DDC 571.8/78--dc23
LC record available at https://lccn.loc.gov/2016036650

Visit the Taylor & Francis Web site at
http://www.taylorandfrancis.com

and the CRC Press Web site at
http://www.crcpress.com

Preface

The inspiration for this book rose from my previous publication "Applications of physiological basis of ageing to forensic sciences. Estimation of age-at-death". During the development of this review, I realized that the basic research in aging and diseases can be applied to other fields and vice versa. In fact, my own research in the forensic sciences with both age-at-death and post-mortem interval estimation is based on the application of my previous knowledge of cell biology and signaling pathways towards these purposes. As a result, this book is intended to expand the review, focusing on four of the hallmarks of aging: aspartic acid racemization, advanced glycation endproducts, telomere shortening and mitochondrial mutations; describing their role in aging and diseases; and their application in age-at-death estimation in forensic sciences in greater depth, displaying the interconnecting pathways among these processes. An additional chapter related to Epigenetics and its role in aging, diseases and forensic age estimation has been included. Since it is a new and growing field, it is worthwhile to introduce the reader to this topic and to present the latest advances. All chapters contain extensive bibliographies, thus allowing the interested readers to explore a given theme further.

This book attempts to reach a broad audience; from students, introducing them to aging, diseases and forensic science research, to basic, biomedical and forensic scientists interested in the different topics, complementing their knowledge in their respective fields and at the same time increasing their knowledge in other disciplines, which hopefully encourages a multidisciplinary approach to their specialties.

I would like to thank the readers for considering this book for their libraries. I hope it fulfills their expectations. This would not have been possible without the authors. I thank the contributing authors for their willingness to undertake the task of writing a review chapter in their corresponding fields and sharing their knowledge and experience with the scientific community. I also thank the Editorial Department of CRC Press, particularly Raju Primlani and Priyanka Primlani, for being patient and helping with the doubts that arose during the development of this book. I would like to acknowledge Christian Thomas for the language revision of the chapters. I would also like to thank Dr. Pablo Bermejo Álvarez for reviewing my draft chapters and his valuable contributions to them. I wish to thank Dr. Vicente Andrés for his suggestion with respect to the title of the book. For providing the cover figures, I would like to thank Dr. Joe Adserias Garriga and Dr. Isaac Antolín González. For encouraging me to develop this book, I would like to thank Dr. Sofía T. Menéndez and Luis Cabo-Pérez and all the people who helped me during this process.

Finally, I dedicate this book to my Spanish and American families, for their continuous support of my career. Thank you very much for everything.

Sara C. Zapico

Contents

List of Abbreviations

A1,	aggrecan
AA,	aplastic anemia
AAI,	amino acid isomerization
AAR,	aspartic acid racemization
Aβ,	amyloid beta
AD,	Alzheimer's Disease
ADAM,	A-Distintegrin and Metalloprotease
ADP,	Adenosine diphosphate
adPEO,	autosomal dominant progressive external ophthalmoplegia
AF,	annulus fibrosus
AGEs,	advanced glycation endproducts
AKR1B1,	Aldo-Keto Reductase Family 1, Member B1
Ala,	alanine
ALE,	advanced lipo-oxidation endproducts
ALS,	amyotrophic lateral sclerosis
ALT,	alternative lengthening of telomeres
ALT-711,	dimethyl-3-phenayl-thiazolium chloride or Alagebrium
AMP,	adenosine monophosphate
AMPK,	adenosyl monophosphate-dependent kinase
ANT1,	adenine nucleotide translocase type 1
APE,	apurinic/apyrimidinic endonuclease
APP,	amyloid precursor protein
apoE-ko,	apolipoprotein E-knockout mice
AR,	aldose reductase
arPEO,	autosomal recessive progressive external ophthalmoplegia
ARTs,	artificial reproductive techniques
Asn,	asparagine
Asp,	Aspartic acid
ASPA,	Aspartoacylase
Asx,	asparagine or aspartic acid
AT,	ataxia telangiectasia
ATM,	ataxia telangiectasia mutated
ATP,	adenosine triphosphate
ATR,	ataxia telangiectasia and Rad3 related
ATRX,	Alpha Thalassemia/Mental Retardation syndrome x-linked

BACE1,	β-secretase cleavage
BBB,	blood-brain barrier
BCC,	basal cell carcinomas
BER,	base excision repair
bFGF,	basic fibroblast growth factor
BNP,	brain natriuretic peptide
BRAF,	B-Raf proto-oncogen, serine/threonine kinase
BRCA1,	Breast Cancer 1
BRCA2,	Breast Cancer 2
BS-seq,	bisulfite treatment
BSCL1,	Berardinelli-Seip congenital lipodystrophy
CAD,	coronary artery disease
CAT,	catalase
CCD-UPLC-MS/MS,	covalent chiral derivatized ultraperformance liquid chromatography tandem mass spectrometry
CD4,	cluster of differentiation 4
CD8,	cluster of differentiation 8
CD36,	cluster of differentiation 36
CEL,	carboxyethyl
CEP,	cartilage endplates
ChIP,	chromatin immunoprecipitation
Chk2,	check-point kinase 2
CI,	confidence interval
CML,	carboxymethyl-lysine
CMT2A,	Charcot-Marie-Tooth Neuropathy Type 2A
CMT4A,	Charcot-Marie-Tooth neuropathy type 4A
CNS,	central nervous system
CoOH,	ubisemiquinone
$CoQH_2$,	ubiquinol
COX,	cytochrome c oxidase
COX10,	cytochrome c oxidase 10 heme A: farnesyltransferase cytochrome c oxidase assembly factor
COX15,	cytochrome c oxidase assembly homolog
CPEO,	chronic progressive external ophtalmoplegia
CpGs,	cytosine/guanine-rich regions
CR,	control region
CREB,	cAMP response element-binding protein
CRISPR,	clustered regularly interspaced short palindromic repeats
CSF,	Cerebrospinal fluid
CTE,	C-terminal extension domain
CTX,	Carboxy-terminal collagen crosslinks
Cu/ZnSOD,	copper/zinc SOD
CVD,	cardiovascular disease
Da,	Dalton
DA,	Dopamine
DAAO,	D-amino acid oxidases

D-amino acid	dextrorotary-amino acid
D-asp,	D-aspartic acid
DC,	Dyskeratosis Congenital
DDR,	DNA-Damage Response
DGUOK,	deoxyguanosine kinase
DKC,	dyskeratosis congenital
DKC1,	dyskerin
DMT,	disease modifying treatment
DNA,	deoxyribonucleic acid
DNMTs,	DNA methyltransferases
DN RAGE,	Dominant negative receptor for advanced glycation end products
DOA,	dominant optic atrophy syndrome
DSB,	double-strand breaks
D-ser,	D-serine
DVI,	Disaster Victim Identification
ECM,	extracellular matrix
ECs,	endothelial cells
EDARADD,	EDAR-associated death domain
EGFR,	epidermal growth factor receptor
EGR1,	Early Growth Response 1
ELISA,	Enzyme-Linked ImmunoSorbent Assay
ELK1,	member of ETS oncogene family
ELK4,	ETS-domain protein SRF accessory protein 1
ELOVL2,	Elongation of very long chain fatty acids-like 2
ELP,	Elongator Protein 3
EPCs,	endothelial progenitor cells
ERK ½,	extracellular signal-regulated kinase ½
esRAGE,	endogenous secretory RAGE
ETC,	electron transport chain
ETS/TCFs,	E-twenty-six/ternary complex factors
FA,	Fanconi anemia
FADH$_2$,	flavin adenine dinucleotide reduced form
FAIRE-Seq,	formaldehyde-assisted isolation of regulatory elements
FALS,	familial amyotrophic lateral sclerosis
FAOXs,	fructosyl-amine oxidases
Fe-S,	ion-sulfur
FHL2,	Four and a Half LIM Domains 2
FN3K,	fructosamine-3-kinase
G2,	gap 2 phase
GABP-α,	GA binding protein transcription factor, alpha subunit
GAG,	glycosaminoglycan
GAPDH,	glyceraldehyde 3-phosphate dehydrogenase
GC,	Gas Chromatography
GC/MS,	gas chromatography/mass spectrometry
GDAP1,	ganglioside-induced differentiation associated protein 1

GH,	growth hormone
Glo-1,	glyoxalase I
Glo-2,	glyoxalase II
Glu,	Glutamic acid
glx,	Glutamic acid/glutamine
GO,	glyoxal
GOLD,	1,3-bis-(5-amino-5-carboxypentyl)-1H-imidazolium
GPx,	glutathione peroxidase
GSH,	reduced glutathione
GSK-3,	glycogen synthase kinase-3
GWAS,	genome-wide association studies
H_2O_2,	hydrogen peroxide
HAT,	histone acetyltransferases
HBV,	Hepatitis B virus
HCC,	hepatocellular carcinoma
HD,	Hungtington's disease
HDACs,	histone deacetylases
HDR,	homologous direct recombination
HGPS,	Hutchinson-Gilford Progeria syndrome
HMGB-1,	high mobility group box 1 protein
HPA,	hybridization protection assay
HPLC,	high performance liquid chromatography
HR,	homologous repair
HSCs,	hematopoietic stem cells
HSP,	heat shock proteins
hTERC,	human telomerase RNA gene
HTT,	Huntingtin
HUMARA,	Human Androgen Receptor
HV1,	hypervariable region 1
HV2,	hypervariable region 2
HV3,	hypervariable region 3
IB,	immunoblotting
ICAM-1,	intracellular adhesion molecule-1
ICSI,	intracytoplasmic sperm injection
IF,	immunofluorescence
IGF-1,	insulin-like growth factor
IHC,	immunohistochemistry
IIS,	insulin/insulin-like growth factor signaling
IL-6,	interleukin-6
I/R:	ischemia/reperfusion
ITG2B,	Integrin subunit alpha 2b
IVD,	intervertebral disc
IVF,	*in vitro* fertilization
JAK,	Janus kinase
JNK,	Jun amino-terminal kinases
KDa,	KiloDalton

KLF14,	Kruppel-like factor 14
KSS,	Kearns-Sayre syndrome
L-asp,	L-aspartic acid
LAP2α,	lamina-associated polypeptide-2α
L-amino acid,	levorotary-amino acid
LC-ESI-TOF-MS,	liquid chromatography-electrospray ionization time-of-flight mass spectrometry
LC,	Liquid chromatography
LC/MS,	liquid chromatography/mass spectrometry
Ldlr-KO,	low-density lipoprotein receptor-knockout mouse
LHON,	Leber's hereditary optic neuropathy
LMNA,	lamin A
lncRNAs,	long ncRNAs
LRP-1,	Low density lipoprotein receptor-related protein 1
LRPPRC,	leucine rich pentatricopeptide repeat containing
LS,	Leigh syndrome
LTD,	long-term depression
LTL,	leukocyte telomere length
LTP,	long-term potentiation
Lys,	lysine
MAD,	mean absolute deviation
MAM,	mitochondria-associated ER membranes
MAPKs,	mitogen-activated protein kinases
MCI,	Mild cognitive impairment
MCO,	metal-catalyzed oxidation
MDS,	myelodysplastic syndrome
MeDIP,	Methylated DNA immunoprecipitation
MELAS,	mitochondrial encephalomyopathy with lactic acidosis and stroke-like episodes
MERRF,	myoclonic epilepsy with ragged red fibers
MFN2,	mitofusin 2
MGO,	methylglyoxal
MH,	methylgyoxal
MI,	myocardial infarction
MILS,	maternally inherited Leigh syndrome
MIRAS,	mitochondrial ataxia syndrome with autosomal recessive inheritance
miRNAs,	microRNAs
MMPs,	matrix metalloproteinases
MNase-Seq,	micrococcal nuclease digestion and sequencing
MNGIE,	Mitochondrial Neurogastrointestinal Encephalopathy disease
MnSOD,	Manganese superoxide dismutase
MOLD,	1,3-bis-(5-amino-5-carboxypentyl)-4-methyl-1H-imidazolium
mPTP,	mitochondrial permeability transition pore
MPV17,	mitochondrial inner membrane protein

mRNA,	messenger Ribonucleic Acid
MS,	multiple sclerosis
MSL,	multiple symmetrical lipomatosis
mtATP,	mitochondrial ATP synthase
mtCytB,	mitochondrial cytochrome c oxidoreductase
mtDNA,	mitochondrial DNA
mtND,	mitochondrially encoded NADH dehydrogenase
mtTFA,	mitochondrial Transcription Factor A
NADH,	nicotinamide adenine dinucleotide (reduced form)
NADPH,	nicotinamide adenine dinucleotide phosphate
NAF1,	nuclear assembly factor 1 ribonucleoprotein
NARP,	neurogenic weakness, ataxia and retinitis pigmentosa
ncRNA,	non-coding RNAs
ND1,	NADH dehydrogenase 1
ND2,	NADH dehydrogenase 2
ND4L,	NADH 4L dehydrogenase
ND6,	NADH dehydrogenase 6
nDNA,	nuclear DNA
NDUFA1,	NADH-Dehydrogenase (Ubiquinone) 1 Alpha
NDUFS,	NADH-Dehydrogenase (Ubiquinone) Fe-S
NDUFV1,	NADH-Ubiquinone-oxidoreductase flavoprotein
NEG,	non-enzymatic glycation
NF-kB,	nuclear factor kappaB
NGS,	next-generation sequencing
NHEJ,	non-homologous end joining repair
NMDA,	N-methyl-D-aspartate
NO,	nitric oxide (NO)
Non-LTR,	non-long-terminal repeat
NP,	nucleus pulposus
Nrf2,	nuclear factor-erythroid 2 p45 subunit-related factor 2
NSCLC,	non-small-cell lung carcinoma
NTH1,	endonuclease III-like protein
OBFC1,	oligonucleotide/oligosaccharide binding fold containing 1
OGG1,	8-oxoguanine DNA glycosylase-1
O-GlcNAc,	β-N-acetylglucosamine
OH,	origin of replication Heavy (H) strand
OL,	origin of replication Light (L) strand
OPA1,	optic atrophy 1
OXPHOS,	oxidative phosphorylation
P38MAPK,	p38 mitogen-activated protein kinase
PAGE,	polyacrylamide gel electrophoresis
PARP,	poly-ADP ribose polymerase
PCR,	Polymerase Chain Reaction
PD,	Parkinson's Disease
PDE4C,	Phosphodiesterase 4C, cAMP-specific
PEO,	progressive external ophthalmoplegia

PGCs,	primordial germ cells
PGD,	preimplantation genetic diagnosis
PGs,	proteoglycans
Pi,	inorganic phosphate
PI3K,	phosphatidyl inositol 3 kinase
PIMT,	protein isoaspartyl methyl transferase
PINK1:	PTEN-induced putative kinase 1
PIWI,	P-element induced WImpy testis
PKC,	protein kinase C
PMI,	Postmortem interval
PML,	promyelocytic leukemia
PMT,	N-phenacyl-4,5-dimethylthiazolium
PNA,	peptide nucleic acid
POLG,	catalytic subunit mitochondrial polymerase γ
POT1,	protection of telomeres 1
PPARγ,	proliferator-activator receptor γ
PS,	Pearson Syndrome
PTB,	N-phenacylthiazolium bromide
PTEN,	Phosphatase and tensin homolog
PTMs,	Post-translational modifications
Q-FISH,	quantitative fluorescence *in situ* hybridization
QPCR,	quantitative polymerase chain reaction
RAGE,	receptor for AGEs
RCCs,	renal cell carcinomas
RNA,	Ribonucleic acid
RNA-Seq,	RNA sequencing
ROS,	reactive oxygen species
RRFS,	ragged red fibers
RRM2B,	ribonucleotide reductase controlled by p53 (p53R2)
rRNA,	ribosomal RNA
RT,	reverse transcriptase domain
RTEL1,	regulator of telomere elongation helicase 1
Ryr,	ryanodine receptors
S,	synthesis phase
SALS,	sporadic amyotrophic lateral sclerosis
SAPK,	Stress-activated protein kinases
SC,	stratum corneum
SCO,	synthesis of cytochrome c oxidase
SDH,	succinate dehydrogenase complex
SDS-PAGE,	sodium dodecyl sulfate polyacrylamide gel electrophoresis
Ser	Serine
SERCA,	sarco-endoplasmic reticulum Ca^{2+} ATPase
SG,	stratum granulosum
SHR,	spontaneously hypertensive rats
SIM,	selected ion monitoring

SIM/GC-MS,	selected ion monitoring gas chromatography-mass spectrometry
siRNAs,	short-interfering RNAs
SIRT1,	Sirtuin 1
sjTREC,	signal join T-cell receptor excision circles
SNCA,	α-synuclein gene
SNpc,	substancia nigra pars compacta
SNPs,	single nucleotide polymorphisms
SOD,	superoxide dismutase
sRAGE,	soluble RAGE
SS,	stratum spinosum
ssDNA,	single-stranded DNA
STAT,	signal transducers and activators of transcription
STE,	stem terminus element
STELA,	single telomere length analysis
SUCLA2,	succinate-CoA ligase ADP-forming beta subunit
SUCLG1,	succinate-CoA ligase alpha subunit
SURF1,	surfeit 1
TALEN,	Transcription activator-like effector nucleases
TCA,	tricarboxylic acid
TEN,	N-terminal extension domain
Terc-KO,	Terc knock-out
TERC,	telomerase RNA component
TERT,	telomerase reverse transcriptase
TERTp,	promoter of the telomerase
TETs,	Ten-eleven translocator enzymes
TK2,	thymidine kinase 2
TL,	telomere length
TNF-α,	tumor necrosis factor α
T-OLA,	Telomeric-oligonucleotide ligation assay
TP,	thymidine phosphorylase
TR,	telomerase
TRAP,	telomeric repeat amplification protocol
TRBD,	Telomerase RNA binding domain
TRF,	telomere restriction fragment
TRF2,	TATA box binding protein-related factor 2
TRIM59,	Tripartite motif containing 59
tRNA,	transfer RNA
TROC,	telomere rate of change
T-SCE,	Telomere-Sister Chromatid Exchange
UQCRB,	ubiquinol-cytochrome c reductase binding protein
UQCRQ,	ubiquinol-cytochrome c reductase complex III subunit VII
UV-B,	Ultraviolet-B
UVR,	ultraviolet radiation
VSMC,	vascular smooth muscle cell
WHO,	World Health Organization

WRN,	Werner syndrome RecQ-like helicase
Wt,	wild-type
yr,	year
ZFN,	Zinc finger nucleases

Section I
Aspartic Acid Racemization

1

Introduction to Aspartic Acid Racemization

Christian Thomas[a] and *Sara C. Zapico**

INTRODUCTION

Before the emergence of life, it is believed that only L-amino acids, and not their specular D-amino acid forms, were chosen for protein formation. Hence, homochirality is essential for life (Motoie et al. 2009). However, in 1936, Kuhn presented kinetic arguments for the inevitable racemization of optically active substances within biological systems and suggested that the accumulation of the wrong stereoisomers could be responsible for the aging process (Helfman et al. 1977). Thus, despite elimination mechanisms and despite the stereospecificity of biosynthetic pathways, some D-amino acids may eventually be incorporated into synthetic enzymes. This would affect the specificity of further synthesis and could culminate in a "quasi-autocatalytical collapse" (Kuhn 1958).

Racemization is a spontaneous post-translational process, which eventually converts optically active compounds into a racemic mixture. Amino acids get converted from the native and common L-form to the relatively rare mirror image, D-form (McCudden and Kraus 2006). The L-amino acids found in living systems are the result of the stereochemical specificity of enzymes which use only L-enantiomeres (Helfman and Bada 1975). Racemization should take place in any metabolically stable protein, which has not turned over during the life-time of long lived organisms and as a consequence of racemization, these proteins will have altered conformations which would probably produce changes in their biological activities or chemical properties

Smithsonian Institution, National Museum of Natural History, Anthropology Department, 10th and Constitution Ave, NW, PO Box 37012, Washington, DC 20560, USA.
[a] Email: crf.thomas@gmail.com
* Corresponding author: saiczapico@gmail.com

(Masters et al. 1977). The resultant alterations in the physicochemical properties of affected proteins may contribute to the progressive changes associated with the aging process (Helfman et al. 1977).

History of Amino Acid Racemization Research

The first studies of amino acid racemization started with geochemistry, based on the assumption that amino acids produced abiotically would adopt an amino acid pool with equal quantities of L and D forms, while biological processes would produce an excess of L-forms; these methods were developed to search for evidence of extraterrestrial life in meteorites (Kvenvolden et al. 1970, Smith and Kvenvolden 1971, Bada and McDonald 1996). After these first studies, racemization of amino acids was used for dating various fossil materials, like marine sediments (Wehmiller and Hare 1971) and fossil bones (Bada and Helfman 1975). Based on these previous works, in 1975, Kuhn's general proposition concerning the inevitability of racemization in living mammals was confirmed in humans (Helfman and Bada 1975). Helfman and Bada analyzed the aspartic acid racemization in tooth enamel from living humans, finding an increase in D/L ratios with age. Furthermore, they demonstrated that the proportion of D-aspartic acid in human hemoglobin does not increase with the age of the individual since hemoglobin is a rapidly turned-over protein. Subsequently, a link between racemization, aging and disease was established, where it was observed that D-amino acid accumulation in the lens of the eye could not be only the result of aging, but also a cause of cataracts (Helfman et al. 1977, Masters et al. 1978, 1977). The most recent development in the use of racemization has been to improve forensic age estimation accuracy (Ohtani and Yamamoto 2005).

Chemistry of D-Amino Acid Formation

Racemization

All amino acids (except glycine) have two optical isomer forms L and D, based on their ability to rotate polarized light. As described above, L-amino acids are found in living systems based on the preference of ribosomal subunits for the L-enantiomer (McCudden and Kraus 2006). D-amino acids are formed in a process called racemization and are accumulated during aging. Although this process occurs in all amino acids, aspartic acid has the fastest racemization rate, hence this amino acid is most commonly analyzed and evaluated.

D-Asp forms via transitional succinimide intermediates originated by Asx (Asn) residues. The initial event in these reactions is the nucleophilic attack of the side-chain carbonyl group by the peptide-bond nitrogen, to produce an unstable five-membered L-succinimide ring. This ring is disposed to hydrolysis, inducing isomerization and racemization, yielding peptides α and β-Asp in both D and L configurations (Cloos and Fledelius 2000) (Fig. 1).

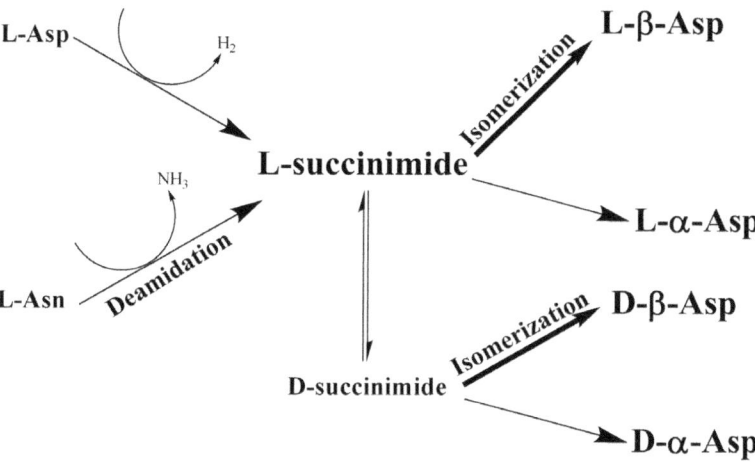

Figure 1. Formation of D-Asp. Involvement of three processes in the formation of D-Asp, racemization, isomerization and deamidation of the peptide bonded Asp and Asn residues.

Isomerization

Isomerization involves a change in the peptide backbone, as a consequence of the carbon chain changes, from the α to β conformation. This process requires a succinimide intermediate. As described above, hydrolysis of this succinimide ring can produce two different conformations, L-α-Asp (native conformation) and L-β-Asp (L-isoAsp) (Fig. 1) (Clarke 2003, Geiger and Clarke 1987).

Deamidation

D-Asp can be generated by Asn deamidation. During this process, the side chain functional amine group of Asn is lost, which can lead to the formation of a succinimide intermediate, which can racemize or isomerize (McCudden and Kraus 2006) (Fig. 1).

Influential factors in racemization

As a first-order chemical reaction, racemization is influenced by numerous factors:

Protein conformation

Protein structure influences all three processes of D-Asp formation: racemization, isomerization and deamidation, although its influence is clearest in this latter process (Robinson and Robinson 2004). However, several studies pointed out an association between protein conformation and the rates of racemization and isomerization; for example in α-crystallin and the insoluble fraction of bone collagen (Capasso et al. 1996, Capasso and Salvadori 1999, Fujii et al. 1999a, Radkiewicz et al. 2001).

Temperature and pH

Deamidation and racemization are influenced by temperature and pH. As temperature increases, these reactions are stimulated (Ohtani et al. 2005). This is of special importance in forensic cases, since this feature makes this technique unsuitable for corpses exposed to high temperatures (Zapico and Ubelaker 2013). In contrast, alkaline pH increases deamidation reactions, while acidic pH accelerates racemization (Brennan and Clarke 1993, Oliyai and Borchardt 1994).

Applications of Aspartic Acid Racemization

Aspartic acid racemization's research has been applied to aging studies, which leads to the main focus of this research: its application in forensic age estimation. Further, taking into account its accumulation with aging, its role in diseases has been also studied. Following chapters will describe in depth the implication of aspartic acid racemization in these processes. Apart from these aforementioned processes, aspartic acid racemization has been used to estimate the long-lived proteins' turnover rates (Maroudas et al. 1998). A first order kinetic equation is used, where k_T = protein turnover rate and k_1 = inversion rate, determined by quantifying the rate of D-amino acid accumulation in a tissue where the protein has essentially no turnover, like collagen in tooth dentin; and D and L are the optical isomers:

$$k_T = k_1[1 - (D/(D + L))]/[D/(D + L)]$$

This information has yielded an estimated half-life of proteins, such as collagen in articular cartilage.

Methods for Quantifying Amino Acid Racemization

There are several methodologies for detecting and quantifying racemization, with applications to aging and disease studies and forensic age estimation.

With respect to chromatographic methods, HPLC has been widely used in studies of α-crystallin racemization and isomerization (Fujii et al. 1994, Fujii et al. 2001, Fujii et al. 1999b, Fujii et al. 1997, Fujii et al. 1989, Fujii et al. 2004). In contrast, the elected method in forensic science is GC (Helfman and Bada 1976, Masters et al. 1977, Man et al. 1983, Ritz and Schutz 1993, Ritz et al. 1994, Ohtani et al. 2002, Dobberstein et al. 2008, Dobberstein et al. 2010).

These chromatography techniques have several advantages like the use of small volumes, their high throughput, repeatability and allowance for simultaneous measurement of both D- and L-amino acids, enabling rapid estimation of the ratio of D/L. However, the most significant disadvantage is that the harsh conditions of acid hydrolysis (the first step in these techniques) can induce racemization, so a correction is required in order to yield accurate results (McCudden and Kraus 2006).

HPLC and GC are commonly used to determine racemization. Alternative methods have been proposed, like Edman degradation and chiral derivatization. This methodology combines sequencing with chiral derivatization reagents, leading to the determination of both sequence and stereo-configuration of a peptide (Iida et al.

2001). This technique is limited by the peptide side and is also prone to the generation of D-forms during hydrolysis. This methodology has been used to characterize a number of racemized peptides including β-amyloid and various neuropeptides (Iida et al. 2001, Iida et al. 1998).

Other methods to quantify racemization are enzymatic methods, based on the use of stereo-selective enzymes. There are two groups of enzymes that specifically recognize racemized amino acids: DAAO, which selectively recognize isolated amino acids; PIMT, which selectively recognize isoaspartate and D-Asp in polypeptides (McCudden and Kraus 2006).

DAAO has not yet been applied in human research settings due to its limitations: isolated free amino acids are required in relatively high concentrations. In contrast PIMT-based assays have been used to quantify L-β-Asp and D-α-Asp residues in a variety of proteins including growth hormones, histones and type I collagen (Weber and McFadden 1997, Young et al. 2001, Young et al. 2005, Johnson et al. 1989). PIMT is an intracellular protein repair enzyme that prevents the accumulation of racemized and isomerized residues, catalyzing the transfer of the methyl group of S-methyl adenosyl-methionine to the free carboxyl group of D-α-Asp and L-β-Asp (Johnson and Aswad 1991). The abundance of these residues can be indirectly measured by either radioisotopic or HPLC-based assays (Carlson and Riggin 2000, Schurter and Aswad 2000). Hence, PIMT can specifically recognize racemized and isomerized residues within a protein. However, PIMT does not recognize D-β-Asp, which can result in an underestimation of the total damage, or an overestimation of the proportion of any one specific form of modification. Further, it does not recognize residues in conformations that are inaccessible to the enzyme (McCudden and Kraus 2006).

The use of immunological methods is one of the most promising approaches for research applications of amino acid racemization. Some investigators have employed enantiomer-specific antibodies for amino acid racemization detection. ELISAs assays have been developed to quantify urinary telopeptide of CTX-1 biomarkers in collagen (Christgau et al. 1998, Cloos et al. 2004, Rosenquist et al. 1994, Rosenquist et al. 1998). These antibody-based methods have several advantages, such as the prevention of hydrolysis-induced racemization; they can be selective for a specific epitope and for quantifying amounts of different biomarker configurations. In an immunoassay format, this technique is more amenable to high-throughput application than other methods, allowing for a rapid screening (McCudden and Kraus 2006). Moreover, immunological methods can be applied for the localization of D-amino acids within a tissue sample, like in cultured astrocytes (Mothet et al. 2005) and D-Asp in the developing rat testes, adrenal and pineal glands (Lee et al. 1997, Sakai et al. 1998, 1997). In spite of these advantages, the most significant drawback of antibody-based methods is the requirement of identifying a racemized epitope with clinical significance. This process is arduous and there are no assurances that a particular fragment will be of clinical utility (McCudden and Kraus 2006).

Keywords: L-amino acids, D-amino acids, racemization, isomerization, deamidation, L-Aspartic acid, D-Aspartic acid, protein conformation, temperature, pH, forensic age estimation, aging, diseases, GC, HPLC

References

Bada, J. L. and P. M. Helfman. 1975. Amino acid racemization dating of fossil bones. World Archaeol 7(2): 160–73. doi: 10.1080/00438243.1975.9979630.

Bada, J. L. and G. D. McDonald. 1996. Detecting amino acids on Mars. Anal Chem 68: 668A–673A.

Brennan, T. V. and S. Clarke. 1993. Spontaneous degradation of polypeptides at aspartyl and asparaginyl residues: effects of the solvent dielectric. Protein Sci 2(3): 331–8. doi: 10.1002/pro.5560020305.

Capasso, S., A. Di Donato, L. Esposito, F. Sica, G. Sorrentino, L. Vitagliano et al. 1996. Deamidation in proteins: the crystal structure of bovine pancreatic ribonuclease with an isoaspartyl residue at position 67. J Mol Biol 257(3): 492–6. doi: 10.1006/jmbi.1996.0179.

Capasso, S. and S. Salvadori. 1999. Effect of the three-dimensional structure on the deamidation reaction of ribonuclease A. J Pept Res 54(5): 377–82.

Carlson, A. D. and R. M. Riggin. 2000. Development of improved high-performance liquid chromatography conditions for nonisotopic detection of isoaspartic acid to determine the extent of protein deamidation. Anal Biochem 278(2): 150–5. doi: 10.1006/abio.1999.4421.

Christgau, S., C. Rosenquist, P. Alexandersen, N. H. Bjarnason, P. Ravn, C. Fledelius et al. 1998. Clinical evaluation of the Serum CrossLaps One Step ELISA, a new assay measuring the serum concentration of bone-derived degradation products of type I collagen C-telopeptides. Clin Chem 44(11): 2290–300.

Clarke, S. 2003. Aging as war between chemical and biochemical processes: protein methylation and the recognition of age-damaged proteins for repair. Ageing Res Rev 2(3): 263–85.

Cloos, P. A. and C. Fledelius. 2000. Collagen fragments in urine derived from bone resorption are highly racemized and isomerized: a biological clock of protein aging with clinical potential. Biochem J 345 Pt 3: 473–80.

Cloos, P. A., N. Lyubimova, H. Solberg, P. Qvist, C. Christiansen, I. Byrjalsen et al. 2004. An immunoassay for measuring fragments of newly synthesized collagen type I produced during metastatic invasion of bone. Clin Lab 50(5-6): 279–89.

Dobberstein, R. C., J. Huppertz, N. von Wurmb-Schwark and S. Ritz-Timme. 2008. Degradation of biomolecules in artificially and naturally aged teeth: implications for age estimation based on aspartic acid racemization and DNA analysis. Forensic Sci Int 179(2-3): 181–91. doi: 10.1016/j.forsciint.2008.05.017.

Dobberstein, R. C., S. M. Tung and S. Ritz-Timme. 2010. Aspartic acid racemisation in purified elastin from arteries as basis for age estimation. Int J Legal Med 124(4): 269–75. doi: 10.1007/s00414-009-0392-1.

Fujii, N., S. Muraoka and K. Harada. 1989. Purification and characterization of a protein containing D-aspartic acid in bovine lens. Biochim Biophys Acta 999(3): 239–42.

Fujii, N., Y. Ishibashi, K. Satoh, M. Fujino and K. Harada. 1994. Simultaneous racemization and isomerization at specific aspartic acid residues in alpha B-crystallin from the aged human lens. Biochim Biophys Acta 1204(2): 157–63.

Fujii, N., Y. Momose, Y. Ishibashi, T. Uemura, M. Takita and M. Takehana. 1997. Specific racemization and isomerization of the aspartyl residue of alphaA-crystallin due to UV-B irradiation. Exp Eye Res 65(1): 99–104. doi: 10.1006/exer.1997.0315.

Fujii, N., Y. Momose, N. Ishii, M. Takita, M. Akaboshi and M. Kodama. 1999a. The mechanisms of simultaneous stereoinversion, racemization, and isomerization at specific aspartyl residues of aged lens proteins. Mech Ageing Dev 107(3): 347–58.

Fujii, N., L. J. Takemoto, Y. Momose, S. Matsumoto, K. Hiroki and M. Akaboshi. 1999b. Formation of four isomers at the asp-151 residue of aged human alphaA-crystallin by natural aging. Biochem Biophys Res Commun 265(3): 746–51. doi: 10.1006/bbrc.1999.1748.

Fujii, N., S. Matsumoto, K. Hiroki and L. Takemoto. 2001. Inversion and isomerization of Asp-58 residue in human alphaA-crystallin from normal aged lenses and cataractous lenses. Biochim Biophys Acta 1549(2): 179–87.

Fujii, N., H. Uchida and T. Saito. 2004. The damaging effect of UV-C irradiation on lens alpha-crystallin. Mol Vis 10: 814–20.

Geiger, T. and S. Clarke. 1987. Deamidation, isomerization, and racemization at asparaginyl and aspartyl residues in peptides. Succinimide-linked reactions that contribute to protein degradation. J Biol Chem 262(2): 785–94.

Helfman, P. M. and J. L. Bada. 1975. Aspartic acid racemization in tooth enamel from living humans. Proc Natl Acad Sci U S A 72(8): 2891–4.

Helfman, P. M. and J. L. Bada. 1976. Aspartic acid racemisation in dentine as a measure of ageing. Nature 262(5566): 279–81.

Helfman, P. M., J. L. Bada and M. Y. Shou. 1977. Considerations on the role of aspartic acid racemization in the aging process. Gerontology 23(6): 419–25.

Iida, T., H. Matsunaga, T. Santa, T. Fukushima, H. Homma and K. Imai. 1998. Amino acid sequence and D/L-configuration determination of peptides utilizing liberated N-terminus phenylthiohydantoin amino acids. J Chromatogr A 813(2): 267–75.

Iida, T., T. Santa, A. Toriba and K. Imai. 2001. Amino acid sequence and D/L-configuration determination methods for D-amino acid-containing peptides in living organisms. Biomed Chromatogr 15(5): 319–27. doi: 10.1002/bmc.80.

Johnson, B. A., J. M. Shirokawa, W. S. Hancock, M. W. Spellman, L. J. Basa and D. W. Aswad. 1989. Formation of isoaspartate at two distinct sites during *in vitro* aging of human growth hormone. J Biol Chem 264(24): 14262–71.

Johnson, B. A. and D. W. Aswad. 1991. Optimal conditions for the use of protein L-isoaspartyl methyltransferase in assessing the isoaspartate content of peptides and proteins. Anal Biochem 192(2): 384–91.

Kuhn, W. 1958. Possible relation between optical activity and aging. Adv Enzymol Relat Subj Biochem 20: 1–29.

Kvenvolden, K., J. Lawless, K. Pering, E. Peterson, J. Flores, C. Ponnamperuma et al. 1970. Evidence for extraterrestrial amino-acids and hydrocarbons in the Murchison meteorite. Nature 228(5275): 923–6.

Lee, J. A., H. Homma, K. Sakai, T. Fukushima, T. Santa, K. Tashiro et al. 1997. Immunohistochemical localization of D-aspartate in the rat pineal gland. Biochem Biophys Res Commun 231(2): 505–8. doi: 10.1006/bbrc.1996.5902.

Man, E. H., M. E. Sandhouse, J. Burg and G. H. Fisher. 1983. Accumulation of D-aspartic acid with age in the human brain. Science 220(4604): 1407–8.

Maroudas, A., M. T. Bayliss, N. Uchitel-Kaushansky, R. Schneiderman and E. Gilav. 1998. Aggrecan turnover in human articular cartilage: use of aspartic acid racemization as a marker of molecular age. Arch Biochem Biophys 350(1): 61–71. doi: 10.1006/abbi.1997.0492.

Masters, P. M., J. L. Bada and J. S. Zigler, Jr. 1977. Aspartic acid racemisation in the human lens during ageing and in cataract formation. Nature 268(5615): 71–3.

Masters, P. M., J. L. Bada and J. S. Zigler, Jr. 1978. Aspartic acid racemization in heavy molecular weight crystallins and water insoluble protein from normal human lenses and cataracts. Proc Natl Acad Sci U S A 75(3): 1204–8.

McCudden, C. R. and V. B. Kraus. 2006. Biochemistry of amino acid racemization and clinical application to musculoskeletal disease. Clin Biochem 39(12): 1112–30. doi: 10.1016/j.clinbiochem.2006.07.009.

Mothet, J. P., L. Pollegioni, G. Ouanounou, M. Martineau, P. Fossier and G. Baux. 2005. Glutamate receptor activation triggers a calcium-dependent and SNARE protein-dependent release of the gliotransmitter D-serine. Proc Natl Acad Sci U S A 102(15): 5606–11. doi: 10.1073/pnas.0408483102.

Motoie, R., N. Fujii, S. Tsunoda, K. Nagata, T. Shimo-oka, T. Kinouchi et al. 2009. Localization of D-beta-aspartyl residue-containing proteins in various tissues. Int J Mol Sci 10(5): 1999–2009. doi: 10.3390/ijms10051999.

Ohtani, S., Y. Matsushima, Y. Kobayashi and T. Yamamoto. 2002. Age estimation by measuring the racemization of aspartic acid from total amino acid content of several types of bone and rib cartilage: a preliminary account. J Forensic Sci 47(1): 32–6.

Ohtani, S. and T. Yamamoto. 2005. Strategy for the estimation of chronological age using the aspartic acid racemization method with special reference to coefficient of correlation between D/L ratios and ages. J Forensic Sci 50(5): 1020–7.

Ohtani, S., R. Ito, S. Arany and T. Yamamoto. 2005. Racemization in enamel among different types of teeth from the same individual. Int J Legal Med 119(2): 66–9. doi: 10.1007/s00414-004-0506-8.

Oliyai, C. and R. T. Borchardt. 1994. Chemical pathways of peptide degradation. VI. Effect of the primary sequence on the pathways of degradation of aspartyl residues in model hexapeptides. Pharm Res 11(5): 751–8.

Radkiewicz, J. L., H. Zipse, S. Clarke and K. N. Houk. 2001. Neighboring side chain effects on asparaginyl and aspartyl degradation: an ab initio study of the relationship between peptide conformation and backbone NH acidity. J Am Chem Soc 123(15): 3499–506.

Ritz, S. and H. W. Schutz. 1993. Aspartic acid racemization in intervertebral discs as an aid to postmortem estimation of age at death. J Forensic Sci 38(3): 633–40.

Ritz, S., A. Turzynski and H. W. Schutz. 1994. Estimation of age at death based on aspartic acid racemization in noncollagenous bone proteins. Forensic Sci Int 69(2): 149–59.

Robinson, N. E. and A. B. Robinson. 2004. Prediction of primary structure deamidation rates of asparaginyl and glutaminyl peptides through steric and catalytic effects. J Pept Res 63(5): 437–48. doi: 10.1111/j.1399-3011.2004.00148.x.

Rosenquist, C., M. Bonde, C. Fledelius and P. Qvist. 1994. A simple enzyme-linked immunosorbent assay of human osteocalcin. Clin Chem 40(7 Pt 1): 1258–64.

Rosenquist, C., C. Fledelius, S. Christgau, B. J. Pedersen, M. Bonde, P. Qvist et al. 1998. Serum CrossLaps One Step ELISA. First application of monoclonal antibodies for measurement in serum of bone-related degradation products from C-terminal telopeptides of type I collagen. Clin Chem 44(11): 2281–9.

Sakai, K., H. Homma, J. A. Lee, T. Fukushima, T. Santa, K. Tashiro et al. 1997. D-aspartic acid localization during postnatal development of rat adrenal gland. Biochem Biophys Res Commun 235(2): 433–6. doi: 10.1006/bbrc.1997.6783.

Sakai, K., H. Homma, J. A. Lee, T. Fukushima, T. Santa, K. Tashiro et al. 1998. Localization of D-aspartic acid in elongate spermatids in rat testis. Arch Biochem Biophys 351(1): 96–105. doi: 10.1006/abbi.1997.0539.

Schurter, B. T. and D. W. Aswad. 2000. Analysis of isoaspartate in peptides and proteins without the use of radioisotopes. Anal Biochem 282(2): 227–31. doi: 10.1006/abio.2000.4601.

Smith, W. T., Jr. and K. A. Kvenvolden. 1971. Racemization and epimerization. Science 172(3981): 403.

Weber, D. J. and P. N. McFadden. 1997. Injury-induced enzymatic methylation of aging collagen in the extracellular matrix of blood vessels. J Protein Chem 16(4): 269–81.

Wehmiller, J. and P. E. Hare. 1971. Racemization of amino acids in marine sediments. Science 173(4000): 907–11. doi: 10.1126/science.173.4000.907.

Young, A. L., W. G. Carter, H. A. Doyle, M. J. Mamula and D. W. Aswad. 2001. Structural integrity of histone H2B *in vivo* requires the activity of protein L-isoaspartate O-methyltransferase, a putative protein repair enzyme. J Biol Chem 276(40): 37161–5. doi: 10.1074/jbc.M106682200.

Young, G. W., S. A. Hoofring, M. J. Mamula, H. A. Doyle, G. J. Bunick, Y. Hu et al. 2005. Protein L-isoaspartyl methyltransferase catalyzes *in vivo* racemization of aspartate-25 in mammalian histone H2B. J Biol Chem 280(28): 26094–8. doi: 10.1074/jbc.M503624200.

Zapico, S. C. and D. H. Ubelaker. 2013. Applications of physiological bases of ageing to forensic sciences. Estimation of age-at-death. Ageing Res Rev 12(2): 605–17. doi: 10.1016/j.arr.2013.02.002.

2

Aspartic Acid Racemization on Aging

Sara C. Zapico,[a,c] *Christian Thomas*[a,d] *and Sofía Tirados Menéndez*[b,*]

INTRODUCTION

Leon Trotsky said, "old age is the most unexpected of all the things that can happen to a man",[1] but it seems difficult to believe as the process of growing old is accompanied by a plethora of physical signs, which can be easily recognised. The progressive appearance of wrinkles and grey hair as well as changes in height, vision or hearing are the detectable consequences of the cellular and molecular changes taking place in our organism as years go by. Different mechanisms are considered hallmarks of aging like genomic instability, stem cell exhaustion or telomere attrition (Lopez-Otin et al. 2013). Among them, proteostasis is defined as a state of dynamic equilibrium in which protein stability and functionality is preserved, all thanks to a complex network of processes designed to control the fate of a protein from synthesis to degradation (Balch et al. 2008). Unfortunately, these mechanisms cannot completely avoid the accumulation of unwanted protein products and it is becoming clear that proteostasis malfunction has a key role in aging and its related disorders (Taylor and Dillin 2011).

[a] Smithsonian Institution, National Museum of Natural History, Anthropology Department, 10th and Constitution Ave, NW, PO Box 37012, Washington, DC 20560, USA.
[c] Email: saiczapico@gmail.com
[d] Email: crf.thomas@gmail.com
[b] King's College London, Randall Division of Cell & Molecular Biophysics, 2nd Floor - New Hunt's House. SE1 1UL London, United Kingdom.
[*] Corresponding author: maria_tirados.menendez@kcl.ac.uk

[1] Leon Trotsky, Diary in Exile (1935).

Protein turnover is an effective strategy that allows the degradation and replacement of old and damaged polypeptides, thus contributing to the maintenance of protein homeostasis (Savas et al. 2012, D'Angelo et al. 2009). The rate of turnover is specific for each protein, ranging from minutes to days, and can even be extended over a cell's lifespan. This is the case for tissues, where maintenance of covalent modifications is crucial for their function (i.e., learning and memory in the brain (Chain et al. 1999)) or those presenting a limited biosynthetic capacity (i.e., red blood cells (Kay et al. 1991) and eye lens (de Vries et al. 1991)). These long-lived proteins become more predisposed to bear PTMs which, in the end, contribute to cell decline and loss of viability (Taylor and Dillin 2011). PTMs are due to both enzymatic and non-enzymatic reactions, which take place after protein maturation and result in structural changes that alter both the biological function and the degradation of the protein (Jaisson and Gillery 2010). About 300 types of PTMs have been described and more than half correspond to enzymatic reactions (Zhao and Jensen 2009), like Glycosylation and Citrullination. Despite being less abundant than their enzymatic counterpart, non-enzymatic modifications are also frequent and varied: Oxidation, Glycation, Isomerization, Deamidation, Nitration, Carbonylation, Carbamylation, and Racemization among others (Jaisson and Gillery 2010). Interestingly, the accumulation of products derived from these reactions seem to increase during aging. The aim of this chapter is to delve into the role of Aspartic Acid Racemization in aging.

Racemization, a World of Possibilities

The term Racemization refers to the interconversion of optical forms of amino acids. The vast majority of proteinogenic amino acids is chiral about its α-carbon, which means that they can exist in two stereoisomeric forms: the L-form and the D-form. Due to steric hindrance, the human organism uses mainly L-amino acids in protein synthesis. In permanent proteins that are synthesized early in life and not subsequently exchanged, D-aspartic acid can accumulate during aging. The study performed by Helfman and Bada in 1975 was the first to demonstrate the potential applicability of the measurement of D-Asp levels in biochronology (Helfman and Bada 1975). Previously, the extent of racemization has been extensively used for dating of fossil materials (Schroeder and Bada 1973, Bada et al. 1970) and human remains (Bada and Protsch 1973) at the geological time scale. But it was not until the publication of this pioneering work in 1975 that the world learned of the usefulness of this technique in living organisms. Since then, Aspartic Acid Racemization has been favourably applied in different disciplines.

In short-lived proteins, Aspartic Acid Racemization provides information about the biological age of the tissue or reflects the metabolic state of the protein in question (Catterall et al. 2015, Sivan et al. 2008, Gineyts et al. 2000), while in long-lived proteins it measures the lifetime of the organism and can then be used in forensic sciences to determine the age at death (Zapico and Ubelaker 2013). A plethora of studies have shown the possibility of estimating the chronological age by analysing the extent of aspartic acid racemization in several tissues, such as dental enamel (Helfman and Bada 1975), dentine (Helfman and Bada 1976), brain (Man et al. 1983), eye (Masters et al.

1978), cartilage (Matzenauer et al. 2014), lung (Shapiro et al. 1991) and bone (Ritz et al. 1994). The information provided by this research sets the basis of our understanding of the role that this uncommon amino acid plays in aging.

Teeth, bone and cartilage

Whole tooth enamel, dentin and cartilage present higher racemization rates than bone and also show an extremely high lineal correlation between D/L ratios and age (Zapico and Ubelaker 2013). On the other hand, whole bone samples can be also correlated with an increase in D-isomer levels but with less precision, probably due to bone remodelling (Pfeiffer et al. 1995b). These works not only demonstrated that the observed correlation with age varied within a tissue, but also suggested the presence of pools of proteins that displayed different turnover patterns (Ritz et al. 1996, Thorpe et al. 2010, Pfeiffer et al. 1995a,b).

Dentin and bone show higher racemization rates in the acid-soluble peptide fraction when compared with the corresponding acid insoluble collagen fraction. But such difference is not detected in cartilage where both fractions share similar rates of racemization (Pfeiffer et al. 1995b). Furthermore, age-related increase on D-Asp was more consistent when only one protein was evaluated, like osteocalcine a universal constituent of bone and dentin (Ritz et al. 1996). It has been proposed that acid soluble fractions include fragments derived from protein degradation (Schulz and Jundt 1989). As the amount and nature of degraded proteins changes with age, the content of this soluble fraction varies with age and that could be the reason for the absence of correlation with age (Ritz et al. 1994).

In addition, aspartic acid racemization studies in horses showed that slow turnover in high strain tendons is responsible for the accumulation of fragments derived from collagen degradation; as the damage is not repaired, the amount of degraded fragments will increase with age, thus possibly affecting the mechanical properties of the tendon (Thorpe et al. 2010). Even though D-Asp levels proved to be a useful tool to assess the consequences of aging in tendons, there is no evidence of the biological significance of this increment with age. Taking into account the rate of racemization of aspartic acid in tooth enamel, Helfman and Bada predicted that only 8% of the total protein will be in its D-Asp form after 60 years (Helfman and Bada 1975). This percentage is very small and probably not enough to affect the stability and function of structural proteins, although more research is needed to clarify this point.

In humans, age-related diseases of the bone and the cartilage occur with altered levels of racemization of Aspartic acid. Racemization can impact protein conformation through reorientation of the angles of peptide bonds, thus affecting electrostatic interactions responsible for protein tertiary structure (Hol et al. 1981). Furthermore, racemization destabilizes the Collagen triple helix (Shah et al. 1999) and modifications in the structure of Collagen type I have been involved in the reduced bone strength seen in Osteoporosis (Boskey et al. 1999). Following this reasoning, studies performed in postmenopausal women demonstrated an association between higher levels of age-related forms of Asp in Collagen type I and an increased risk of osteoporotic fracture (Garnero et al. 2002). In addition, the levels of D-Asp in Collagen from cartilage diagnosed with osteoarthrosis are reduced due to the synthesis of new proteins in an attempt to repair the damage caused by the disease (Stabler et al. 2009). Interestingly, a recent study questioned whether protein turnover in osteoarthritic cartilage depends on joint site. The authors reported a higher anabolic metabolism in the knee when compared with the hip (Catterall et al. 2015), although the results cannot clarify if this difference indicates distinct mechanisms of disease progression or if it is a result of different cartilage biology.

Eye lens

Mammalian eyes are a complex design that has nothing to envy from photographic cameras. The iris regulates the amount of light entering the eye while the crystalline lens and the cornea refract light to be focused on the retina, thus allowing the formation of a sharp and precise image of the object in sight. Briefly, the eye lens is a transparent and biconvex structure that can modulate the focal distance of the eye by changes in its curvature. This shape adjustment is called accommodation and permits the eye to focus on objects at different distances from it.

Crystallines (Hoehenwarter et al. 2006) constitutes approximately 90% of eye lens composition; they are water-soluble proteins that tend to form soluble, high-molecular weight aggregates that pack tightly in lens fibres and contribute to increase the index of refraction while maintaining its transparency. Crystallines from vertebrate eye lenses can be classified in three main types: alpha, beta and gamma. Alpha-Crystalline is the major soluble protein in the human lens and it is composed by covalent aggregates of two different subunits, A and B. The two subunits have a sequence homology of 57% (de Jong et al. 1975, Dubin et al. 1990), a molecular weight of approximately 20 kDa (αA contains 173 amino acids and αB 175) and can function as chaperones, thus preventing other proteins from aggregation (Klemenz et al. 1991, Horwitz 1992).

Racemization of L-Asp residues in α-crystalline starts with an intramolecular cyclation that yields L-succinimide, and which will be subsequently converted to its D-isoform (Fujii et al. 1999b, de Jong et al. 1975, Nakamura et al. 2008, Fujii et al. 2004). Since racemization is a reversible first order reaction, D- and L-succinimide are in equilibrium and both isomers can undergo hydrolysis. As a result, four by-products can be formed: D-α-Asp, D-β-Asp, L-α-Asp and L-β-Asp. However, succinimide intermediates prefer to open to β-Asp and the majority of the product of the reaction is D-β-Asp (Geiger and Clarke 1987). The use of mutant A-crystalline subunits and peptides corresponding to A and B-crystalline sequences demonstrated that the process is dominated by steric hindrance of both the size of the amino acid preceding the Asp residue (Nakamura et al. 2006, Fujii et al. 1999a) and the higher order structure surrounding the residue (Nakamura et al. 2008).

D-Asp residues have been detected in specific positions of both polypeptides, Asp36 and Asp62 of αB-crystalline, as well as Asp58 and Asp151 of αA-crystalline are highly inverted to the D-isomer in eye lenses from aged subjects. Interestingly, Asp racemization can also be found in younger subjects but to a lesser extent (Fujii et al. 1999a). Consequently, it has been proposed that the formation of this biologically uncommon isomer starts shortly after birth (Fujii et al. 1999b), reaching the loss of approximately half of the initial amount of the normal L-Asp by the age of 30. And its decrease continues until it reaches the 42% observed in 80-year-old range subjects (Kaji et al. 2007, Fujii et al. 1999b, Fujii et al. 2004). Strikingly, the antipeptide 3R antibody, designed to specifically recognize D-β-Asp containing peptides, showed no reactivity against human eye samples from donors younger than 18 years old (Kaji et al. 2007).

During aging, thickening of basement membranes and loss of elasticity are noticeable changes at the histological level (Dunlevy and Rada 2004, Ramrattan et al. 1994, Ida et al. 2004). The eye is not an exception. An increased thickness of eye membranes has been reported not only in animal models of aging, but also in natural aging (Ramrattan et al. 1994, Ida et al. 2004). Reduced elasticity as well as increase in high molecular proteins have been shown in the lamina cribosa or the sclera (Dunlevy and Rada 2004). Immunostaining of D-β-Asp was detected in the lamina cribosa, the sclera and different types of basement membranes of the eye (Kaji et al. 2007).

As the accumulation of D-β-Asp residues leads to an elongation of the main chain of the protein and also modifies the orientation of the amino acid side chain, proteins containing this uncommon residue tend to aggregate. Subsequently, aggregates can deposit in the tissue, affecting its thickness and also modifying its function (Nakamura et al. 2008, Kaji et al. 2007). Similar changes are described for such age-related diseases of the eye as cataracts, pterygium or macular degeneration where the increase of D-Asp has also been reported (Kaji et al. 2007, Kaji et al. 2015, Kaji et al. 2010).

Skin

Typical signs of aged skin like laxity, sagging and wrinkling, have been associated with the degradation of elastic fibres (Frances and Robert 1984, Imayama and Braverman 1989). Approximately 90% of the elastic fibres are elastin, one of the most important structural proteins in the human organism. Elastin exhibits a very low turnover rate. Under healthy conditions, the vast majority of the protein is synthesized during development and its production finishes after birth. Even more, elastin synthesis may be supressed in adult tissues by a post-transcriptional mechanism mediating a rapid decay of tropoelastin mRNA (Parks 1997, Zhang and Parks 1999).

In UVR protected skin from healthy donors, analysis of racemization rates in purified elastin showed a rapid increase in D-Asp levels that was highly correlated with age (Ritz-Timme et al. 2003b). As sun exposure is one of the bigger contributors to skin aging, it should be noted that D-Asp residues have been also detected in degraded elastin from UVR damaged skin (Fujii et al. 2002). In this case, synthesis of new elastin has been observed but it does not result in the formation of normal elastic fibres, thus modifying the elasticity or resilience of normally aged skin (Bernstein et al. 1994, Uitto and Bernstein 1998). Furthermore, it has been reported that UV-B irradiation also enhances D-Asp generation in keratin-1, keratin-6B, keratin-10, and keratin-14 of epidermis (Mori et al. 2011).

Due to its lack of aromatic residues, Asp is not a photosensitive amino acid. Interestingly, recent *in vitro* studies with synthesized peptides demonstrated that the inclusion of a photosensitive amino acid, like tryptophan or tyrosine, close to the Asp residue induced its racemization after UVR exposure (Cai et al. 2013). This could not only explain the link between D-Asp and photo-aging, but also the D-Asp association with UV-B-irradiated lens proteins and eye diseases where UVR is a developing factor (i.e., cataracts, pterygium) (Nolan et al. 2003, Di Girolamo et al. 2005).

Elastin

Aged aortas display increased calcification and reduced elasticity (Mohiaddin et al. 1989), which are responsible for the loss of elastic recoil of the young aorta. These physical changes are accompanied by biochemical changes, like the accumulation of proteoglycans or the decrease of the elastic connective tissue (Nejjar et al. 1990). Elastin and collagen are the main proteic components of aorta (30% and 60% respectively). The amount of D-Asp in collagen from aorta samples was almost constant with age, but the levels of the isomer increased linearly with age (Powell et al. 1992).

A linear relationship between D-Asp in elastin and the age of the subjects analysed has also been found in the lungs (Shapiro et al. 1991) and yellow ligaments (Ritz-Timme et al. 2003a). Both tissues present considerable amounts of the protein, which plays a key role in the maintenance of their elasticity. Because racemization changes the overall protein structure, it was hypothesized that tissues with accumulation of molecules with low turnover rates, such as elastin, collagen, crystalline or aggrecan would be subject to damage accumulation. Thus D-Asp racemization of elastin fibres may be responsible for the decline of lung function with age, through structural changes that could negatively impact the normal lung elastic recoil (Shapiro et al. 1991).

Erythrocytes

Despite the fact that the erythrocyte's half-life is about 120 days (Shemin and Rittenberg 1946, McFadden and Clarke 1982), many physiological parameters have been correlated with its aging (Cohen et al. 1976, Bocci 1981). As the erythrocyte becomes older, it experiences modifications in the deformability properties of the membrane that increases its rigidity until passage through capillaries is impaired and the cell has to be removed from circulation (Shiga et al. 1979, LaCelle 1970). Measurements of Aspartic acid racemization in the membranes of erythrocytes showed that the amount of D-Asp increases as a function of cell age (Shapiro et al. 1991), while remaining steady in the cytosolic fraction (Brunauer and Clarke 1986).

Although there is no evidence that protein carboxylmethyl transferases can recognize D-β-Asp residues, experimental results suggest that methylation of acidic residues on erythrocyte membrane proteins occurs exclusively on D-Asp residues (McFadden and Clarke 1982). Thus methylation has been proposed as part of a cellular mechanism that recognizes aged proteins to promote their degradation or repair (Barber and Clarke 1983). Interestingly, the same evidence has been reported in brain tissue (O'Connor et al. 1984). Increased racemization of Aspartic acid residues in amyloid-β-peptide enhances its aggregation by affecting the solubility of the peptide (Mori et al. 1994), thus suggesting a role of this modification in the development of the amyloid plaques in Alzheimer's disease.

Conclusions

In summary, the amount of D-Asp shows a linear correlation with age in different tissues. Furthermore, altered levels of this modified amino acid can be detected in age-related diseases such as cataracts, Alzheimer's disease or osteoarthritis, suggesting a role of this alteration in aging. However, none of the evidence indicates the nature of this role. Despite the time elapsed since the first studies about Aspartic Acid Racemization and aging, it is still unclear if this process is a driver or merely a consequence of the process of growing older. On the other hand, Aspartic Acid racemization has revealed itself as a very useful tool for the assessment of the time of residence of proteins in a certain tissue. Hence, determination of changes in D-Asp levels can provide information about the repair or synthesis status of proteins in both health and disease.

Keywords: Aspartic acid, racemization, aging, protein turnover, eye lens, elastin, erythrocytes

References

Bada, J. L., B. P. Luyendyk and J. B. Maynard. 1970. Marine sediments: dating by the racemization of amino acids. Science 170(3959): 730–2.

Bada, J. L. and R. Protsch. 1973. Racemization reaction of aspartic acid and its use in dating fossil bones. Proc Natl Acad Sci U S A 70(5): 1331–4.

Balch, W. E., R. I. Morimoto, A. Dillin and J. W. Kelly. 2008. Adapting proteostasis for disease intervention. Science 319(5865): 916–9. doi: 10.1126/science.1141448.

Barber, J. R. and S. Clarke. 1983. Membrane protein carboxyl methylation increases with human erythrocyte age. Evidence for an increase in the number of methylatable sites. J Biol Chem 258(2): 1189–96.

Bernstein, E. F., Y. Q. Chen, K. Tamai, K. J. Shepley, K. S. Resnik, H. Zhang et al. 1994. Enhanced elastin and fibrillin gene expression in chronically photodamaged skin. J Invest Dermatol 103(2): 182–6.

Bocci, V. 1981. Determinants of erythrocyte ageing: a reappraisal. Br J Haematol 48(4): 515–22.

Boskey, A. L., T. M. Wright and R. D. Blank. 1999. Collagen and bone strength. J Bone Miner Res 14(3): 330–5. doi: 10.1359/jbmr.1999.14.3.330.

Brunauer, L. S. and S. Clarke. 1986. Age-dependent accumulation of protein residues which can be hydrolyzed to D-aspartic acid in human erythrocytes. J Biol Chem 261(27): 12538–43.

Cai, S., N. Fujii, T. Saito and N. Fujii. 2013. Simultaneous ultraviolet B-induced photo-oxidation of tryptophan/tyrosine and racemization of neighboring aspartyl residues in peptides. Free Radic Biol Med 65: 1037–46. doi: 10.1016/j.freeradbiomed.2013.08.171.

Catterall, J. B., R. D. Zura, M. P. Bolognesi and V. B. Kraus. 2015. Aspartic acid racemization reveals a high turnover state in knee compared with hip osteoarthritic cartilage. Osteoarthritis Cartilage. doi: 10.1016/j.joca.2015.09.003.

Chain, D. G., J. H. Schwartz and A. N. Hegde. 1999. Ubiquitin-mediated proteolysis in learning and memory. Mol Neurobiol 20(2-3): 125–42. doi: 10.1007/bf02742438.

Cohen, N. S., J. E. Ekholm, M. G. Luthra and D. J. Hanahan. 1976. Biochemical characterization of density-separated human erythrocytes. Biochim Biophys Acta 419(2): 229–42.

D'Angelo, M. A., M. Raices, S. H. Panowski and M. W. Hetzer. 2009. Age-dependent deterioration of nuclear pore complexes causes a loss of nuclear integrity in postmitotic cells. Cell 136(2): 284–95. doi: 10.1016/j.cell.2008.11.037.

de Jong, W. W., E. C. Terwindt and H. Bloemendal. 1975. The amino acid sequence of the A chain of human alpha-crystallin. FEBS Lett 58(1): 310–3.

de Vries, A. C., M. A. Vermeer, A. L. Hendriks, H. Bloemendal and L. H. Cohen. 1991. Biosynthetic capacity of the human lens upon aging. Exp Eye Res 53(4): 519–24.

Di Girolamo, N., M. Coroneo and D. Wakefield. 2005. Epidermal growth factor receptor signaling is partially responsible for the increased matrix metalloproteinase-1 expression in ocular epithelial cells after UVB radiation. Am J Pathol 167(2): 489–503. doi: 10.1016/s0002-9440(10)62992-6.

Dubin, R. A., A. H. Ally, S. Chung and J. Piatigorsky. 1990. Human alpha B-crystallin gene and preferential promoter function in lens. Genomics 7(4): 594–601.

Dunlevy, J. R. and J. A. Rada. 2004. Interaction of lumican with aggrecan in the aging human sclera. Invest Ophthalmol Vis Sci 45(11): 3849–56. doi: 10.1167/iovs.04-0496.

Frances, C. and L. Robert. 1984. Elastin and elastic fibers in normal and pathologic skin. Int J Dermatol 23(3): 166–79.

Fujii, N., K. Harada, Y. Momose, N. Ishii and M. Akaboshi. 1999a. D-amino acid formation induced by a chiral field within a human lens protein during aging. Biochem Biophys Res Commun 263(2): 322–6. doi: 10.1006/bbrc.1999.1279.

Fujii, N., L. J. Takemoto, Y. Momose, S. Matsumoto, K. Hiroki and M. Akaboshi. 1999b. Formation of four isomers at the asp-151 residue of aged human alphaA-crystallin by natural aging. Biochem Biophys Res Commun 265(3): 746–51. doi: 10.1006/bbrc.1999.1748.

Fujii, N., S. Tajima, N. Tanaka, N. Fujimoto, T. Takata and T. Shimo-Oka. 2002. The presence of D-beta-aspartic acid-containing peptides in elastic fibers of sun-damaged skin: a potent marker for ultraviolet-induced skin aging. Biochem Biophys Res Commun 294(5): 1047–51. doi: 10.1016/s0006-291x(02)00597-1.

Fujii, N., N. Takeuchi, N. Fujii, T. Tezuka, K. Kuge, T. Takata et al. 2004. Comparison of post-translational modifications of alpha A-crystallin from normal and hereditary cataract rats. Amino Acids 26(2): 147–52. doi: 10.1007/s00726-003-0050-8.

Garnero, P., P. Cloos, E. Sornay-Rendu, P. Qvist and P. D. Delmas. 2002. Type I collagen racemization and isomerization and the risk of fracture in postmenopausal women: the OFELY prospective study. J Bone Miner Res 17(5): 826–33. doi: 10.1359/jbmr.2002.17.5.826.

Geiger, T. and S. Clarke. 1987. Deamidation, isomerization, and racemization at asparaginyl and aspartyl residues in peptides. Succinimide-linked reactions that contribute to protein degradation. J Biol Chem 262(2): 785–94.

Gineyts, E., P. A. Cloos, O. Borel, L. Grimaud, P. D. Delmas and P. Garnero. 2000. Racemization and isomerization of type I collagen C-telopeptides in human bone and soft tissues: assessment of tissue turnover. Biochem J 345 Pt 3: 481–5.

Helfman, P. M. and J. L. Bada. 1975. Aspartic acid racemization in tooth enamel from living humans. Proc Natl Acad Sci U S A 72(8): 2891–4.

Helfman, P. M. and J. L. Bada. 1976. Aspartic acid racemisation in dentine as a measure of ageing. Nature 262(5566): 279–81.

Hoehenwarter, W., J. Klose and P. R. Jungblut. 2006. Eye lens proteomics. Amino Acids 30(4): 369–89. doi: 10.1007/s00726-005-0283-9.

Hol, W. G., L. M. Halie and C. Sander. 1981. Dipoles of the alpha-helix and beta-sheet: their role in protein folding. Nature 294(5841): 532–6.

Horwitz, J. 1992. Alpha-crystallin can function as a molecular chaperone. Proc Natl Acad Sci U S A 89(21): 10449–53.

Ida, H., K. Ishibashi, K. Reiser, L. M. Hjelmeland and J. T. Handa. 2004. Ultrastructural aging of the RPE-Bruch's membrane-choriocapillaris complex in the D-galactose-treated mouse. Invest Ophthalmol Vis Sci 45(7): 2348–54.

Imayama, S. and I. M. Braverman. 1989. A hypothetical explanation for the aging of skin. Chronologic alteration of the three-dimensional arrangement of collagen and elastic fibers in connective tissue. Am J Pathol 134(5): 1019–25.

Jaisson, S. and P. Gillery. 2010. Evaluation of nonenzymatic posttranslational modification-derived products as biomarkers of molecular aging of proteins. Clin Chem 56(9): 1401–12. doi: 10.1373/clinchem.2010.145201.

Kaji, Y., T. Oshika, Y. Takazawa, M. Fukayama, T. Takata and N. Fujii. 2007. Localization of D-beta-aspartic acid-containing proteins in human eyes. Invest Ophthalmol Vis Sci 48(9): 3923–7. doi: 10.1167/iovs.06-1284.

Kaji, Y., T. Oshika, Y. Takazawa, M. Fukayama and N. Fujii. 2010. Accumulation of D-beta-aspartic acid-containing proteins in age-related ocular diseases. Chem Biodivers 7(6): 1364–70. doi: 10.1002/cbdv.200900329.

Kaji, Y., T. Oshika, R. Nejima, S. Mori, K. Miyata and N. Fujii. 2015. Immunohistochemical localization of d-beta-aspartic acid-containing proteins in pterygium. J Pharm Biomed Anal 116: 86–9. doi: 10.1016/j.jpba.2015.01.057.

Kay, M. M., J. J. Marchalonis, S. F. Schluter and G. Bosman. 1991. Human erythrocyte aging: cellular and molecular biology. Transfus Med Rev 5(3): 173–95.

Klemenz, R., E. Frohli, R. H. Steiger, R. Schafer and A. Aoyama. 1991. Alpha B-crystallin is a small heat shock protein. Proc Natl Acad Sci U S A 88(9): 3652–6.

LaCelle, P. L. 1970. Alteration of membrane deformability in hemolytic anemias. Semin Hematol 7(4): 355–71.

Lopez-Otin, C., M. A. Blasco, L. Partridge, M. Serrano and G. Kroemer. 2013. The hallmarks of aging. Cell 153(6): 1194–217. doi: 10.1016/j.cell.2013.05.039.

Man, E. H., M. E. Sandhouse, J. Burg and G. H. Fisher. 1983. Accumulation of D-aspartic acid with age in the human brain. Science 220(4604): 1407–8.

Masters, P. M., J. L. Bada and J. S. Zigler, Jr. 1978. Aspartic acid racemization in heavy molecular weight crystallins and water insoluble protein from normal human lenses and cataracts. Proc Natl Acad Sci U S A 75(3): 1204–8.

Matzenauer, C., A. Reckert and S. Ritz-Timme. 2014. Estimation of age at death based on aspartic acid racemization in elastic cartilage of the epiglottis. Int J Legal Med 128(6): 995–1000. doi: 10.1007/s00414-013-0940-6.

McFadden, P. N. and S. Clarke. 1982. Methylation at D-aspartyl residues in erythrocytes: possible step in the repair of aged membrane proteins. Proc Natl Acad Sci U S A 79(8): 2460–4.

Mohiaddin, R. H., S. R. Underwood, H. G. Bogren, D. N. Firmin, R. H. Klipstein, R. S. Rees et al. 1989. Regional aortic compliance studied by magnetic resonance imaging: the effects of age, training, and coronary artery disease. Br Heart J 62(2): 90–6.

Mori, H., K. Ishii, T. Tomiyama, Y. Furiya, N. Sahara, S. Asano et al. 1994. Racemization: its biological significance on neuropathogenesis of Alzheimer's disease. Tohoku J Exp Med 174(3): 251–62.

Mori, Y., K. Aki, K. Kuge, S. Tajima, N. Yamanaka, Y. Kaji et al. 2011. UV B-irradiation enhances the racemization and isomerizaiton of aspartyl residues and production of nepsilon-carboxymethyl lysine (CML) in keratin of skin. J Chromatogr B Analyt Technol Biomed Life Sci 879(29): 3303–9. doi: 10.1016/j.jchromb.2011.05.010.

Nakamura, T., Y. Sadakane and N. Fujii. 2006. Kinetic study of racemization of aspartyl residues in recombinant human alphaA-crystallin. Biochim Biophys Acta 1764(4): 800–6. doi: 10.1016/j.bbapap.2006.02.005.

Nakamura, T., M. Sakai, Y. Sadakane, T. Haga, Y. Goto, T. Kinouchi et al. 2008. Differential rate constants of racemization of aspartyl and asparaginyl residues in human alpha A-crystallin mutants. Biochim Biophys Acta 1784(9): 1192–9. doi: 10.1016/j.bbapap.2008.04.008.

Nejjar, I., M. T. Pieraggi, J. C. Thiers and H. Bouissou. 1990. Age-related changes in the elastic tissue of the human thoracic aorta. Atherosclerosis 80(3): 199–208.

Nolan, T. M., N. DiGirolamo, N. H. Sachdev, T. Hampartzoumian, M. T. Coroneo and D. Wakefield. 2003. The role of ultraviolet irradiation and heparin-binding epidermal growth factor-like growth factor in the pathogenesis of pterygium. Am J Pathol 162(2): 567–74. doi: 10.1016/s0002-9440(10)63850-3.

O'Connor, C. M., D. W. Aswad and S. Clarke. 1984. Mammalian brain and erythrocyte carboxyl methyltransferases are similar enzymes that recognize both D-aspartyl and L-isoaspartyl residues in structurally altered protein substrates. Proc Natl Acad Sci U S A 81(24): 7757–61.

Parks, W. C. 1997. Posttranscriptional regulation of lung elastin production. Am J Respir Cell Mol Biol 17(1): 1–2. doi: 10.1165/ajrcmb.17.1.f135.

Pfeiffer, H., H. Mornstad and A. Teivens. 1995a. Estimation of chronologic age using the aspartic acid racemization method. I. On human rib cartilage. Int J Legal Med 108(1): 19–23.

Pfeiffer, H., H. Mornstad and A. Teivens. 1995b. Estimation of chronologic age using the aspartic acid racemization method. II. On human cortical bone. Int J Legal Med 108(1): 24–6.

Powell, J. T., N. Vine and M. Crossman. 1992. On the accumulation of D-aspartate in elastin and other proteins of the ageing aorta. Atherosclerosis 97(2-3): 201–8.

Ramrattan, R. S., T. L. van der Schaft, C. M. Mooy, W. C. de Bruijn, P. G. Mulder and P. T. de Jong. 1994. Morphometric analysis of Bruch's membrane, the choriocapillaris, and the choroid in aging. Invest Ophthalmol Vis Sci 35(6): 2857–64.

Ritz, S., A. Turzynski and H. W. Schutz. 1994. Estimation of age at death based on aspartic acid racemization in noncollagenous bone proteins. Forensic Sci Int 69(2): 149–59.

Ritz, S., A. Turzynski, H. W. Schutz, A. Hollmann and G. Rochholz. 1996. Identification of osteocalcin as a permanent aging constituent of the bone matrix: basis for an accurate age at death determination. Forensic Sci Int 77(1-2): 13–26.

Ritz-Timme, S., I. Laumeier and M. Collins. 2003a. Age estimation based on aspartic acid racemization in elastin from the yellow ligaments. Int J Legal Med 117(2): 96–101. doi: 10.1007/s00414-002-0355-2.

Ritz-Timme, S., I. Laumeier and M. J. Collins. 2003b. Aspartic acid racemization: evidence for marked longevity of elastin in human skin. Br J Dermatol 149(5): 951–9.

Savas, J. N., B. H. Toyama, T. Xu, J. R. Yates, 3rd and M. W. Hetzer. 2012. Extremely long-lived nuclear pore proteins in the rat brain. Science 335(6071): 942. doi: 10.1126/science.1217421.

Schroeder, R. A. and J. L. Bada. 1973. Glacial-postglacial temperature difference deduced from aspartic acid racemization in fossil bones. Science 182(4111): 479–82. doi: 10.1126/science.182.4111.479.

Schulz, A. and G. Jundt. 1989. Immunohistological demonstration of osteonectin in normal bone tissue and in bone tumors. Curr Top Pathol 80: 31–54.

Shah, N. K., B. Brodsky, A. Kirkpatrick and J. A. Ramshaw. 1999. Structural consequences of D-amino acids in collagen triple-helical peptides. Biopolymers 49(4): 297–302. doi: 10.1002/(sici)1097-0282(19990405)49:4<297::aid-bip4>3.0.co;2-q.

Shapiro, S. D., S. K. Endicott, M. A. Province, J. A. Pierce and E. J. Campbell. 1991. Marked longevity of human lung parenchymal elastic fibers deduced from prevalence of D-aspartate and nuclear weapons-related radiocarbon. J Clin Invest 87(5): 1828–34. doi: 10.1172/jci115204.

Shemin, D. and D. Rittenberg. 1946. The life span of the human red blood cell. J Biol Chem 166(2): 627–36.

Shiga, T., N. Maeda, T. Suda, K. Kon and M. Sekiya. 1979. The decreased membrane fluidity of *in vivo* aged, human erythrocytes. A spin label study. Biochim Biophys Acta 553(1): 84–95.

Sivan, S. S., E. Wachtel, E. Tsitron, N. Sakkee, F. van der Ham, J. Degroot et al. 2008. Collagen turnover in normal and degenerate human intervertebral discs as determined by the racemization of aspartic acid. J Biol Chem 283(14): 8796–801. doi: 10.1074/jbc.M709885200.

Stabler, T. V., S. S. Byers, R. D. Zura and V. B. Kraus. 2009. Amino acid racemization reveals differential protein turnover in osteoarthritic articular and meniscal cartilages. Arthritis Res Ther 11(2): R34. doi: 10.1186/ar2639.

Taylor, R. C. and A. Dillin. 2011. Aging as an event of proteostasis collapse. Cold Spring Harb Perspect Biol 3(5). doi: 10.1101/cshperspect.a004440.

Thorpe, C. T., I. Streeter, G. L. Pinchbeck, A. E. Goodship, P. D. Clegg and H. L. Birch. 2010. Aspartic acid racemization and collagen degradation markers reveal an accumulation of damage in tendon collagen that is enhanced with aging. J Biol Chem 285 (21): 15674–81. doi: 10.1074/jbc.M109.077503.

Uitto, J. and E. F. Bernstein. 1998. Molecular mechanisms of cutaneous aging: connective tissue alterations in the dermis. J Investig Dermatol Symp Proc 3(1): 41–4.

Zapico, S. C. and D. H. Ubelaker. 2013. Applications of physiological bases of ageing to forensic sciences. Estimation of age-at-death. Ageing Res Rev 12(2): 605–17. doi: 10.1016/j.arr.2013.02.002.

Zhang, M. and W. C. Parks. 1999. Posttranscriptional regulation of lung elastin expression involves binding of a developmentally regulated cytosolic protein to an open-reading frame cis-element in the messenger RNA. Chest 116(1 Suppl): 74S.

Zhao, Y. and O. N. Jensen. 2009. Modification-specific proteomics: strategies for characterization of post-translational modifications using enrichment techniques. Proteomics 9(20): 4632–41. doi: 10.1002/pmic.200900398.

3

Aspartic Acid Racemization and Aging in Cartilaginous Tissue

Sarit Sara Sivan[a],* *Alice Maroudas*[b] and *Ellen Wachtel*[c]

INTRODUCTION

Ultrastructure of Cartilaginous Tissue: The Intervertebral Disc and Articular Cartilage

Depending on the composition of the extracellular matrix, cartilage in the human body is classified as elastic, fibrous, fibroelastic or hyaline. In this chapter, we will review what has been documented concerning AAR in two of these forms of cartilage: in fibroelastic cartilage, as exemplified by the IVD, and in articular cartilage, a specialized type of hyaline cartilage which provides a gliding surface for synovial joints. Both IVD and articular cartilage are aneural, avascular and alymphatic tissues, containing very few cells embedded in an extracellular matrix composed of diverse proteins which are described below.

The IVD, the largest avascular cartilaginous structure in the human body (Maroudas et al. 1975), plays a primary mechanical role in transmitting loads through the spine and providing flexibility to the spinal column. The overall shape of the IVD is that of a truncated cylinder with macromolecular components: collagen, elastin and proteoglycans—aggregating (aggrecan, which forms aggregates in association with hyaluronan) and non-aggregating, both containing numerous negatively charged GAG chains.

[a] Department of Biotechnology Engineering, ORT Braude College, Karmiel, 21982 Israel.
[b] Department of Biomedical Engineering, Technion-Israel Institute of Technology, 32000, Israel.
   Email: alice.maroudas@gmail.com
[c] Faculty of Chemistry, Weizmann Institute of Science, Rehovot, 76100, Israel.
   Email: ellen.wachtel@weizmann.ac.il
* Corresponding author: ssivan@braude.ac.il

The disc has three structurally and functionally distinct regions: the central, collagen-poor NP; the outer, collagen-rich AF; and the CEP, which encase the complete structure at the interface with the vertebrae. The biochemical heterogeneity of the mature IVD reflects its developmental origins as well as its metabolic history, which is shaped by mechanical influences experienced during growth and differentiation. In contrast to other connective tissues, the macromolecules in IVD are commonly found in a particularly fragmented form (Donohue et al. 1988). Because of the large size and unique structure of the disc, which restricts diffusion, these partially degraded fragments accumulate, with amounts increasing as a function of age and state of degeneration.

Articular cartilage, in addition to providing a low friction, gliding surface for synovial joints, must also demonstrate compressive strength and wear-resistance. Morphologically, the matrix of articular cartilage can be divided into three zones: the outermost or superficial zone, poor in proteoglycans and in which thin collagen fibrils lie parallel to the joint surface; the transition (or intermediate) zone which is richer in aggrecan and in which the direction of the collagen fibrils changes from parallel to perpendicular to the joint surface; and the deep zone, which is richest in proteoglycans and in which thick collagen fibrils are radial to the joint surface.

In both IVD and articular cartilage, the small population of cells has the essential function of maintaining and repairing the matrix, both by synthesizing matrix macromolecules and by producing degradative enzymes, which are involved in tissue metabolism and turnover. Thus, the normal functioning of cartilaginous tissue is dependent on a balance between synthesis and matrix breakdown. When this balance is maintained, damaged tissue can be restored by cellular repair response. Otherwise, the matrix composition and organization are altered. Hence, the degraded matrix can no longer function effectively, leading to tissue degeneration. Human IVD is graded 1–5 (1–2, normal or near-normal, 3–5 increasingly degenerated), on the basis of histological assessment (Thompson et al. 1990). In characterizing tissue capacity for self-repair, an important parameter is the lifetime of the constituent macromolecules or, equivalently, the rate of molecular turnover.

Aspartic Acid Racemization as a Tool for Molecular Age Estimation in Human Tissues

A reliable, widely accepted method for assessing protein age and turnover in human tissues is the measurement of the accumulation of the D-isomer of amino acids. Amino acids are synthesized in nature as the L-isomers. However, during a lifetime, spontaneous, non-enzymatic racemization converts L-amino acids into a racemic mixture of L- and D-forms. The characteristic racemization rate for each amino acid depends on the local protein conformation as well as on the temperature, pH and ionic strength of the environment (Bada 1984). Of all the stable amino acids, aspartic acid has one of the highest racemization rates (Bada 1984). This enables measurement of D-isomer accumulation in human subjects in proteins that are not renewed or that turn over slowly (Bada 1984, Helfman and Bada 1975).

Measurements of age-dependent racemization in a number of different human and animal tissues have been published, e.g. (Arany et al. 2004, Bada 1981, Bada

1984, Dobberstein et al. 2010, Helfman and Bada 1975, 1976, Man et al. 1983, Masters 1983, Masters et al. 1977, 1978, Powell et al. 1992, Ritz and Schutz 1993, Ritz-Timme et al. 2003, Sivan et al. 2006a, Sivan et al. 2012, Sivan et al. 2008, Thorpe et al. 2010, Verzijl et al. 2001, Verzijl et al. 2000, Bada and Protsch 1973, Ohtani 1998, Ohtani et al. 1998). Since the relationship between the age of a protein and its D-Asp content depends on both racemization and turnover rates, and since, with independent knowledge of the former, racemization can be correlated with the age of long-lived proteins, it can be used as a molecular clock of protein aging and turnover. Under normal physiological conditions, protein longevity is observed to vary widely. Three distinct patterns of D-Asp accumulation have been observed: (i) Proteins with rapid turnover, such as hemoglobin, with *in vivo* lifetime of 120 days, are exchanged before a measurable accumulation of D-Asp can occur. (ii) Proteins with longer half-life can exhibit measurable accumulation of D-Asp. However, due to protein turnover, the lifetime relationship between D-Asp accumulation and age may not be linear; age-related equilibrium between accumulated D-Asp residues and L-Asp of newly synthesized proteins can prevent additional net accumulation of the D-isomer. (iii) Proteins with very slow (if any) turnover exhibit an approximately linear relationship between accumulation of D-Asp and age. Slow turnover implies long residence, suggesting that it is more difficult for cells to repair defects, which therefore may accumulate during the human life span (Bada 1984, Helfman and Bada 1976). This is the basis for the identification of long-lived proteins that age in parallel with the human being.

Calculation of protein turnover rates

Accumulation of D-Asp depends upon competition between two linear processes described by the rate constants of racemization (k_i) and of protein turnover (k_T). When the amount of D-isomer is small (D/L < 0.15), racemization may be considered irreversible. This is generally the case in the tissues discussed here; a reverse reaction would be subsumed in protein turnover. The time rate of change of the amount of D-isomer can be written as (Maroudas et al. 1998):

$$d(C_{Asp}(D)/(D + L))/dt = k_i C_{Asp}(L)/(D + L) - k_T C_{Asp}(D)/(D + L) \qquad (1)$$

where C_{Asp} is the molar concentration of Asp in the protein and D/(D+L) and L/(D+L) represent the fractions of the D- and L-isomers, respectively. Assuming that the content of the D-isomer is so small that D/(D+L) ≈ (D/L) and L/(D+L) ≈ 1, which is usually the case *in vivo*, Equation (1) can be rewritten as:

$$d(D/L)/dt = k_i - k_T(D/L) \qquad (2)$$

When the amount of protein changes with age (rate α), Equation 2 is modified as:

$$d(D/L)/dt = k_i - D/L\,(k_T + \alpha/(1 + \alpha t)) \qquad (3)$$

Although the amount of aggrecan in IVD is indeed not constant during a lifetime ($\alpha \approx 0.008$ year^{-1}) (Sivan et al. 2006b), the resulting effective change in k_T is < 5% and is not considered here. When the ratio D/L and/or the turnover rate (k_T) are very small, the rate of change of D/L will be approximately linear with time. Then k_i can be approximated by the accumulation rate. When the amount of D/L becomes relatively large and/or the turnover rate is significant, the time rate of change of D/L will no longer be constant, but will rather decrease with age. However, if the turnover rate itself is not constant but decreases with age, then it is possible that the accumulation of the D-isomer might still appear to be approximately linear. When k_T can be determined, then the molecular half-life $t_{1/2}$ (i.e., the necessary time for half of the molecules to be replenished) is:

$$t_{1/2} = \ln (2)/k_T \tag{4}$$

In this work, age groupings of $t_{1/2}$ values will be derived from regression curves rather than individual data points as previously reported (Sivan et al. 2014a, Sivan et al. 2014b). What values of k_i are used in studies of protein racemization? For collagen, the accepted value of k_i comes from the accumulation rate of collagen in dentin, where during a human lifetime there is little or no protein turnover (Helfman and Bada 1976). Since the temperatures experienced by dentin and by IVD can both be assumed to be 37°C, no correction for temperature needs to be made in the value of k_i for IVD collagen, i.e., 7.6×10^{-4}/year (Bada 1984). This value is also used for collagen in hip joint articular cartilage. Applying the temperature correction (Bada 1981)

$$k_i(T) = k_i(310) \times exp(-(16{,}890 \times (310 - T)/(T \times 310))) \tag{5}$$

where temperature T is in units of Kelvin, for collagen in articular cartilage of the knee joint (average temperature ~33°C (Haimovici 1982)), we find that k_i is approximately half as large as that of dentin. In the case of aggrecan, there is currently no tissue which may be used to extract a standard racemization rate analogous to dentin for collagen. The work of Maroudas et al. (Maroudas et al. 1998) on aggrecan fractions from human knee joint cartilage permitted estimation of k_i. It was found to be 1.873×10^{-3} per year and assumed to apply to all aggrecan fractions.

Aspartic Acid racemization in human intervertebral disc

Accumulation of D-Asp with age for the major structural proteins of the normal IVD matrix, i.e., aggrecan (A1, which includes the intact molecule and its fragments), collagen, and elastin, is shown in Fig. 1. It is evident that the concentration of D-Asp varies widely among the three proteins, with elastin at donor age 70 years having 2.3 times more D-Asp than collagen and 3.5 times more than aggrecan A1. No significant differences were observed for NP as compared to AF (p > 0.05) for both normal and degenerate tissue (Sivan et al. 2014b).

Aggrecan

In human IVD aggrecan (A1), the racemization pattern is similar to that of the type ii proteins described above. The amount of D-Asp increases with donor age until ~60 years (Fig. 1). The average accumulation rates during this time, defined as the rate of change of the ratio of the D-isomer to total Asp content are $4.74 \pm 0.44 \times 10^{-4}$ and $2.78 \pm 0.43 \times 10^{-4}$ year^{-1} for aggrecan from normal and degenerate IVD tissue, respectively (Sivan et al. 2014b, Sivan et al. 2006a). Beyond 60 years, accumulation

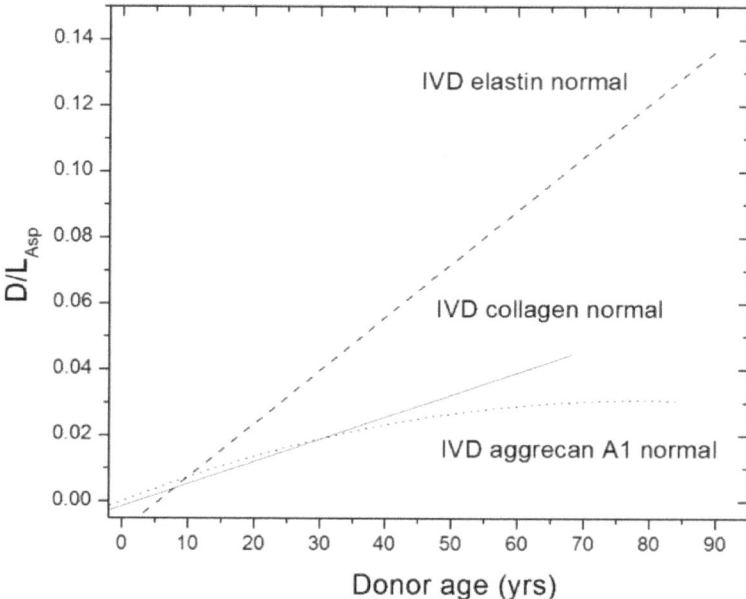

Figure 1. Increase of D/L$_{Asp}$ with age of elastin (Sivan et al. 2012), collagen (Sivan et al. 2008) and aggrecan (A1) (Sivan et al. 2006a), all isolated from normal adult IVD tissue.

appears to level off. From this, we infer that in aggrecan, equilibrium between accumulated D-Asp and newly synthesized L-Asp and/or the preferential loss of D-Asp is established. From D-Asp accumulation rates and an independent evaluation of k_i, as described above, the rate at which human IVD aggrecan turns over can be determined. A1 turnover is not constant during a lifetime. The calculated change in turnover rate with age is shown for aggrecan from both normal and degenerate IVD in Fig. 2A.

Turnover is a decreasing function of age and is consistently higher for degenerate IVD as compared to normal. The mean turnover rate for aggrecan from normal tissue is 0.138 year^{-1} and the mean half-life, 5.2 years. Calculating aggrecan half-life (defined by Equation 4) for different age groups gives Fig. 3A.

Collagen and elastin

Unlike aggrecan, for longer-lived proteins in IVD, i.e., collagen and elastin, a steady accumulation of D-Asp (similar to type iii proteins described above) is observed. The accumulation rates for collagen from normal and degenerate IVD are 6.74 ± 0.44 × 10^{-4} and 5.18 ± 0.44 × 10^{-4} year^{-1}, respectively (Sivan et al. 2008) and for elastin from normal and degenerate IVD (until the mid 50's), 16.2 ± 3.1 × 10^{-4} and 11.7 ± 3.1 × 10^{-4} year^{-1}, respectively (Sivan et al. 2012). In neither case is the normal/degenerate difference statistically significant (p > 0.05, T-test), a fact that influences the accuracy of turnover and lifetime calculations. Using the racemization rate found previously for collagen molecules in dentin, as described above, we found that the rate of collagen turnover (k_T) in disc is not constant but rather decreases with age (Fig. 2B). The average

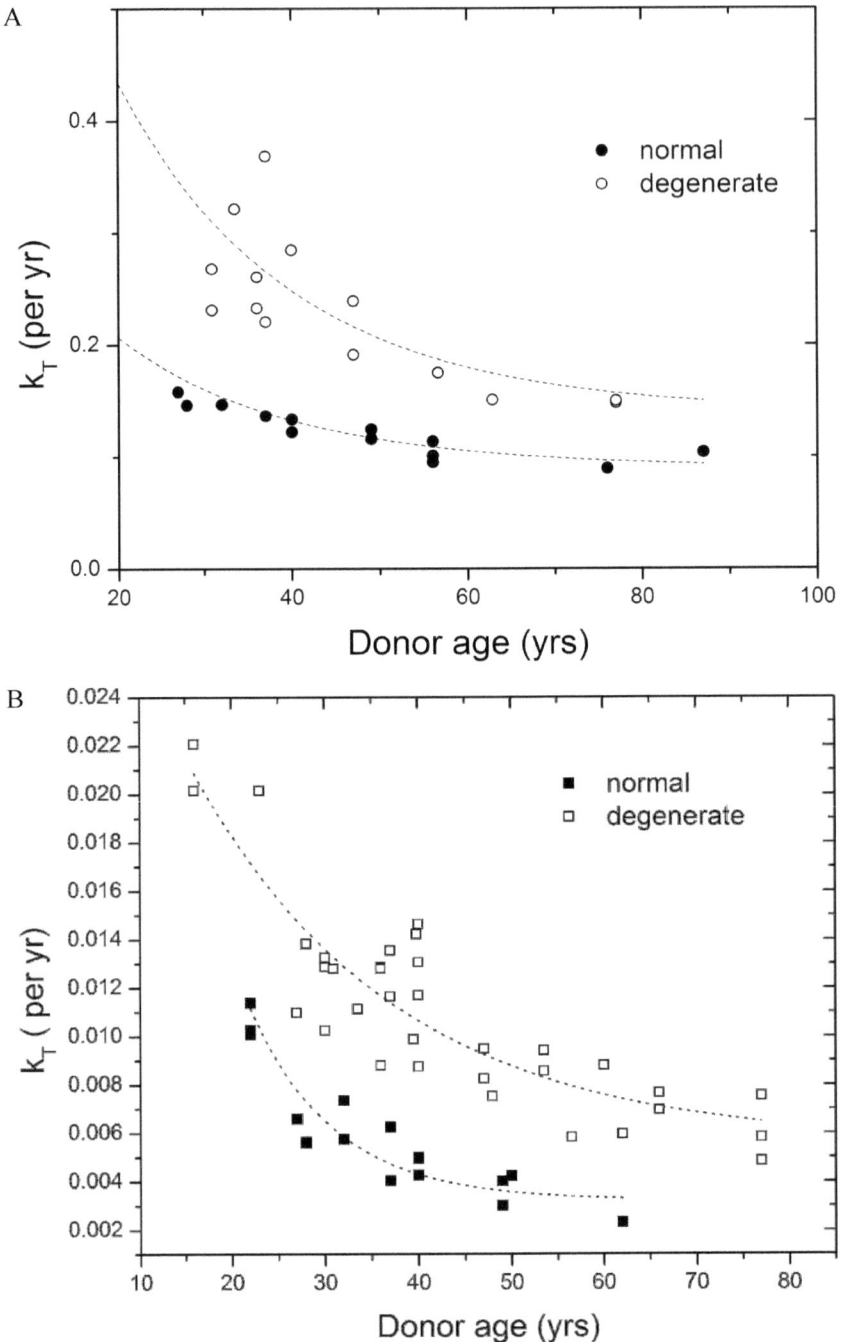

Figure 2. Age dependent change in the rate of turnover (k_T) of (A) human IVD aggrecan (Sivan et al. 2014b) and (B) collagen (Sivan et al. 2008) as determined by the accumulation of the D-isomer of aspartic acid. The values of k_i used in Equation 2 are 3.82×10^{-3} year^{-1} for aggrecan and for collagen, 7.6×10^{-4} year^{-1}.

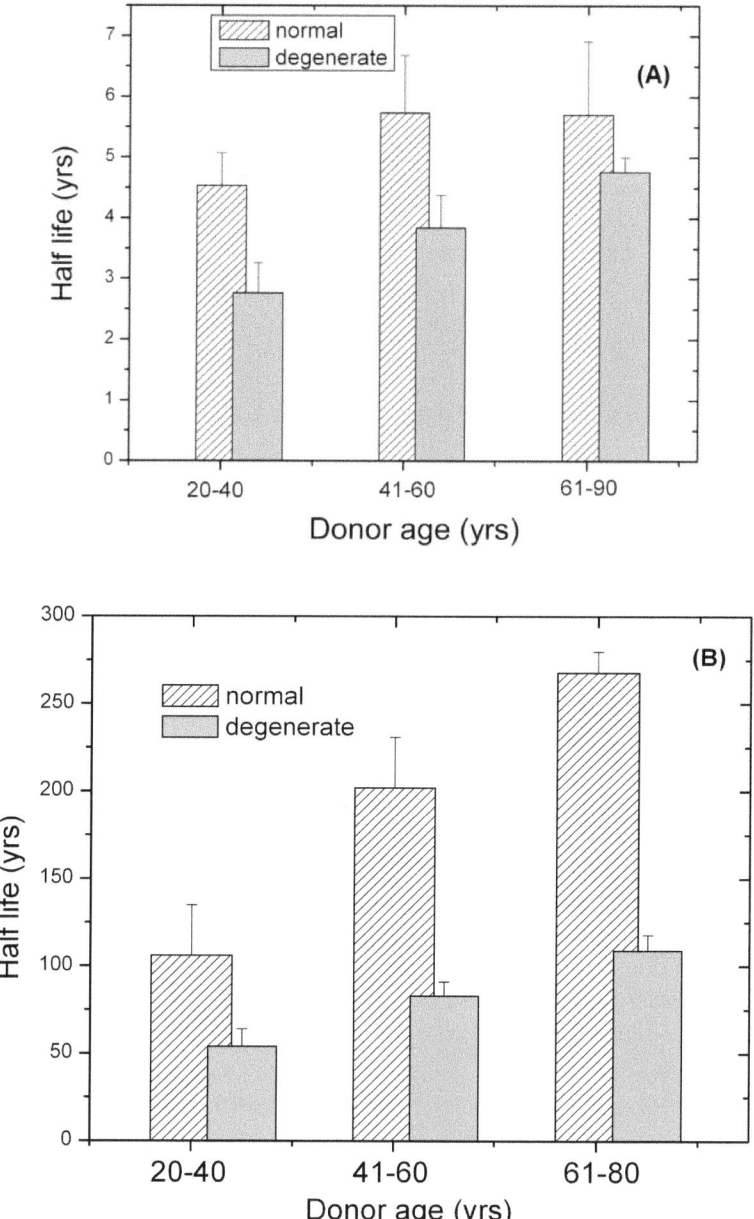

Figure 3. Half-life of IVD aggrecan (A) and IVD collagen (B) grouped by age. Data for (A) are taken from Sivan et al. 2014b and for (B) from Sivan et al. 2008.

turnover rate in normal disc between ages 20 to 40 is 0.00728 ± 0.00275 year^{-1} and between ages 50 to 65, 0.00323 ± 0.000947 year^{-1}. Half lives, grouped by age, vary from 106 to 268 years (Fig. 3B).

Turnover of collagen from degenerate discs appears to be more rapid than in normal discs. The much shorter mean half-life of aggrecan in normal IVD as compared to that of collagen (Sivan et al. 2008), is partly due to aggrecan's greater susceptibility to proteolysis; the fibrillar packing of collagen likely increases its stability. That half-life is an increasing function of age reflects changes in the rate of protein synthesis. Shorter half-life for degenerate IVD tissue than for normal tissue, suggests protein synthesis is more rapid in degenerate tissue. It should be noted that in none of the human IVD samples studied (Sivan et al. 2012, Sivan et al. 2008) did the accumulation of D-Asp in the NP and the AF differ significantly. However, in the case of whole, healthy human IVD tissue, Ritz and Schutz (1993) did find a higher D-Asp accumulation rate in the aggrecan-rich NP (23×10^{-4} year^{-1}) that decreased towards the peripheral collagen-rich AF (10.8×10^{-4} year^{-1}). This difference can be attributed to the presence of long lived, minor macromolecular components.

The Relationship between Aspartic Acid Racemization and other Post Translational Modifications in IVD

In addition to racemization, the most frequently investigated post-translational modifications of proteins are oxidation, glycation and deamidation (McKerrow 1979, Rattan et al. 1992, Robinson and Robinson 2001, Rosenberger 1991, Sajdok et al. 2001). These processes may occur in parallel. It was shown that both aspartyl residue racemization and the formation, via the Maillard reaction, of pentosidine, a non-enzymatic AGE, correlate well with disc age in IVD proteins; the amount of pentosidine increases with age in both normal and degenerate tissue (Sivan et al. 2006b, Verzijl et al. 2001, Verzijl et al. 2000). A strong correlation is obtained between the levels of D-Asp and pentosidine in IVD proteins (Fig. 4).

The fact that there is no statistical significance ($p > 0.05$) to the difference in slopes between normal and degenerate tissue suggests that these two markers change at a similar rate during aging and/or degeneration. These findings support the previously noted relevance of both these markers to the biology of aging (Geiger and Clarke 1987).

Aspartic acid racemization in human articular cartilage

As in the case of IVD, major structural proteins of the matrix of human articular cartilage are collagen and aggrecan. Elastin is also present but to the best of our knowledge, there have not been any studies of racemization in elastin from articular cartilage. AAR has been quantitated in both femoral joint and knee joint cartilage (Maroudas et al. 1998, Verzijl et al. 2000), the former in both native and degenerate tissue. Since, as described above, the racemization constant k_i is dependent on the body temperature at the site of the tissue, we have to note the difference in temperature between these two joints: knee joint ~33°C (Haimovici 1982), and femoral head, ≥ 37°C (Bergmann et al. 2001).

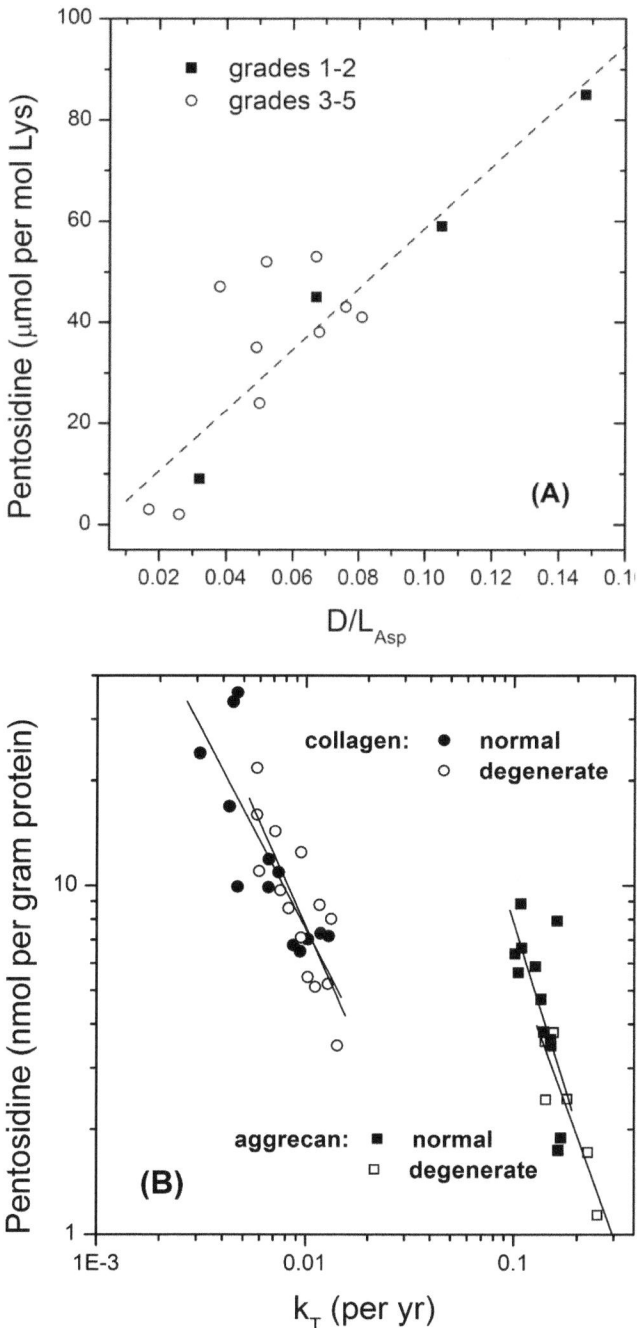

Figure 4. Correlation of pentosidine accumulation: (A) in IVD elastin as a function of D/L$_{Asp}$ (Sivan et al. 2012); and (B) in IVD collagen (Sivan et al. 2008) and aggrecan (Sivan et al. 2014b) as a function of turnover rate (k_T). For elastin, the accumulation rate is 581 µmol/mol Lys per D/L$_{Asp}$ when normal (grades 1–2) and degenerate (grades 3–5) tissues are considered together (R = 0.87, P < 0.05, solid line).

Collagen

The accumulation of D-Asp in collagen from normal knee joint cartilage appears to be linear with age (similar to type iii proteins described above) until approximately 80 years with the maximum content of D-Asp reaching ~2.3% (Verzijl et al. 2000). Using Equation 2 with $k_i = 3.8 \times 10^{-4}$ year^{-1}, the turnover rate k_T is found to be very small, decreasing from 0.00027 year^{-1} at age 20 to 0.000065 year^{-1} at age 80. Few data points are available for the age dependence of the accumulation of D-Asp in collagen from normal human femoral joint cartilage and these are from proteoglycan-depleted tissue (Table 1) (Maroudas et al. 1992). Nevertheless, the values are similar to those observed for knee joint cartilage, in spite of the expected temperature based difference in racemization rates. On this basis, we would expect to find slower turnover of collagen in the knee joint than in the femoral joint (Equation 2).

Table 1. Racemization of articular cartilage collagen from human femoral head.[a]

Age (years)	Corrected D/L Asp × 100	
	PG-depleted*	Native**
30	0.4 ± 0.1	1.0 ± 0.2
75	2.1 ± 0.2	5.3 ± 0.3
86	2.0 ± 0.2	4.8 ± 0.2

[a] (Maroudas et al. 1992)
Procedural Corrections:
*1.28 (blank: young depleted bovine cartilage and fresh rat tail collagen)
**I .6 (blank: undepleted young bovine cartilage)

Aggrecan

AAR in aggrecan (A1) from normal human knee joint cartilage (Maroudas et al. 1998) produces an apparently linear dependence of D-Asp content on donor age (Fig. 5). However, also in this case, the linear behavior does not mean that the protein is not undergoing turnover. Using $k_i = 1.873 \times 10^{-3}$ for the knee joint, as described above, k_T is in fact found (using Equation 2) to be non-zero, decreasing from 0.055 year^{-1} at age 20 to 0.025 year^{-1} at age 80. The half-life therefore increases accordingly (Equation 3).

However, for the A1D1 fraction (enriched in the large aggrecan monomer), there is no obvious accumulation of D-Asp with age. Beginning at ~18 years, the amount of D-Asp is approximately constant at ~2.4%. These results are very similar to those obtained for IVD aggrecan (Sivan et al. 2006b) (data not shown). Since aggrecan is heterogeneous with respect to molecular size and composition, both in its protein core and in its covalently attached carbohydrate chains, the difference in D-Asp accumulation between the two samples indicates that the differently sized molecules which constitute A1 display differential turnover.

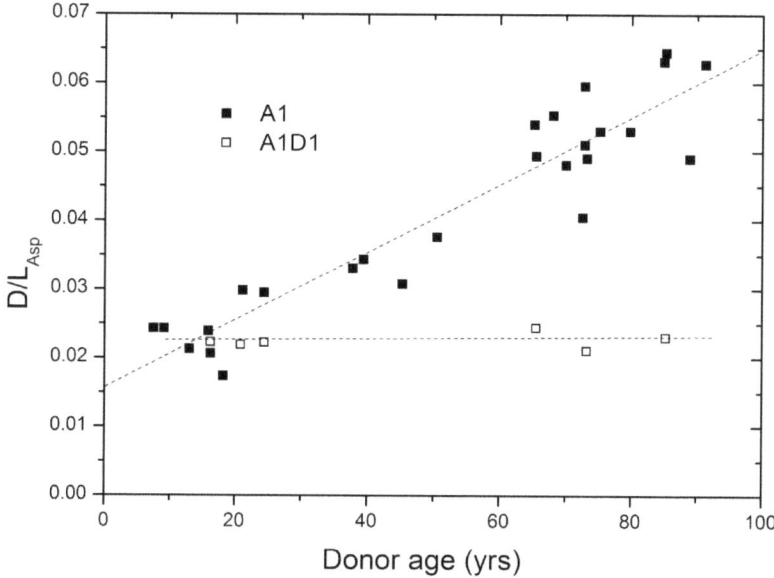

Figure 5. D/L_{Asp} ratios for the A1 and A1D1 preparations obtained from human knee joint as a function of age. Linear regression for A1: R = 0.95, P < 0.0001. Data are from Maroudas et al. 1998.

Correlation between aspartic acid racemization and pentosidine formation in collagen from articular cartilage

As for IVD, there is strong correlation between the extent of AAR and the formation of pentosidine in collagen from human knee joint cartilage (Verzijl et al. 2000).

Comparing IVD and articular cartilage on the same graph shows that this correlation appears to be preserved across tissue types, in spite of the fact that the maximum degree of racemization in knee joint collagen is significantly lower than that in IVD (Fig. 6).

Physiological significance of aspartic acid racemization in human intervertebral disc and articular cartilage

The physiological significance of AAR is two-fold. On the one hand, AAR has been identified as a non-enzymatic modification of covalent bonds that leads to age-dependent accumulation of abnormal protein in human tissue. On the other hand, since *in vivo* racemization correlates with the age of long-lived proteins, AAR can be used as a molecular indicator of protein ageing and turnover; this is especially true for proteins which have turnover rates too slow to be determined by conventional labelling techniques.

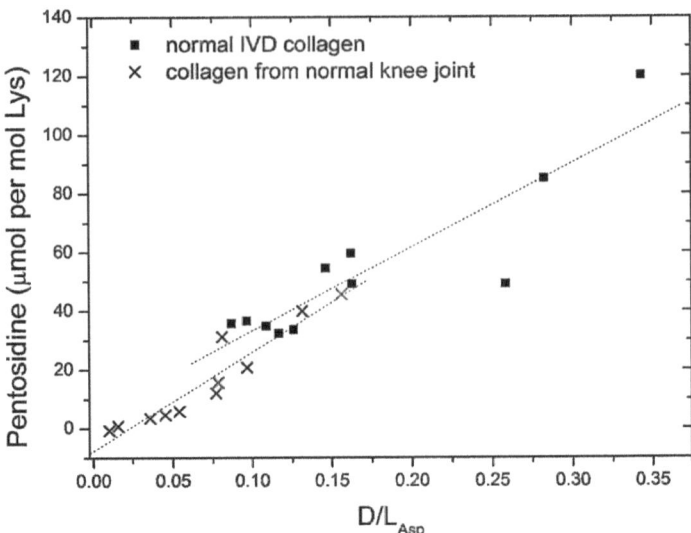

Figure 6. Comparison of pentosidine formation and D-Asp accumulation in collagen from normal IVD and human knee joint cartilage (Sivan et al. 2008). The difference in slopes is not statistically significant ($p > 0.05$, T-test).

The AAR data reviewed here for human IVD and articular cartilage suggest that the structural integrity of IVD collagen and the co-located elastin (Yu et al. 2005, Yu et al. 2002) and collagen from articular cartilage may be at particular risk due to rapid accumulation of D-isomer and/or slow turnover. There is, however, no unambiguous identification of the nature of that risk. While it is true that aggrecan turns over more rapidly than collagen in IVD, and that degenerate IVD tissue experiences more rapid turnover than normal tissue, yet aggrecan on the whole is also relatively long-lived so that in articular cartilage, it even accumulates more D-Asp than collagen. In the normal functioning of both IVD and articular cartilage matrix, proteoglycan molecules must resist compressive forces while embedded within a fibrillar collagen network that provides tensile strength (Kempson et al. 1973, Schmidt et al. 1990). The finely tuned physiological balance of collagen, proteoglycan, and water is dependent on a minimum supply of native protein: isomerization and turnover must of course affect that supply. The correlation between AAR and pentosidine formation in both IVD and articular cartilage may exclude the possibility of considering Asp racemization as a simple molecular event independent of other macromolecular processes, e.g., glycation. The origin of the correlation is not clear; however, it has been suggested that racemization of aspartyl residues may induce posttranslational stereochemical modifications, altering structure-function relationships of proteins (Bada 1984), which in turn could result in reactions such as crosslinking (Helfman et al. 1977), higher susceptibility to degradation (Chen et al. 2002, Geiger and Clarke 1987) as well as to conformational instability (Masters et al. 1978). Aggregation of long-lived, structural proteins could also occur (Masters et al. 1978, Roher et al. 1993, Shimizu et al. 2000, Tomiyama et al. 1994). *In vitro* amplification of pentosidine content has been shown to result in increased cartilage stiffness (Verzijl et al. 2002), while accumulation of crosslinked

AGEs may render the cartilage collagen network more brittle (Chen et al. 2001). Both changes may, in turn, contribute to the age-related failure of cartilage to resist mechanical damage, and may thus be a factor that predisposes to the development of osteoarthritis.

Acknowledgments

The authors wish to thank Dr. Peter Roughley (Shriners Hospital for Children, Montreal, Canada) for his critical reading of the manuscript and most helpful suggestions.

Keywords: Aspartic acid racemization, advanced glycation endproduct (AGE), turnover, age estimation, articular cartilage, intervertebral disc

References

Arany, S., S. Ohtani, N. Yoshioka and K. Gonmori. 2004. Age estimation from aspartic acid racemization of root dentin by internal standard method. Forensic Sci Int 141(2-3): 127–30. doi: 10.1016/j.forsciint.2004.01.017.

Bada, J. L. and R. Protsch. 1973. Racemization reaction of aspartic acid and its use in dating fossil bones. Proc Natl Acad Sci U S A 70(5): 1331–4.

Bada, J. L. 1981. Racemization of amino acids in fossil bones and teeth from Olduvai Gorge region, Tanzania, East Africa. Earth and Planetary Science Letters 55: 292–298.

Bada, J. L. 1984. *In vivo* racemization in mammalian proteins. Methods Enzymol 106: 98–115.

Bergmann, G., F. Graichen, A. Rohlmann, N. Verdonschot and G. H. van Lenthe. 2001. Frictional heating of total hip implants. Part 1: measurements in patients. J Biomech 34(4): 421–8.

Chen, A. C., M. M. Temple, D. M. Ng, C. D. Richardson, J. DeGroot and N. Verzijl. 2001. Age-related crosslinking alters tensile properties of articular cartilage. Transact Orthop Res Soc 26: 128.

Chen, A. C., M. M. Temple, D. M. Ng, N. Verzijl, J. DeGroot, J. M. TeKoppele et al. 2002. Induction of advanced glycation end products and alterations of the tensile properties of articular cartilage. Arthritis Rheum 46(12): 3212–7. doi: 10.1002/art.10627.

Dobberstein, R. C., S. M. Tung and S. Ritz-Timme. 2010. Aspartic acid racemisation in purified elastin from arteries as basis for age estimation. Int J Legal Med 124(4): 269–75. doi: 10.1007/s00414-009-0392-1.

Donohue, P. J., M. R. Jahnke, J. D. Blaha and B. Caterson. 1988. Characterization of link protein(s) from human intervertebral-disc tissues. Biochem J 251(3): 739–47.

Geiger, T. and S. Clarke. 1987. Deamidation, isomerization, and racemization at asparaginyl and aspartyl residues in peptides. Succinimide-linked reactions that contribute to protein degradation. J Biol Chem 262(2): 785–94.

Haimovici, N. 1982. Three years experience in direct intraarticular temperature measurement. Prog Clin Biol Res 107: 453–61.

Helfman, P. M. and J. L. Bada. 1975. Aspartic acid racemization in tooth enamel from living humans. Proc Natl Acad Sci U S A 72(8): 2891–4.

Helfman, P. M. and J. L. Bada. 1976. Aspartic acid racemisation in dentine as a measure of ageing. Nature 262(5566): 279–81.

Helfman, P. M., J. L. Bada and M. Y. Shou. 1977. Considerations on the role of aspartic acid racemization in the aging process. Gerontology 23(6): 419–25.

Kempson, G. E., H. Muir, C. Pollard and M. Tuke. 1973. The tensile properties of the cartilage of human femoral condyles related to the content of collagen and glycosaminoglycans. Biochim Biophys Acta 297(2): 456–72.

Man, E. H., M. E. Sandhouse, J. Burg and G. H. Fisher. 1983. Accumulation of D-aspartic acid with age in the human brain. Science 220(4604): 1407–8.

Maroudas, A., R. A. Stockwell, A. Nachemson and J. Urban. 1975. Factors involved in the nutrition of the human lumbar intervertebral disc: cellularity and diffusion of glucose *in vitro*. J Anat 120(Pt 1): 113–30.

Maroudas, A., G. Palla and E. Gilav. 1992. Racemization of aspartic acid in human articular cartilage. Connect Tissue Res 28(3): 161–9.

Maroudas, A., M. T. Bayliss, N. Uchitel-Kaushansky, R. Schneiderman and E. Gilav. 1998. Aggrecan turnover in human articular cartilage: use of aspartic acid racemization as a marker of molecular age. Arch Biochem Biophys 350(1): 61–71. doi: 10.1006/abbi.1997.0492.

Masters, P. M., J. L. Bada and J. S. Zigler, Jr. 1977. Aspartic acid racemisation in the human lens during ageing and in cataract formation. Nature 268(5615): 71–3.

Masters, P. M., J. L. Bada and J. S. Zigler, Jr. 1978. Aspartic acid racemization in heavy molecular weight crystallins and water insoluble protein from normal human lenses and cataracts. Proc Natl Acad Sci U S A 75(3): 1204–8.

Masters, P. M. 1983. Stereochemically altered noncollagenous protein from human dentin. Calcif Tissue Int 35(1): 43–7.

McKerrow, J. H. 1979. Non-enzymatic, post-translational, amino acid modifications in ageing. A brief review. Mech Ageing Dev 10(6): 371–7.

Ohtani, S. 1998. Rate of aspartic acid racemization in bone. Am J Forensic Med Pathol 19(3): 284–7.

Ohtani, S., Y. Matsushima, Y. Kobayashi and K. Kishi. 1998. Evaluation of aspartic acid racemization ratios in the human femur for age estimation. J Forensic Sci 43(5): 949–53.

Powell, J. T., N. Vine and M. Crossman. 1992. On the accumulation of D-aspartate in elastin and other proteins of the ageing aorta. Atherosclerosis 97(2-3): 201–8.

Rattan, S. I., A. Derventzi and B. F. Clark. 1992. Protein synthesis, posttranslational modifications, and aging. Ann N Y Acad Sci 663: 48–62.

Ritz, S. and H. W. Schutz. 1993. Aspartic acid racemization in intervertebral discs as an aid to postmortem estimation of age at death. J Forensic Sci 38(3): 633–40.

Ritz-Timme, S., I. Laumeier and M. J. Collins. 2003. Aspartic acid racemization: evidence for marked longevity of elastin in human skin. Br J Dermatol 149(5): 951–9.

Robinson, N. E. and A. B. Robinson. 2001. Prediction of protein deamidation rates from primary and three-dimensional structure. Proc Natl Acad Sci U S A 98(8): 4367–72. doi: 10.1073/pnas.071066498.

Roher, A. E., J. D. Lowenson, S. Clarke, C. Wolkow, R. Wang, R. J. Cotter et al. 1993. Structural alterations in the peptide backbone of beta-amyloid core protein may account for its deposition and stability in Alzheimer's disease. J Biol Chem 268(5): 3072–83.

Rosenberger, R. F. 1991. Senescence and the accumulation of abnormal proteins. Mutat Res 256(2-6): 255–62.

Sajdok, J., A. Kotrbova-Kozak and A. Pilin. 2001. Age determination using peptide mapping of non-collagenous proteins in human dentin. Soud Lek 46(1): 5–8.

Schmidt, M. B., V. C. Mow, L. E. Chun and D. R. Eyre. 1990. Effects of proteoglycan extraction on the tensile behavior of articular cartilage. J Orthop Res 8(3): 353–63. doi: 10.1002/jor.1100080307.

Shimizu, T., A. Watanabe, M. Ogawara, H. Mori and T. Shirasawa. 2000. Isoaspartate formation and neurodegeneration in Alzheimer's disease. Arch Biochem Biophys 381(2): 225–34. doi: 10.1006/abbi.2000.1955.

Sivan, S. S., E. Tsitron, E. Wachtel, P. J. Roughley, N. Sakkee, F. van der Ham et al. 2006a. Aggrecan turnover in human intervertebral disc as determined by the racemization of aspartic acid. J Biol Chem 281(19): 13009–14. doi: 10.1074/jbc.M600296200.

Sivan, S. S., E. Tsitron, E. Wachtel, P. Roughley, N. Sakkee, F. van der Ham et al. 2006b. Age-related accumulation of pentosidine in aggrecan and collagen from normal and degenerate human intervertebral discs. Biochem J 399(1): 29–35. doi: 10.1042/BJ20060579.

Sivan, S. S., E. Wachtel, E. Tsitron, N. Sakkee, F. van der Ham, J. Degroot et al. 2008. Collagen turnover in normal and degenerate human intervertebral discs as determined by the racemization of aspartic acid. J Biol Chem 283(14): 8796–801. doi: 10.1074/jbc.M709885200.

Sivan, S. S., B. Van El, Y. Merkher, C. E. Schmelzer, A. M. Zuurmond, A. Heinz et al. 2012. Longevity of elastin in human intervertebral disc as probed by the racemization of aspartic acid. Biochim Biophys Acta 1820(10): 1671–7. doi: 10.1016/j.bbagen.2012.06.010.

Sivan, S. S., A. J. Hayes, E. Wachtel, B. Caterson, Y. Merkher, A. Maroudas et al. 2014a. Biochemical composition and turnover of the extracellular matrix of the normal and degenerate intervertebral disc. Eur Spine J 23 Suppl 3: S344–53. doi: 10.1007/s00586-013-2767-8.

Sivan, S. S., E. Wachtel and P. Roughley. 2014b. Structure, function, aging and turnover of aggrecan in the intervertebral disc. Biochim Biophys Acta 1840(10): 3181–9. doi: 10.1016/j.bbagen.2014.07.013.

Thompson, J. P., R. H. Pearce, M. T. Schechter, M. E. Adams, I. K. Tsang and P. B. Bishop. 1990. Preliminary evaluation of a scheme for grading the gross morphology of the human intervertebral disc. Spine (Phila Pa 1976) 15(5): 411–5.

Thorpe, C. T., I. Streeter, G. L. Pinchbeck, A. E. Goodship, P. D. Clegg and H. L. Birch. 2010. Aspartic acid racemization and collagen degradation markers reveal an accumulation of damage in tendon collagen that is enhanced with aging. J Biol Chem 285(21): 15674–81. doi: 10.1074/jbc.M109.077503.

Tomiyama, T., S. Asano, Y. Furiya, T. Shirasawa, N. Endo and H. Mori. 1994. Racemization of Asp23 residue affects the aggregation properties of Alzheimer amyloid beta protein analogues. J Biol Chem 269(14): 10205–8.

Verzijl, N., J. DeGroot, S. R. Thorpe, R. A. Bank, J. N. Shaw, T. J. Lyons et al. 2000. Effect of collagen turnover on the accumulation of advanced glycation end products. J Biol Chem 275(50): 39027–31. doi: 10.1074/jbc.M006700200.

Verzijl, N., J. DeGroot, R. A. Bank, M. T. Bayliss, J. W. Bijlsma, F. P. Lafeber et al. 2001. Age-related accumulation of the advanced glycation endproduct pentosidine in human articular cartilage aggrecan: the use of pentosidine levels as a quantitative measure of protein turnover. Matrix Biol 20(7): 409–17.

Verzijl, N., J. DeGroot, Z. C. Ben, O. Brau-Benjamin, A. Maroudas, R. A. Bank et al. 2002. Crosslinking by advanced glycation end products increases the stiffness of the collagen network in human articular cartilage: a possible mechanism through which age is a risk factor for osteoarthritis. Arthritis Rheum 46(1): 114–23. doi: 10.1002/1529-0131(200201)46:1<114::AID-ART10025>3.0.CO;2-P.

Yu, J., P. C. Winlove, S. Roberts and J. P. Urban. 2002. Elastic fibre organization in the intervertebral discs of the bovine tail. J Anat 201(6): 465–75.

Yu, J., J. C. Fairbank, S. Roberts and J. P. Urban. 2005. The elastic fiber network of the anulus fibrosus of the normal and scoliotic human intervertebral disc. Spine (Phila Pa 1976) 30(16): 1815–20.

4

Racemization and Isomerization of Aspartyl Residues in Amyloid Peptides Involved in the Development of Alzheimer's Disease

Koichi Inoue[1] and *Toshimasa Toyo'oka*[2,*]

INTRODUCTION

Neurodegenerative disease is currently the most demanding disorder in our progressively aging society. A major concern is the development of clinical strategies not only for the proper utilization of medicines and/or treatments related to the central nervous system agents, but also for risk assessments derived from exercise, anxiolytic activity, nutrition and unknown effects that might be triggered by other complicating factors in health care. Regarding the definitive therapy for each neurodegenerative disease, we will seek a generalization about the pathological changes to brain tissues. However, it is very difficult for the resolution to have excellent controlled and/or stopping advanced neurodegenerative effects for the most serious patients using only pathology. Thus, we can overcome the challenges facing neurodegenerative systems caused by various molecular biological factors inducing expanded nerve damage, chronic syndrome, aggregation and fibrous response. Further studies should go on to investigate the aging brain in detail, which has not received sufficient attention from molecular biological research and discussion as of now.

Dementia in particular is a very important neurodegenerative disorder in the current global community. In 2008, the WHO declared dementia as a priority condition

[1] Laboratory of Clinical and Analytical Chemistry, College of Pharmaceutical Sciences, Ritsumeikan University, 1-1-1 Nojihigashi, Kusatsu, Shiga 525-8577, Japan.
Email: kinoue@fc.ritsumei.ac.jp
[2] Laboratory of Analytical and Bio-Analytical Chemistry, School of Pharmaceutical Sciences, University of Shizuoka, 52-1 Yada, Suruga-ku, Shizuoka 422-8526, Japan.
* Corresponding author: toyooka@u-shizuoka-ken.ac.jp

through the Mental Health Gap Action Programme (WHO Mental Health Gap Action Programme (mhGAP) 2008). The prevalence and incidence prediction indicated that the number of dementia patients will continue to rapidly increase, particularly among the aging population. The total number of dementia patients worldwide in 2010 was estimated at 35.6 million, which will then double every twenty years and thus expected to be 65.7 million in 2030 and 115.4 million in 2050. This report indicated that much of the increase will be in developing countries, the fastest growth in the elderly population taking place in China, India, South Asia and Western Pacific countries. Moreover, Alzheimer's disease (AD) has become a major public health concern as the world's population ages. It is projected that by 2050, people aged 60 and above will account for 22% of the world's population with four-fifths living in Asia, Latin America or Africa. It is apparent that for the AD problem, there is an urgent need for action and expeditious research.

AD is diagnosed by the progressive decline in cognitive function, and substantially increases among people aged 65 years or more, with a progressive decline in memory, cognitive function, and emotional capacity as compared to normal age-related decline. The focal AD reaction is related to the injury and depredation of neurons that are involved with memory and learning in a specific brain region such as the hippocampus, frontal and temporal lobes. Many AD researchers are hopeful that they can find a way of stopping the destruction of neuronal cells altogether or at least to drastically slow it down. Research will be needed to identify, understand and define, and then prevent the pathogenesis of AD in the aging population. The veritable cause of AD is not yet understood, although a one-molecule analysis based on pathology has come to dominate explanations for the damage that occurs to the brain. This chapter is a review of how research into one hypothesis, namely, the racemization and isomerization of aspartyl residues in amyloid peptides, is enabling researchers to identify ways to develop early and effective treatments for AD.

Amyloid Peptides from AD Pathology

Amyloid beta are short peptides that are the abnormal proteolytic byproducts of the transmembrane-type protein APP. This function is unclear but thought to be involved in neuronal development and signaling nexus that transfers neuronal information about a range of extracellular conditions, including neuronal damage, to induction of intracellular signaling events (van der Kant and Goldstein 2015). On the other hand, decomposed Aβ monomers are soluble peptides that undergo a dramatic conformational change to form a beta sheet-a rich tertiary structure that aggregates to form amyloid fibrils, known as a principal component of senile plaques in AD pathology. These fibrils deposit outside the neurons in dense formations known as senile plaques and sometimes in the walls of small cerebral blood vessels via a process called amyloid angiopathy (Sugiyama and Tanaka 2014, Canobbio et al. 2015). This senile plaque is a very important pathological finding in the postmortem specimen (Fig. 1). In recent discussions, this senile plaque, based on decomposed Aβ monomers, has no known cause, consequence or byproducts. However, we cannot negate the results from the pathological findings in the postmortem specimen. Thus, the molecular

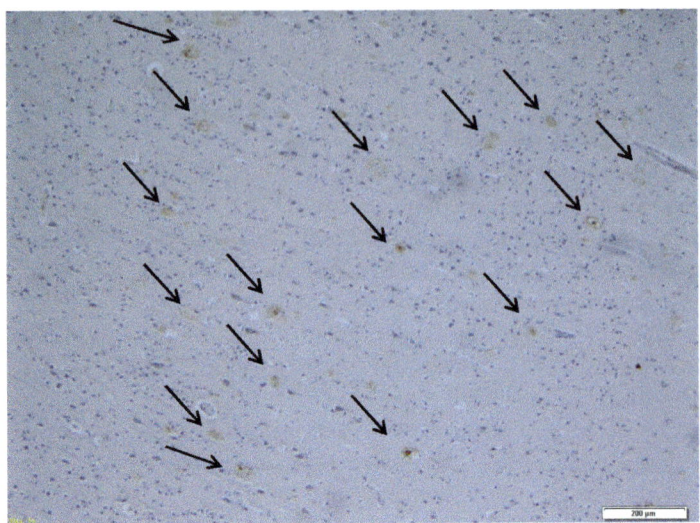

Figure 1. Senile plaque from pathological findings in the postmortem specimen. Reference: Inoue et al. 2006.

biology of Aβ peptides is needed using various investigations for the identification of modifications and structural variations in the brain tissues.

Post-translational modifications of Aβ peptides

Based on the traditional amyloid hypothesis, most AD cases are sporadic without changes in the production of the Aβ peptides. However, the propensity to form aggregates and toxic species in senile plaque from the pathology of AD may be driven by factors other than changes in the production of Aβ peptides from the APP function. Thus, several PTMs have been discovered in the increased aggregation rate of Aβ from the pathology of AD. Figure 2 shows some of these modifications such as oxidation, truncated reaction, nitration and racemization/isomerization that are obviously induced by the results of protein aging with physiological reactions.

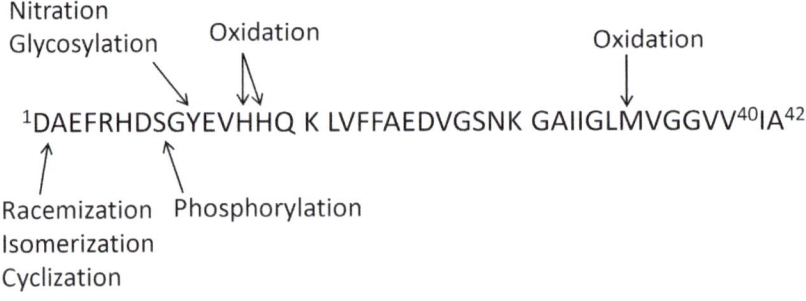

Figure 2. Post-translational modification of Aβ peptides such as oxidation, truncated reaction, nitration and racemization/isomerization.

During the oxidation of Aβ peptides by PTMs, the most prominent amino acid site of oxidative changes is methionine at position 35 (Met^{35}). The oxidation of Met^{35} in the Aβ peptides by MCO has been observed (Hong and Schöneich 2001). Thus, Met^{35} in Aβ peptides could potentially be a target of oxidation by ROS (Butterfield 2002, Butterfield and Kanski 2002). However, it has been suggested that oxidation of the histidine at position 13 (His^{13}) and position 14 (His^{14}) most likely precedes the oxidation of Met^{35} by ROS generated from ascorbic acid/copper ion (Schöneich and Williams 2002, Schöneich 2004). In addition, our study suggested that two tryptic peptides containing Met^{35} were obtained from the MCO reaction of the Aβ peptide. The peptides were identical to those obtained by tryptic digestion of the untreated Aβ peptide. Therefore, oxidation of the full length Aβ peptide oxidation proceeded in an analogous manner to truncated forms of the Aβ peptide that lacked Met^{35}, but contained His^{13} and His^{14}. These findings demonstrated that the MCO mediated ROS generation exclusively occurs at His^{13} and His^{14} and that Met^{35} oxidation does not occur in the Aβ peptide under these conditions (Inoue et al. 2006). Moreover, this MCO occurred such that the N-terminal Aβ peptides were truncated corresponding to a decrease in mass of 45 and 89 Da compared to the model peptide and cyclization by the ROS reaction (Fig. 3) (Inoue et al. 2009, Inoue et al. 2010, Lee et al. 2014).

For the other reactions of Aβ peptides by PTMs, nitration, phosphorylation and glycosylations were reported. Based on the theory of phosphorylation, Aβ peptides possess three potential reaction sites at the serine residues of position 8 (Ser^{8}) and position 26 (Ser^{26}), and tyrosine at position 10 (Tyr^{10}). In particular, for the phosphorylation of the Aβ peptides, the Ser^{8} site has been studied by phospho-serine-8-specific Aβ antibodies that revealed localized aggregation of the Aβ from the pathology of the brains in transgenic mice and human AD brains (Kumar et al. 2011). On the other hand, nitration was reported such that Aβ peptides can be nitrated *in vitro* by binding to heme in the presence of nitrite and hydrogen peroxide (Zhao et al. 2015). For the glycosylations from the PTMs of Aβ peptides, Zhu et al. reported *O*-GlcNAc in the Aβ peptides (Zhu et al. 2014). As just described, various PTMs of the Aβ peptides were discussed, however, no common perception of the essential foundation of the AD pathology, was determined.

Detections of Racemization and Isomerization of Amino Acids in Aβ Peptides

In 1988, Shapira et al. investigated the extent of amino acid racemization in Aβ peptides from the AD pathology (Shapira et al. 1988). This study suggested that the purified core Aβ was found to contain relatively large proportions of D-Asp (5%) and D-Ser (2%) (Shapira et al. 1988). In addition, Fisher et al. reported that the preliminary results showed that these neurofibrillary tangle preparations contained a greater number of these racemized D-Asp residues from normal brain tissues (Fisher et al. 1992). On the other hand, Mori et al. showed that the amino acid racemization of D-Asp (4.54%/7.31%, soluble/insoluble fractions), Glu (2.25%/2.15%), Ala (1.07%/1.52%) and Lys (4.76%/16.8) were detected in the supernatant fraction of the AD brain homogenate (Mori et al. 1994). Based on these reports regarding

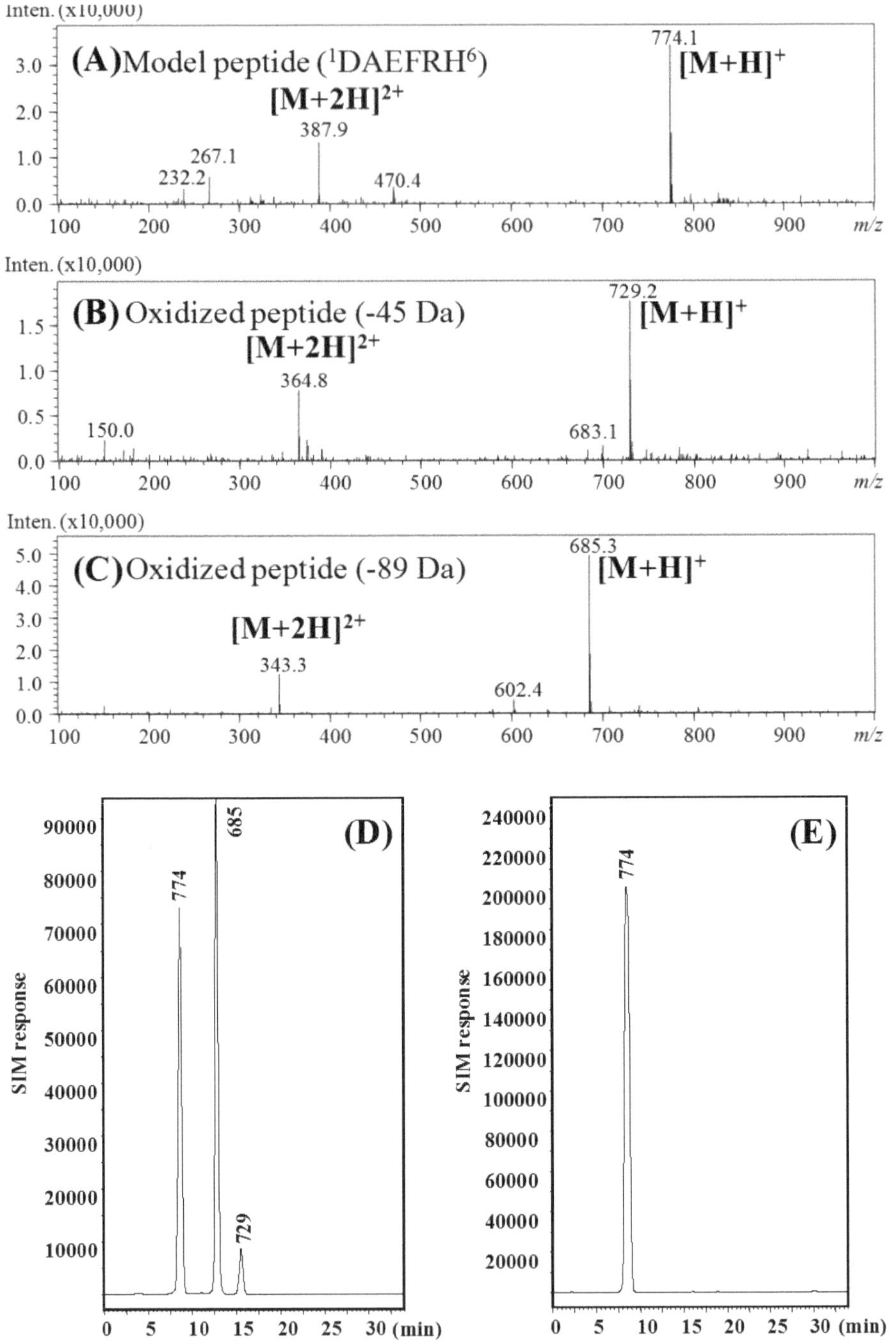

Figure 3. LC-MS analysis of N-terminal truncated modification of Aβ peptides. (A) MS spectrum of native Aβ$_{1-6}$ sequence. (B) MS spectrum of modified Aβ$_{1-6}$ sequence (–45 Da). (C) MS spectrum of modified Aβ$_{1-6}$ sequence (–89 Da). (D) LC-MS chromatogram of metal catalyzed modification of Aβ$_{1-6}$ sequence. (E) LC-MS chromatogram of H$_2$O$_2$ oxidation of Aβ$_{1-6}$ sequence. Reference: Inoue et al. 2009.

the amino acid racemization of Aβ peptides, the Edman degradation was applied using the enantiomeric separation of the produced amino acids by a LC technique (Iida et al. 1998). This technique makes it very difficult to suppress the amino acid residue racemization based on the sample preparation and experimental procedure. In addition, it is impossible to evaluate the isomerization and sequence of the native peptides in the biological samples. Correspondingly, a panel of antibodies was used that specifically discriminate the terminal structures and modifications at the amino and carboxyl termini of the Aβ peptides (Iwatsubo et al. 1996). This result showed that diffuse plaques found in the cerebral and cerebellar cortex, neostriatum, and hypothalamus of Down's syndrome, AD, and control brains were strongly immunoreactive for Aβ peptides with the N-terminal L-Asp, L-isomerized (iso) Asp, and D-Asp (Iwatsubo et al. 1996). Recently, the covalent CCD-UPLC-MS/MS technique was developed for the sensitive and selective analysis of N-terminal tryptic Aβ peptides in AD brain tissues (Fig. 4) (Inoue et al. 2014). In this result, the detection levels of the N-terminal Aβ sequence from AD patients with the predictable L-Asp of about $59.0 \pm 26.0\%$ based on the D/L-iso/D-iso Asp ratio ($4.8 \pm 5.7/25.4 \pm 15.0/10.8 \pm 11.2\%$, AD patients, n = 10) were observed (Inoue et al. 2014). The presence of racemized and/or isomerized Aβ peptides in AD brain tissues has long been described by various techniques. Thus, this existence in the AD pathology is believed to be certain. Although it is not clear whether these reactions of the Aβ peptides occurred before or after the progressive disease, these observations led us to the hypothesis that the AD racemization/isomerization proteins would be increased and the mechanism to repair the racemized/isomerized Aβ peptides may aggregate more quickly.

Reactive Property of N-terminal Asp Racemization and Isomerization in Aβ Peptides

The N-terminal structure of the Aβ peptides, extracted from senile plaques such neuritic deposits, showed the cleaved Asp of about 8%, the formation of the isoaspartate form iso-Asp of about 20%, pyroglutamate-3 (p3Glu), the cyclization of the N-terminal glutamyl residue of about 51% and the native form of only about 20% by amino acid analysis and MS analysis (Kuo et al. 1997). These pathways of the post-translational N-terminal modification of Aβ are characterized to represent the most frequent type of aging protein damage. These reactions proceed through the formation of a cyclic succinimide intermediate, which rapidly undergoes spontaneous chemical modifications, such as AAR and AAI, during the degradation of Asp included peptides (Fig. 5) (Shimizu et al. 2000). During the spontaneous chemical modification of the Aβ peptides, the functionality of the AAR and AAI of Asp is to act as a specific repair system based on the PIMT (Shimizu et al. 2005). However, the PIMT repair system is not completely efficient due to the fact that the D-Asp residue is not recognized and is not equally functional in all tissues (Lowenson and Clarke 1991, 1992). It has been reported that the modified Asp sequence dramatically increased with aging in the brain (David et al. 1998). The AAR and AAI represent the major non-enzymatic and chemical modifications affecting the Aβ folding and degradation in the pathology of aging dementia. Recently, an interesting study was reported that when iso-Asp is

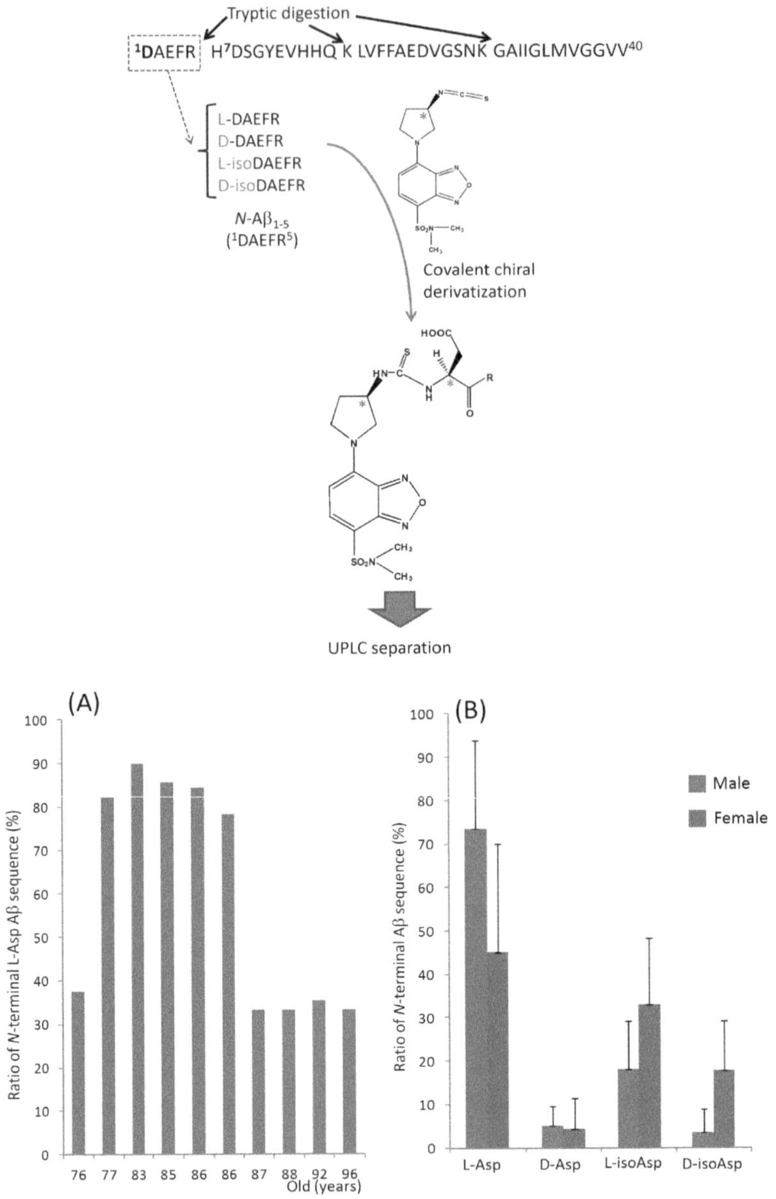

Figure 4. CCD-UPLC-MS/MS analysis of N-terminal tryptic Aβ peptides in AD brain tissues. (A) Pattern of the post-translational AAR and AAI formation of N-terminal Aβ sequence for AD brain (n = 10). (B) Pattern of gender difference of average AD patient (male, n = 5 and female, n = 5) for the decreased L-Asp. Reference: Inoue et al. 2014.

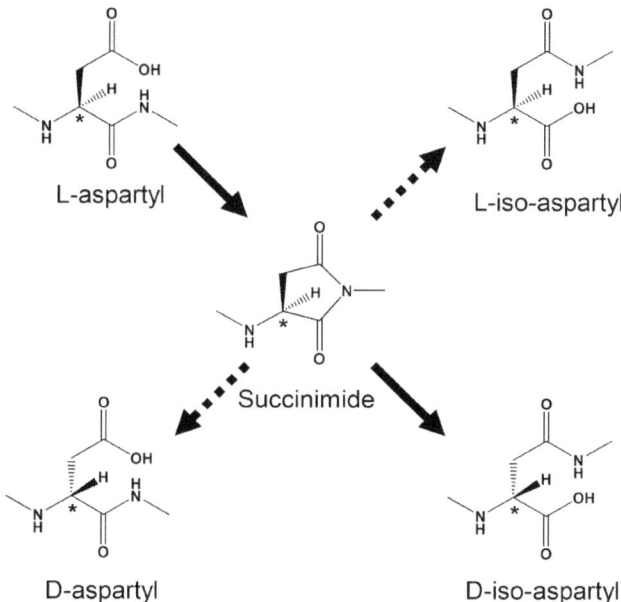

L-aspartyl

L-iso-aspartyl

Succinimide

D-aspartyl

D-iso-aspartyl

Figure 5. The formation of a cyclic succinimide intermediate that occurred by spontaneous chemical modifications such as AAR and AAI during degradation of Asp included peptides.

present at position 672 in the APP, cathepsin B can catalyze the cleavage between methionine (Met) at 671 and isoAsp at 672 with a high efficiency (Böhme et al. 2008). Since the spontaneous AAI cannot readily take place in the post-translational N-terminal modification, the iso-Asp formation in the native Aβ can only occur as an early event in the Aβ production before the BACE1. This means that the detection of AAR and AAI in the N-terminal sequence has two aspects used to evaluate the production and aggregation of the Aβ peptides. In any case, the consequences of these chemical modifications are site-specific and can affect the age-related accumulation according to the structural alterations produced by the site-specific incorporation of the modified N-terminal residues. There have been many studies of the AAR and AAI events of Aβ peptides *in vivo* and/or *vitro*, but its role in the natural Aβ peptides pathogenesis in the focal brain tissues is still unclear.

Isomerization and racemization of L-aspartate spontaneously occurs in proteins and proceeds through a common transient cyclic succinimidyl intermediate. Succinimide is immediately hydrolyzed to form two more stable compounds: L-isoaspartate and L-aspartate. Alternatively, L-succinimide racemizes to D-succinimide, which is rapidly hydrolyzed to D-isoaspartate and D-aspartate.

Diagnostic Markers of Asp Racemization and Isomerization for AD Pathology

Asp racemization/isomerization is a natural process that occurs under physiological conditions. Recent reports suggest that the water molecules or radical reaction with ions would effect the succinimide formation of the Asp residues (Takahashi et al. 2014, Tambo et al. 2013). Moreover, the technical practicability is comparable to that of age estimation based on AAR in dentin (Matzenauer et al. 2014). Based on the relationship between age and the racemization/isomerization of Asp, a variety of human tissues containing metabolically stable proteins were investigated (Zapico and Ubelaker 2013). If this reaction can take place in any metabolically-activated protein, these proteins would have altered conformations which produce changes in their biological activities and/or chemical properties. The physicochemical properties of the affected proteins may contribute to the progressive changes associated with the aging process related to the homeostatic mechanism. Thus, there is the possibility that some of this Asp racemization/isomerization in the Aβ peptides might serve as useful diagnostic markers for the AD progress. Actually, an interesting study suggested that antibodies recognized either oligomeric or post-translationally modified Aβ peptides in the plasma and CSF that might be a relevant target for passive or active immunotherapy (Britschgi et al. 2009). In addition, Aβ peptides can be easily excreted from the normal brain into the blood (Ueno et al. 2014). The research of the mechanisms that control the Aβ folding/aggregation and the balance between the Aβ production and clearance is a central step in the comprehension of the AD pathology. However, the mechanism for the onset of AD is unknown, thus to capture the conversion from normal aging to dementia and the most appropriate biomarkers for the AD progress will be a significant challenge in future studies. Although the role of the Asp racemization/isomerization of the Aβ peptides in AD pathology is still unclear, evidence has demonstrated that the Aβ peptides isolated from AD brain tissues contain high levels of Asp that have undergone racemization/isomerization. On the other hand, the accumulation of the Asp racemization/isomerization in stable proteins is linked to aging. Roher et al. suggested that these alterations may contribute to the production and stability of the Aβ peptide deposits in the AD brain of dementia patients (Roher et al. 1993). These data were considered indicative of the different times of formation of the two kinds of Aβ deposits: plaque deposition seems to be older and consequently an earlier pathological event that precedes the formation of vascular deposits in the AD pathology (Roher et al. 1993). This may be a preliminary screening of the Asp racemization/isomerization in biological samples of AD patients and these modifications can be considered as novel AD biomarkers between aging and pathology.

Keywords: Alzheimer's disease, amyloid-β, aspartic acid, racemization, isomerization, liquid chromatography mass spectrometry

References

Böhme, L., T. Hoffmann, S. Manhart, R.Wolf and H. U. Demuth. 2008. Isoaspartate-containing amyloid precursor protein-derived peptides alter efficacy and specificity of potential beta-secretases. Biol Chem 389: 1055–1066.
Britschgi, M., C. E. Olin, H. T. Johns, Y.Takeda-Uchimura, M. C. LeMieux, K. Rufibach et al. 2009. Neuroprotective natural antibodies to assemblies of amyloidogenic peptides decrease with normal aging and advancing Alzheimer's disease. Proc Natl Acad Sci USA 106: 12145–12150.
Butterfield, D. A. 2002. Amyloid beta-peptide (1-42)-induced oxidative stress and neurotoxicity: implications for neurodegeneration in Alzheimer's disease brain. A review. Free Radic Res 36: 1307–1313.

Butterfield, D. A. and J. Kanski. 2002. Methionine residue 35 is critical for the oxidative stress and neurotoxic properties of Alzheimer's amyloid beta-peptide 1-42. Peptides 23: 1299–1309.

Canobbio, I., A. A. Abubaker, C. Visconte, M. Torti and G. Pula. 2015. Role of amyloid peptides in vascular dysfunction and platelet dysregulation in Alzheimer's disease. Front Cell Neurosci 9: 65.

David, C. L., J. Orpiszewski, X. C. Zhu, K. J. Reissner and D. W. Aswad. 1998. Isoaspartate in chrondroitin sulfate proteoglycans of mammalian brain. J Biol Chem 273: 32063–32070.

Fisher, G. H., I. L. Payan, S. J. Chou, E. H. Man, S. Cerwinski, T. Martin et al. 1992. Racemized D-aspartate in Alzheimer neurofibrillary tangles. Brain Res Bull 28: 127–131.

Hong, J. and C. Schöneich. 2001. The metal-catalyzed oxidation of methionine in peptides by Fenton systems involves two consecutive one-electron oxidation processes. Free Radic Biol Med 31: 1432–1441.

Iida, T., H. Matsunaga, T. Santa, T. Fukushima, H. Homma and K. Imai. 1998. Amino acid sequence and D/L-configuration determination of peptides utilizing liberated N-terminus phenylthiohydantoin amino acids. J Chromatogr A 813: 267–275.

Inoue, K., C. Garner, B. L. Ackermann, T. Oe and I. A. Blair. 2006. Liquid chromatography/tandem mass spectrometry characterization of oxidized amyloid beta peptides as potential biomarkers of Alzheimer's disease. Rapid Commun Mass Spectrom 20: 911–918.

Inoue, K., A. Nakagawa, T. Hino and H. Oka. 2009. Screening assay for metal-catalyzed oxidation inhibitors using liquid chromatography-mass spectrometry with an N-terminal beta-amyloid peptide. Anal Chem 81: 1819–1825.

Inoue, K., M. Kaneko, T. Hino and H. Oka. 2010. Simple and novel screening assay of natural antioxidants for Cu(II) ion/adrenaline-mediated oxidation of N-terminal amyloid beta by liquid chromatography/ mass spectrometry. J Agric Food Chem 58: 9413–9417.

Inoue, K., H. Tsutsui, H. Akatsu, Y. Hashizume, N. Matsukawa, T. Yamamoto et al. 2013. Metabolic profiling of Alzheimer's disease brains. Sci Rep 3: 2364.

Inoue, K., D. Hosaka, N. Mochizuki, H. Akatsu, K. Tsutsumiuchi, Y. Hashizume et al. 2014. Simultaneous determination of post-translational racemization and isomerization of N-terminal amyloid-β in Alzheimer's brain tissues by covalent chiral derivatized ultraperformance liquid chromatography tandem mass spectrometry. Anal Chem 86: 797–804.

Iwatsubo, T., T. C. Saido, D. M. Mann, V. M. Lee and J. Q. Trojanowski. 1996. Full-length amyloid-beta (1-42(43)) and amino-terminally modified and truncated amyloid-beta 42(43) deposit in diffuse plaques. Am J Pathol 149: 1823–1830.

Kumar, S., N. Rezaei-Ghaleh, D. Terwel, D. R. Thal, M. Richard, M. Hoch et al. 2011. Extracellular phosphorylation of the amyloid β-peptide promotes formation of toxic aggregates during the pathogenesis of Alzheimer's disease. EMBO J 30: 2255–2265.

Kuo, Y. M., M. R. Emmerling, A. S. Woods, R. J. Cotter and A. E. Roher. 1997. Isolation, chemical characterization, and quantitation of A beta 3-pyroglutamyl peptide from neuritic plaques and vascular amyloid deposits. Biochem Biophys Res Commun 237: 188–191.

Lee, S. H., H. Kyung, R. Yokota, T. Goto and T. Oe. 2014. Hydroxyl radical-mediated novel modification of peptides: N-terminal cyclization through the formation of α-ketoamide. Chem Res Toxicol 28: 59–70.

Lowenson, J. D. and S. Clarke. 1991. Structural elements affecting the recognition of L-isoaspartyl residues by the L-isoaspartyl/D-aspartyl protein methyltransferase. Implications for the repair hypothesis. J Biol Chem 266: 19396–19406.

Lowenson, J. D. and S. Clarke. 1992. Recognition of D-aspartyl residues in polypeptides by the erythrocyte L-isoaspartyl/D-aspartyl protein methyltransferase. Implications for the repair hypothesis. J Biol Chem 267: 5985–5995.

Matzenauer, C., A. Reckert and S. Ritz-Timme. 2014. Estimation of age at death based on aspartic acid racemization in elastic cartilage of the epiglottis. Int J Legal Med 128: 995–1000.

Mori, H., K. Ishii, T. Tomiyama, Y. Furiya, N. Sahara, S. Asano et al. 1994. Racemization: its biological significance on neuropathogenesis of Alzheimer's disease. Tohoku J Exp Med 174: 251–262.

Roher, A. E., J. D. Lowenson, S. Clarke, A. S. Woods, R. J. Cotter, E. Gowing et al. 1993. beta-Amyloid-(1-42) is a major component of cerebrovascular amyloid deposits: implications for the pathology of Alzheimer disease. Proc Natl Acad Sci USA 90: 10836–10840.

Schöneich, C. and T. D. Williams. 2002. Cu(II)-catalyzed oxidation of beta-amyloid peptide targets His[13] and His[14] over His[6]: Detection of 2-Oxo-histidine by HPLC-MS/MS. Chem Res Toxicol 15: 717–722.

Schöneich, C. 2004. Selective Cu[2+]/ascorbate-dependent oxidation of Alzheimer's disease beta-amyloid peptides. Ann NY Acad Sci 1012: 164–170.

Shapira, R., G. E. Austin and S. S. Mirra. 1988. Neuritic plaque amyloid in Alzheimer's disease is highly racemized. J Neurochem 50: 69–74.

Shimizu, T., A. Watanabe, M. Ogawara, H. Mori and T. Shirasawa. 2000. Isoaspartate formation and neurodegeneration in Alzheimer's disease. Arch Biochem Biophys 381: 225–234.

Shimizu, T., Y. Matsuoka and T. Shirasawa. 2005. Biological significance of isoaspartate and its repair system. Biol Pharm Bull 28: 1590–1596.

Sugiyama, S. and M. Tanaka. 2014. Self-propagating amyloid as a critical regulator for diverse cellular functions. J Biochem 155: 345–351.

Takahashi, O., R. Kirikoshi and N. Manabe. 2014. Roles of intramolecular and intermolecular hydrogen bonding in a three-water-assisted mechanism of succinimide formation from aspartic acid residues. Molecules 19: 11440–11452.

Tambo, K., T. Yamaguchi, K. Kobayashi, E. Terauchi, I. Ichi and S. Kojo. 2013. Racemization of the aspartic acid residue of amyloid-β peptide by a radical reaction. Biosci Biotechnol Biochem 77: 416–418.

Ueno, M., Y. Chiba, K. Matsumoto, T. Nakagawa and H. Miyanaka. 2014. Clearance of beta-amyloid in the brain. Curr Med Chem 21: 4085–4090.

van der Kant, R. and L. S. Goldstein. 2015. Cellular functions of the amyloid precursor protein from development to dementia. Dev Cell 32: 502–515.

World Health Organization. WHO Mental Health Gap Action Programme (mhGAP) 2008. http://www.who.int/mental_health/mhgap/en/

Zapico, S. C. and D. H. Ubelaker. 2013. Applications of physiological bases of aging to forensic sciences. Estimation of age-at-death. Aging Res Rev 12: 605–617.

Zhao, J., P. Wang, H. Li and Z. Gao. 2015. Nitration of Y10 in Aβ1-40: is it a compensatory reaction against oxidative/nitrative stress and Aβ aggregation? Chem Res Toxicol 28: 401–407.

Zhu, Y., X. Shan, S. A. Yuzwa and D. J. Vocadlo. 2014. The emerging link between O-GlcNAc and Alzheimer disease. J Biol Chem 289: 34472–34481.

5

Aspartic Acid Racemization
Applications to Forensic and Archaeological Age Estimation

Rebecca C. Griffin[1],* and *Matthew Collins*[2]

INTRODUCTION

Determining the age-at-death of human remains provides essential information for both forensic and archaeological scientists. For juveniles, it is possible to obtain highly accurate age estimates based on the known patterns of development of the skeleton. However, for adult remains estimating age is problematic. The majority of existing methods rely heavily on degenerative changes that take place in the skeleton as part of the aging process. These changes do not occur at the same rate in all individuals, leading to very broad age estimates (Aykroyd et al. 1999, Wittwer-Backofen et al. 2008). There is, therefore, a real need for the development of new methods for estimating age that are able to provide more accurate age estimates needed by forensic and archaeological scientists.

Aspartic Acid Racemization (AAR) could provide a potential solution to this problem, as in long-lived proteins within the body, the extent of racemization increases with age. AAR occurs at a constant rate at any given temperature. Because human body temperature is relatively constant, racemization should therefore occur at a steady rate in the body during life and should also occur at a similar rate in different individuals. AAR, therefore, has the potential to provide more reliable age estimates than traditional age estimation techniques.

[1] University of Sydney, Discipline of Anatomy and Histology, Anderson Stuart Building (F13), University of Sydney, NSW, Australia, 2006.

[2] University of York, BioArch Environment Building, Wentworth Way, York, North Yorkshire. United Kingdom, YO10 5DD.
Email: matthew.collins@york.ac.uk

* Corresponding author: rebecca.griffin@sydney.edu.au

The application of AAR for forensic age estimation relies upon the fact that the temperature experienced by the body usually drops significantly post-mortem, with below ground temperatures in cool temperate climates (e.g., UK, South Island New Zealand) estimated to be approximately 10 degrees, 27°C below body temperature. This results in a 100 fold slower rate of racemization post-mortem. In warmer burial environments, the differences are smaller (20 fold at 20°C, threefold at 30°C), and therefore the resolution is worse. As a result, over short periods of burial, little racemization should occur post-mortem, so the relative proportions of D and L amino acids (D/L value) in long-lived proteins within the body can be used to estimate age directly. However, for remains from warmer climates, or with a longer post-mortem interval, significant amounts of racemization occur post-mortem and this must be taken into account when estimating age using this technique.

Methodologies for AAR Age Estimation

The application of racemization to forensic age estimation has largely focused on Asx, as this is the most rapidly racemizing amino acid. Its higher rate of racemization is a result of its ability to racemize in-chain. For most other residues, racemization rates are suppressed within the protein chain, but are much higher when (as a consequence of hydrolysis) the residue finds itself in an N-terminal position. However, there are a number of other amino acids that can also provide useful information, which can complement the results obtained using Asx. Like Asx, Ser racemizes rapidly as it can also isomerise in-chain (Demarchi et al. 2013) but shows a weaker correlation with age (Carolan et al. 1997, Griffin et al. 2008). Glx and Ala also show a relationship between D/L and age, but with a weaker correlation than for Asx or Ser (Ohtani et al. 2004). Given the number of amino acids that racemize at an appreciable rate, there is scope for developing new methods which utilize multiple amino acids simultaneously. However to do so, there will need to be better comprehension of the rates of amino acid racemization in different secondary structures. For example Collins et al. (1999) argue that the pattern of racemization observed in collagen is largely controlled by the tertiary structure of the collagen molecule. This avenue has yet to be fully explored, but may potentially increase the accuracy of the age estimates produced.

The majority of work on AAR age estimation has focussed on dental tissues, as these are formed during childhood and undergo minimal remodelling during life. Dentine is the most frequently used tissue for AAR analysis, as it is relatively protein-rich. The correlation of Asx D/L with age in dentine has been found to be higher than that of traditional age estimation methods in adults; $R^2 = 0.92$ in Ritz et al. (1990) and $R^2 = 99$ in Ohtani and Yamamoto (1992). Tooth enamel also shows a higher correlation between AAR and age; $R^2 = 0.85$ in Helfman and Bada (1975), $R^2 = 0.86$ in Ohtani and Yamamoto (1992) and $R^2 = 0.92$ in Griffin et al. (2008), though the relationship between Asx D/L and age is weaker than for dentine, possibly due to the difficulty in measuring the much lower amino acid concentrations in this tissue. Some limited work has also been conducted on AAR age estimation using (collagen-rich) tooth cementum (Ohtani et al. 1995), in which Asx D/L again shows a high correlation with age though the relationship is not as strong as in dentine ($R^2 = 0.98$). More recently,

Sakuma et al. (2012) have explored the potential for analysing whole teeth rather than individual tissues, in order to maximize the protein content while also making sampling more straightforward. The kinetics of racemization in such a situation will be much more complicated than in studies utilising a single tissue, given that each of the dental tissues racemizes at a different rate (Ohtani et al. 1995). These variations would need to be taken into account when using a whole tooth to estimate age at death in order to ensure the accuracy of the results.

While much work on AAR age estimation focuses on dental tissues, a number of studies have also explored the potential for using bone for AAR analysis (Ritz et al. 1994, Ohtani et al. 1998). Bone is often the best-preserved tissue in forensic contexts, making it an ideal target for analysis. However, Cunha et al. (2009) recommend against using bone for AAR age estimation, due to the potential for remodelling of the bone during life as explored by Hedges et al. (2007). This may be responsible for the weaker relationship between Asx D/L and age observed for bone ($R^2 = 0.85$ in Ohtani et al. (1998)). However, it has been claimed that it is possible to reduce this effect by purifying and analysing specific proteins within the bone such as osteocalcin that are less susceptible to replacement during remodelling (Ritz et al. 1996). In addition to dental tissues and bones, other tissues which can be used for AAR age estimation include intervertebral discs (Ritz and Schutz 1993), rib cartilage (Pfeiffer et al. 1995), the elastic cartilage of the epiglottis (Matzenauer et al. 2014), and elastin from arteries (Dobberstein et al. 2010). However, these tissues are not always well preserved over longer post-mortem intervals. It is therefore often preferable to use dental tissues such as dentine for AAR age estimation, although this frequently involves invasive sampling strategies, which may raise ethical issues when applied in forensic practice (Cunha et al. 2009).

For AAR age estimation, D/L values are usually determined using either GC/MS or HPLC. HPLC is increasingly the method of choice as it allows faster separation and measurement of multiple amino acids in the same run, is less expensive (Fu et al. 1995, Powell et al. 2013) and can detect amino acids at much lower concentrations (Benesova et al. 2004), a critical consideration when working with samples where the D/L value is likely to be quite low. However, the greater sensitivity of HPLC analysis renders the method more susceptible to inaccuracies due to contamination of samples prior to or during analysis. It is therefore important while using this technique to take precautions to minimize contamination of samples during the sample preparation process.

Factors influencing racemization rates

Although time and temperature are the main factors which determine the extent of racemization, there are a number of environmental factors that must be taken into account when using this approach to estimate the age of forensic remains. The burial environment of the deceased could potentially affect the rate of post-mortem racemization, particularly over longer periods of burial. For example, the presence of water or acidic or basic conditions (Masters 1984) or of copper or zinc ions (Gillard and O'Brien 1978) may affect the rate of racemization. This leads to a situation where the immediate micro-environment surrounding the body can have a significant impact upon the D/L values obtained from remains with a long post-mortem interval (Dobberstein et al. 2008). As the rate of racemization of an amino acid is affected by its position in the protein chain (Liardon and Friedman 1987), with N-terminal residues racemizing the most rapidly, diagenetic processes post-mortem such as microbial cleavage or degradation of the proteins analysed, can also have an impact upon the D/L values observed. However, the extent to which these factors can affect the results of AAR age estimation has not yet been quantified, and it is possible that they may have only a minor impact on the use of the AAR technique. The loss of amino acids to the burial environment by leaching can also have an impact on the D/L values observed if either L or D amino acids are preferentially lost, as observed by Masters (1986). The nature of the burial environment must therefore be taken into account when interpreting the results of AAR age estimation.

The treatment of the remains post-mortem must also be considered when determining whether or not to apply the AAR technique. For example, maceration is sometimes used by forensic anthropologists to remove any remaining flesh from the bones prior to analysis. This technique has been shown to impact the AAR results obtained (Offele et al. 2006), due to the high temperatures used, the altered pH conditions that result, and the potential for the process to increase the rate of protein degradation. Likewise, remains which have been burnt peri- or post-mortem are likely to have artificially elevated D/L values due to the high temperatures experienced, although there is some evidence to suggest that burning may not always prevent accurate age estimation using AAR (Ohtani et al. 1997). AAR age estimation should therefore be used with caution when there is evidence that the remains have been subjected to high temperatures post-mortem, and the forensic team should avoid exposing the remains to high temperatures after recovery if they wish to use this age estimation technique.

Variations have also been observed in the rate of racemization between different individuals. For example, Mornstad et al. (1994) have observed variations in the racemization rate that equate to a difference in the age estimate of up to 20 years in older individuals. Furthermore, variations have also been observed in the extent of racemization between different teeth from the same individual (Ohtani et al. 2005, Griffin et al. 2010). While some of this variation is due to differences in the timing of the formation of the various teeth of the dentition, even when this is taken into account a high level of variation appears to be present in the proteins of teeth from any given individual (Griffin et al. 2010). It is therefore important to sample multiple teeth from each individual in order to determine their age accurately. Sampling must also be consistent as the extent of racemization may vary within the tissue being studied (Ohtani and Yamamoto 2011).

Application of AAR to Age Estimation in Forensic Practice

Studies using extracted teeth from living individuals have shown that AAR can estimate age with a high level of accuracy (e.g., Ohtani and Yamamoto 1991, Helfman and Bada 1975, 1976, Griffin et al. 2008, Ritz et al. 1990, Carolan et al. 1997). The age estimates produced have a 95% confidence interval ranging from ± 4.5 years (Ohtani and Yamamoto 1991) to ± 11.4 years (Ritz et al. 1990) in dentine. Its application to forensic cases has confirmed its potential as a forensic age estimation method (Masters 1986, Ogino et al. 1985, Ohtani et al. 1988, Ohtani 1995). AAR has been found to be one of the most accurate age estimation techniques for cases involving adult remains which have not been burnt or heated (Ritz-Timme et al. 2000, Zapico and Ubelaker 2013), and it is currently considered more useful for forensic practice than other molecular techniques (Meissner and Ritz-Timme 2010). AAR also has the advantage that it can be applied to a range of different tissues, giving it greater flexibility than many other techniques (Zapico and Ubelaker 2013), a particular advantage in forensic practice, where remains are often incomplete.

While the cause of death has been shown to have limited impact upon the accuracy of the AAR technique (Ohtani et al. 1988), there are some circumstances in which AAR should only be used with caution. One such circumstance is the analysis of burnt remains, where the high temperature experienced by the body during burning is likely to influence the extent of racemization. Likewise, if the body was left exposed rather than buried post-mortem, the higher temperatures experienced by the body during the post-mortem interval may result in a higher rate of racemisation after death than for buried remains. If the body was buried in a location susceptible to groundwater leaching, this can also affect the D/L value, by removing soluble protein fragments from the body, which (on account of their higher relative concentration of N-terminal residues) may be most highly racemized (Masters 1986). This process can also be accelerated by diagenetic changes to the proteins during the burial period. The impact of these effects accumulates over time following death, and it has therefore been suggested that AAR may not be useful for age estimation over long post-mortem intervals (Dobberstein et al. 2008). However, limited work has been done to date to assess the impact of these various factors on the age estimates produced.

Application of AAR to Age Estimation in Archaeological Populations

In spite of this caveat, a number of studies have explored the application of AAR age estimation to archaeological populations. The principal challenge in adapting the method from forensics to archaeological practice lies in the much longer burial period of archaeological remains. Over longer periods of burial, racemization occurring post-mortem will make a substantial contribution to the overall D/L value of the individual. The rate of racemization is very sensitive to temperature (Shimoyama and Harada 1984), so the racemization rate of the remains post-mortem will also be affected by both the climate in which they are buried and the depth of the burial. Furthermore, studies of below ground temperature have indicated that this may be more variable than is often assumed (Wehmiller et al. 2000), so even within the same site, burials may be exposed to varying temperatures post-mortem. As with forensic remains, diagenetic changes to the proteins during the burial period may also affect the results of AAR analysis (Carolan et al. 1997, Ritz-Timme et al. 2000). Burial in an environment susceptible to leaching may have an impact upon the D/L value obtained (Masters 1986), while variations in the burial environment such as differing pH and soil hydrology may also influence the results. However, the extent to which these factors may affect the ability of this technique to estimate age has not yet been determined. The strong effect of temperature on racemization also means that the AAR age estimation method cannot be applied to heated or burned remains, so this approach cannot be applied to cremated burials. AAR age estimation has therefore to date largely been used on archaeological populations from cooler, temperate climates and cemeteries where the remains were interred without cremation.

The application of AAR age estimation to archaeological populations has had mixed results to date. Comparison of AAR age estimates with ages determined using morphological techniques has shown that there is a close relationship between the ages produced by these two approaches (Gillard et al. 1992, Griffin et al. 2009, Masters and Zimmerman 1978, Csapo et al. 2004, Masters 1984, Carolan et al. 1997). However, when the AAR method has been blind tested on remains with documented ages at death, the correlation between the known ages and the AAR age estimates was found to be poor (Gillard et al. 1992, Griffin et al. 2009, Carolan et al. 1997), although the estimates produced were no less accurate than those obtained using standard morphological techniques. This may be partly due to the higher proportion of older individuals in populations with documented ages at death, as these populations tend to be from more recent time periods, when life expectancy was higher. As any variations in the rate of racemization during life will be compounded over time, the errors due to these variations are likely to be much greater in older individuals than in younger individuals. However, in all three studies, the relationship of AAR with age was different to that observed in modern populations even for the younger individuals. It therefore seems likely that the inability of AAR to provide accurate age estimates in these populations is due to changes taking place in the burial environment, rather than variations in the rate of racemization during life.

The nature of these changes and their causes are yet to be determined conclusively. Both Carolan et al. (1997) and Dobberstein (2005) have suggested that the primary cause may be post-mortem degradation of collagen, the main protein within bone and dentine. However, recent work by Griffin et al. (2009) has shown that a similar pattern is observed in dental enamel, which contains no collagen. Griffin has suggested that the cause may instead be the loss of free amino acids into the burial environment over time, which will have an impact upon the D/L value of the sample. However, further research is needed to explore this hypothesis further, both in enamel and in dentine and bone.

Conclusion and Future Directions

AAR shows strong potential as a highly accurate age estimation technique for use in forensic practice. There is considerable evidence to indicate that it can provide more accurate age estimates than traditional methods based on morphological changes in the skeleton. However, to date there has been limited uptake of this approach into forensic practice. Among the reasons suggested for this are the scarcity of laboratories which offer this technique, the cost of the analysis and questions about its applicability to remains with longer post-mortem intervals (Cunha et al. 2009). AAR analysis may be more expensive than morphological examination of the skeleton, due to the costs of the chemicals and equipment for the analysis. However, these costs are not substantial, so an AAR analysis should not be significantly more expensive than a standard morphological analysis. The AAR technique will be particularly useful in forensic practice for cases where identification is problematic, and where a more precise and accurate age estimate may help in the identification process.

The shortage of laboratories that are able to conduct AAR is a key limitation on its adoption which clearly indicates that there is a real opportunity for further laboratories to take up this method and explore the questions which remain regarding its application to buried remains. Key issues which still need to be addressed include quantification of the impact of the burial environment, the post-mortem treatment of remains and the length of the post-mortem interval on the D/L values obtained. Furthermore, there is scope to explore the potential for using multiple amino acids to obtain more accurate results, as has already been achieved for amino acid racemization dating (e.g., Penkman et al. 2013). In this way, by developing a better understanding of the chemistry of AAR post-mortem, it will be possible to apply this age estimation technique to a wider range of forensic and archaeological contexts.

Keywords: Aspartic acid racemization, age estimation, forensic applications, archaeology, dentine, enamel, bone, burial environment, diagenesis, post mortem interval

References

Aykroyd, R. G., D. Lucy, A. M. Pollard and C. A. Roberts. 1999. Nasty, brutish, but not necessarily short: A reconsideration of the statistical methods used to calculate age at death from adult human skeletal and dental age indicators. Am Antiq 64(1): 55–70.

Benesova, T., A. Honzatko, A. Pilin, J. Votruba and M. Flieger. 2004. A modified HPLC method for the determination of aspartic acid racemization in collagen from human dentin and its comparison with GC. J Sep Sci 27(4): 330–334.

Carolan, V. A., M. L. G. Gardner, D. Lucy and A. M. Pollard. 1997. Some considerations regarding the use of amino acid racemization in human dentine as an indicator of age at death. J Forensic Sci 42: 10–16.

Collins, M. J., E. R. Waite and A. C. T. van Duin. 1999. Predicting protein decomposition: the case of aspartic-acid racemization kinetics. Phil Trans R Soc B 354(1379): 51–64.

Csapo, J., M. Collins, Z. Csapo-Kiss, E. Varga-Visi, G. Pohn and J. Csapo. 2004. Use of amino acids and amino acid racemization for age determination in archaeometry. pp. 65–78. In: G. Palyi, C. Zucchi and L. Caglioti (eds.). Progress in Biological Chirality. Amsterdam: Elsevier.

Cunha, E., E. Baccino, L. Martrille, F. Ramsthaler, J. Prieto, Y. Schuliar et al. 2009. The problem of aging human remains and living individuals: A review. Forensic Sci Int 193: 1–13.

Demarchi, B., M. Collins, E. Bergstrom, A. Dowle, K. Penkman, J. Thomas-Oates et al. 2013. New experimental evidence for in-chain amino acid racemization of serine in a model peptide. Anal Chem 85(12): 5835–5842.

Dobberstein, R. C. 2005. Ist eine Revision der Aminosäurendatierung möglich? Erlangung des Doktorgrades, der Matematisch-Naturwissenschaftlichen Fakultät, der Christian-Albrechts-Universität, Kiel.

Dobberstein, R. C., J. Huppertz, N. von Wurmb-Schwark and S. Ritz-Timme. 2008. Degradation of biomolecules in artificially and naturally aged teeth: Implications for age estimation based on aspartic acid racemization and DNA analysis. Forensic Sci Int 179: 181–191.

Dobberstein, R. C., S. -M. Tung and S. Ritz-Timme. 2010. Aspartic acid racemisation in purified elastin from arteries as basis for age estimation. Int J Legal Med 124: 269–275.

Fu, S. J., C. C. Fan, H. W. Song and F. Q. Wei. 1995. Age estimation using a modified HPLC determination of ratio of aspartic-acid in dentin. Forensic Sci Int 73(1): 35–40.

Gillard, R. D. and P. O'Brien. 1978. The isomers of alpha-amino acids with Copper (II). Part 4. Catalysis of the racemization of optically active alanine by copper (II) and pyruvate in alkaline solution. J Chem Soc, Dalton Trans 11: 1444–1447.

Gillard, R. D., A. M. Pollard, P. A. Sutton and D. K. Whittaker. 1992. An improved method for age at death determination from the measurement of D-aspartic acid in dental collagen. Archaeometry 32(1): 61–70.

Griffin, R.C., H. Moody, K. E. H. Penkman and M. J. Collins. 2008. The application of amino acid racemization in the acid soluble fraction of enamel to the estimation of the age of human teeth. Forensic Sci Int 175(1): 11–16.

Griffin, R. C., A. T. Chamberlain, G. Hotz, K. E. H. Penkman and M. J. Collins. 2009. Age estimation of archaeological remains using amino acid racemization in dental enamel: A comparison of morphological, biochemical and known ages-at-death. Am J Phys Anthropol 140(2): 244–252.

Griffin, R.C., K. E. H. Penkman, H. Moody and M. J. Collins. 2010. The impact of random natural variability on aspartic acid racemization ratios in enamel from different types of human teeth. Forensic Sci Int 200: 148–152.

Hedges, R. E., J. G. Clement, C. D. Thomas and T. C. O'Connell. 2007. Collagen turnover in the adult femoral mid-shaft: modeled from anthropogenic radiocarbon tracer measurements. Am J Phys Anthropol 133(2): 808–816.

Helfman, P. M. and J. L. Bada. 1975. Aspartic acid racemization in tooth enamel from living humans. Proc Natl Acad Sci USA 72: 2891–2894.

Helfman, P. M. and J. L. Bada. 1976. Aspartic acid racemization in dentine as a measure of ageing. Nature 262: 279–281.

Liardon, R. and M. Friedman. 1987. Effect of peptide-bond cleavage on the racemization of amino- acid-residues in proteins. J Agric Food Chem 35(5): 661–667.

Masters, P. M. and M. R. Zimmerman. 1978. Age determination of an Alaskan mummy: morphological and biochemical correlation. Science 201: 811–812.

Masters, P. M. 1984. Stereochemical age-determinations for the Barrow Eskimo remains. Arctic Anthropol 21(1): 77–82.

Masters, P. M. 1986. Age at death determinations for autopsied remains based on aspartic-acid racemization in tooth dentin—importance of postmortem conditions. Forensic Sci Int 32(3): 179–184.

Matzenauer, C., A. Reckert and S. Ritz-Timme. 2014. Estimation of age at death based on aspartic acid racemization in elastic cartilage of the epiglottis. Int J Legal Med 128: 995–1000.

Meissner, C. and S. Ritz-Timme. 2010. Molecular pathology and age estimation. Forensic Sci Int 203: 34–43.

Mornstad, H., H. Pfeiffer and A. Teivens. 1994. Estimation of dental age using HPLC-technique to determine the degree of aspartic-acid racemization. J Forensic Sci 39(6): 1425–1431.

Offele, D., M. Harbeck, R. C. Dobberstein, N. von Wurmb-Schwark and S. Ritz-Timme. 2006. Soft tissue removal by maceration and feeding of Dermestes sp.: impact on morphological and biomolecular analyses of dental tissues in forensic medicine. Int J Legal Med 121: 341–348.

Ogino, T., H. Ogino and B. Nagy. 1985. Application of aspartic-acid racemization to forensic odontology—postmortem designation of age at death. Forensic Sci Int 29(3-4): 259–267.

Ohtani, S., S. Kato, H. Sugeno, H. Sugimoto, T. Marunco, M. Yamazaki et al. 1988. A study on the use of the amino-acid racemization method to estimate the ages of unidentified cadavers from their teeth. Bull Kanagawa Dent Coll 16(1): 11–21.

Ohtani, S. and K. Yamamoto. 1991. Age estimation using the racemization of amino-acid in human dentin. J Forensic Sci 36(3): 792–800.

Ohtani, S. and K. Yamamoto. 1992. Estimation of age from a tooth by means of racemization of an amino-acid, especially aspartic-acid—comparison of enamel and dentin. J Forensic Sci 37(4): 1061–1067.

Ohtani, S. 1995. Estimation of age from the teeth of unidentified corpses using the amino-acid racemization method with reference to actual cases. Am J Forensic Med Pathol 16(3): 238–242.

Ohtani, S., H. Sugimoto, H. Sugeno, S. Yamamoto and K. Yamamoto. 1995. Racemization of aspartic-acid in human cementum with age. Arch Oral Biol 40(2): 91–95.

Ohtani, S., Y. Yamada and I. Yamamoto. 1997. Age estimation from racemization rate using heated teeth. J Forensic Odontostomatol 15: 9–12.

Ohtani, S., Y. Matsushima, Y. Kobayashi and K. Kishi. 1998. Evaluation of aspartic acid racemization ratios in the human femur for age estimation. J Forensic Sci 43(5): 949–953.

Ohtani, S., Y. Yamada, T. Yamamoto, S. Arany, K. Gonmori and N. Yoshioka. 2004. Comparison of age estimated from degree of racemization of aspartic acid, glutamic acid and alanine in the femur. J Forensic Sci 49(3): 441–445.

Ohtani, S., R. Ito, S. Arany and T. Yamamoto. 2005. Racemization in enamel among different types of teeth from the same individual. Int J Legal Med 119(2): 66–69.

Ohtani, S. and T. Yamamoto. 2011. Comparison of age estimation in Japanese and Scandinavian teeth using amino acid racemization. J Forensic Sci 56(1): 244–247.

Penkman, K. E. H., R. C. Preece, D. R. Bridgland, D. H. Keen, T. Meijer, S. A. Parfitt et al. 2013. An aminostratigraphy for the British Quaternary based on Bithynia opercula. Quat Sci Rev 61: 111–134.

Pfeiffer, H., H. Mornstad and A. Teivens. 1995. Estimation of chronological age using the aspartic-acid racemization method. 1. On human rib cartilage. Int J Legal Med 108(1): 19–23.

Powell, J., M. J. Collins, J. Cussens, N. McLeod and K. E. H. Penkman. 2013. Results from an amino acid racemization inter-laboratory proficiency study: design and performance evaluation. Quat Geochronol 16: 183–197.

Ritz-Timme, S., C. Cattaneo, M. J. Collins, E. R. Waite, H. W. Schutz, H. J. Kaatsch et al. 2000. Age estimation: The state of the art in relation to the specific demands of forensic practise. Int J Legal Med 113(3): 129–136.

Ritz, S., H. W. Schutz and B. Schwarzer. 1990. The extent of aspartic-acid racemization in dentin—a possible method for a more accurate determination of age at death. Z Rechtsmed 103(6): 457–462.

Ritz, S. and H. W. Schutz. 1993. Aspartic-acid racemization in intervertebral disks as an aid to postmortem estimation of age at death. J Forensic Sci 38(3): 633–640.

Ritz, S., A. Turzynski and H. W. Schutz. 1994. Estimation of age at death based on aspartic acid racemization in non-collagenous bone proteins. Forensic Sci Int 69: 149–159.

Ritz, S., A. Turzynski, H. Schutz, A. Hollmann and G. Rochholz. 1996. Identification of osteocalcin as a permanent aging constituent of the bone matrix: Basis for an accurate age at death determination. Forensic Sci Int 77(1-2): 13–26.

Sakuma, A., S. Ohtani, H. Saitoh and H. Iwase. 2012. Comparative analysis of aspartic acid racemization methods using whole-tooth and dentin samples. Forensic Sci Int 223: 198–201.

Shimoyama, A. and K. Harada. 1984. An age-determination of an ancient burial mound man by apparent racemization reaction of aspartic-acid in tooth dentin. Chem Lett 10: 1661–1664.

Wehmiller, J. F., H. A. Stecher, L. L. York and I. Friedman. 2000. The thermal environment of fossils: effective ground temperatures at aminostratigraphic sites on the U.S. Atlantic Coastal Plain. pp. 219–250. *In*: G. A. Goodfriend, M. J. Collins, M. L. Fogel, S. A. Macko and J. F. Wehmiller (eds.). Perspectives in Amino Acid and Protein Geochemistry. Oxford: Oxford University Press.

Wittwer-Backofen, U., J. Buckberry, A. Czarnetzki, S. Doppler, G. Grupe, G. Hotz et al. 2008. Basics in paleodemography: A comparison of age indicators applied to the early medieval skeletal sample of Laucheim. Am J Phys Anthropol 137(4): 384–396.

Zapico, S. C. and D. H. Ubelaker. 2013. Applications of physiological bases of ageing to forensic sciences. Estimation of age-at-death. Ageing Res Rev 12: 605–617.

Section II
Advanced Glycation
Endproducts

6

Advanced Glycation Endproducts
An Introduction

Andreas Simm* and Alexander Navarrete Santos[a]

INTRODUCTION

Advanced Glycation Endproducts (AGEs) are the result of the Maillard reaction. AGEs arise from the non-enzymatic reaction of reactive carbohydrates (sugars) with DNA, proteins or lipids. There is some confusion in the terminology of sugar modified proteins. This was nicely summarized by Rabbini and Thornalley 2012 (Rabbani and Thornalley 2012): In 1985, the Nomenclature Committee of International Union of Biochemistry and the International Union of Pure and Applied Chemistry recommended the term glycation for all reactions that link a sugar to a protein or a peptide, whether or not catalyzed by an enzyme. In 1993, a discrimination between enzymatic modification (glycosylation) and non-enzymatic modification (glycation) of proteins with sugars was proposed (Lis and Sharon 1993). Thereafter, the mix-up of the nomenclature still persisted, so for a long time many names existed for non-enzymatic modified proteins, such as glycated, glycosylated, or non-enzymatic glycosylated and so on. Nowadays, "*glycated protein*" is the accepted terminology for the products of the non-enzymatic reaction of proteins with sugars.

History

Since the detection of fire, food was heated, leading to browning and making it easily digestible nutrition with better taste and flavor. While the effect of heating on taste and color was known, the chemistry behind it was unclear for a long time. It was

Martin-Luther-University Halle-Wittenberg, Department of Cardiac Surgery, Ernst-Grube Str. 40, Halle (Saale), Saxony-Anhalt Germany, D-06120.
[a] Email: alexander.navarrete@uk-halle.de
* Corresponding author: andreas.simm@uk-halle.de

the French chemist Louis-Camille Maillard, who first described in 1912, a reaction between glucose and glycine during a heating process (Maillard 1912). This non-enzymatic browning reaction (using temperatures around 150°C) was later named as the Maillard reaction. But it was only more than 40 years later, in 1953, that the chemist, John Edward Hodge, published the mechanism of the Maillard reaction (Hodge 1953). During the 1920s, the Italian chemist, Mario Amadori, concentrated on the condensation reaction between carbohydrates like glucose and amines like p-anisidine. This rearrangement after building of the first reaction product within the Maillard reaction, the Schiff-base, leading to a more stable product is now known as the Amadori rearrangement. The following reactions are more complex ending in the so called AGEs. Beside the "classical reaction" of a sugar with the amino group of amino acids, it is now well known that small oxidative degradation products of glucose, reactive α-oxoaldehydes like GO and MGO, are important inducers of AGEs. As early as 1913, the enzymatic system which converts methylglyoxal to lactate (the glyoxalase system) was discovered as methylglyoxalase (shortly glyoxalase) (Neuberg 1913). MGO and the regulating glyoxalase system was thought to regulate cell division and the reaction of MGO with proteins to form a Schiff-base ("the protein transfer of one of its electrons onto an acceptor of small size and incorporate this acceptor") was called doping of proteins (Szent-Gyorgyi 1977). The first *in vivo* glycated protein identified was hemoglobin. Samuel Rahbar screened for hemoglobin variants and found in 1968, a minor "abnormally fast-moving hemoglobin band" in diabetic patients, the subfraction HbA1c. This is now the basis of the procedure for evaluating long-term control of diabetes (Rahbar 2005). Thereafter, glycation was detected in a variety of human tissues and organs and discussed to play a major pathophysiological role during induction of degenerative diseases.

Beside glycation *in vivo* and their possible pathophysiological role, there were parallel investigations to understand glycation in the context of nutrition (Fig. 1). Depending on the conditions like sugar content, time and temperature, heating of food can induce AGE formation, resulting in low to very high yields of AGEs. A diversity of products are formed in the Maillard reaction, which seem to be important for color and flavor. Therefore, this reaction is important for the food industry. Interestingly, this was known for a long time, for example, it was of interest that the Maillard reaction could be the cause of the browning that occurs during the manufacture and storage of foods (e.g., review by Stadtman ER, Non-enzymatic browning in fruit products, Advances in Food Research 1(1948) 325–372).

The Maillard Reaction: A Short Overview

Within the classical Maillard reaction, a reducing sugar or aldehyde, like glucose, reacts with a free amino group, like the side chain of a lysine or arginine residue, of a protein. The resulting Schiff's base typically undergoes a rearrangement to form a more stable Amadori product. Further rearrangements, oxidations and eliminations are needed to finally form the members of the highly heterogenous group of AGEs. AGEs are stable at physiological pH, and their rate of accumulation in tissues depends especially on the turn-over rate of the affected proteins (Stitt 2010). Therefore,

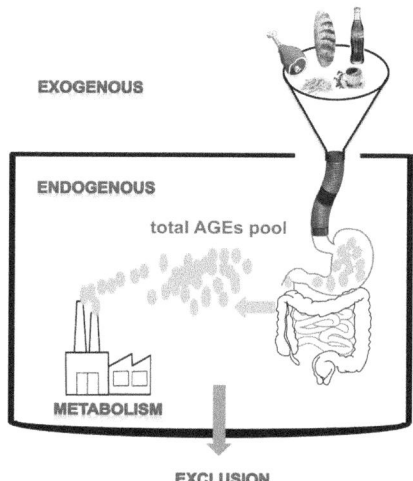

Figure 1A. AGEs not only reach the body with the food intake (exogenously) but are also produced as a result of the metabolism (endogenously). The exogenous and endogenous AGEs build the total AGEs pool in the body. Most of the AGEs are then excluded from the body.

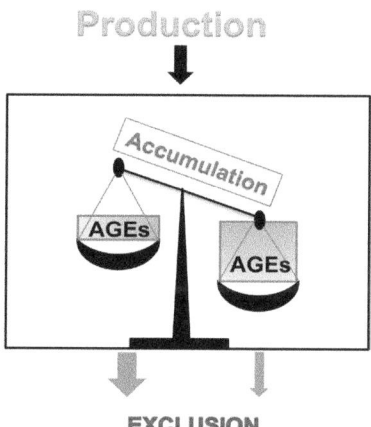

Figure 1B. When there is a misbalance between the production and exclusion (e.g., diseases), AGEs accumulate.

especially extracellular matrix proteins like elastin and collagen are modified due to their normally long half-life *in vivo*.

Sugars differ in their ability to react with amino groups. Generally, smaller sugar molecules (oxoaldehydes) are more reactive than sugars with a higher number of carbon atoms (Bunn and Higgins 1981). Reactivity also depends on the proportion of the sugars in the more reactive open chain conformations (Franks 1987). For example, fructose, which is to a higher degree in the open form than glucose, is about 10-fold more reactive in terms of glycation of proteins than glucose. In line, fructose-supplemented growth of *Saccharomyces cerevisiae* cells as compared to growth on glucose resulted in a more pronounced age-related decline in yeast reproductive ability and higher cell mortality. This was mainly explained by the higher reactivity of fructose in glycation and ROS generation *in vivo* (Semchyshyn et al. 2011). D-ribose participates in protein glycation both inside (endogenous glycation) and outside the body (exogenous glycation). Syrovy studied glycation of albumin using four different reducing sugars. Results showed that D-ribose was more active in glycation than the other sugars used (Syrovy 1994). The glycating ability of reducing sugars seems to increase as follows: D-glucose < D-mannose < D-galactose < D-xylose < D-fructose < D-arabinose < D-ribose < 2-deoxy-D-ribose (Monnier 1990).

Important acceptors of the sugars are amino acids like lysine or arginine or amino groups on nucleobases and lipids. Besides monosaccharides, small α-oxocarbonylic compounds (especially dicarbonyls) contribute significantly to AGE formation as they are far more reactive than sugars (Lo et al. 1994, Thornalley 2007). High levels of these reactive compounds such as 3-deoxyglucosone, glyoxal or methylglyoxal induces the so-called "carbonyl stress". Different reactions lead to the formation of dicarbonyls, from a byproduct of the glycolysis (methylglyoxal) to a product of lipid peroxidation (glyoxal). Depending on the amino acid sequence of the proteins, specific amino acids can be modified by sugars, resulting in some kind of unexpected specificity (Munch et al. 1999).

Biological Action of Glycated Proteins

AGEs can have impact on the function of biological systems by several means. Modifications on proteins can clearly alter structure, enzymatic activity and biological half-life (Bulteau et al. 2001, Friguet et al. 1994). If DNA is modified, the consequence can be mutations, and if membrane lipids are hit, this might affect transport and signalling processes. Last, but not the least, AGEs can act through specific receptor molecules.

For several enzymes it has been shown that the presence of an AGE-modification alters, if not destroys their activity. Glycation of the Na,K-ATPase *in vitro* in the presence of ATP, results in a shift in the steady state kinetics of ATP hydrolysis, whereas in the absence of ATP, the enzyme activity is irreversibly inhibited (Garner et al. 1990).

Extracellular proteins are well known targets for AGE-modifications. Proteins such as collagen have a relatively long biological half-life and are, especially in the case of diabetes, directly exposed to high levels of glucose outside the cell. Interestingly, modified collagen becomes more resistant to degradation by MMPs which causes accumulation of AGE-modified collagens in the ECM (Bartling et al. 2007). The occurrence of AGE cross links such as GOLD or MOLD result in stiffening of the ECM, which often compromises organ function and is associated with several chronic diseases such as diabetes, vascular diseases, retinopathy, arthritis and also Alzheimer's syndrome.

Although glycation seems to be a rather unspecific reaction, it seems that some proteins are prone to becoming the major modified protein in a cell. It is not clear whether this is simply a consequence of exposure of reactive residues on the surface or in the catalytic site of the proteins, or whether it reflects physiological functions such as a protective mechanism. Examples of predominantly modified proteins include several HSPs. In case of HSP-27, a higher anti-apoptotic effect became evident after methylglyoxal modification. In stem cells, the constitutive heat shock protein HSP-70 protein was described as by far the most modified protein (Hernebring et al. 2006). In yeast cells, only a few proteins were found to be modified by methylglyoxal (Gomes et al. 2006, Gomes et al. 2005). These proteins include enolase, phosphoglyceratemutase and aldolase, all involved in glycolysis. Additionally, very similar to mammalians, three heat shock proteins, involved in protein salvage were glycated (HSP 71/72, HSP26). Although enolase was inhibited by methylglyoxal modification, the glycolytic flux was not affected (Gomes et al. 2006) showing that yeast glycolytic machinery is resistant to glycation that results from the high glycolytic activity in these cells.

According to an early hypothesis, the formation of AGE-modifications might have regulatory functions as it reflects physiological states, such as glycolytic activity of cells. This is clearly in contrast to the modern view that focuses on the association of AGE-formation with pathophysiological processes. However, it could be shown that high glycolytic flow caused increased modification of the transcriptional corepressor mSIN3a by methylglyoxal, which resulted in coupling of glycolytic activity to changes in gene expression, namely angiopoetin-2 transcription (Yao et al. 2006, 2007a, Yao et al. 2007b).

The finding that embryonic stem cells contain large amounts of glycated proteins that are rapidly eliminated when differentiation processes occur (Hernebring et al. 2006) showed that glycation per se is not generally deleterious to the cells and may even be involved in maintaining the undifferentiated state.

Cells possess specific binding and receptor molecules for AGEs (Ott et al. 2014). The major and best known receptor for AGEs is the "Receptor for AGEs", RAGE or synonymous AGER. But other binding proteins have also been described. These are oligosaccharyltransferase (OST48, AGE-R1), 80K-H (AGE-R2) (Li et al. 1996), galectin-3 (AGE-R3) (Vlassara et al. 1995), CD36 (Kuniyasu et al. 2003, Ohgami et al. 2001, Ohgami et al. 2002), and scavenger receptors IIa and –b (Takata et al. 1988, 1989). RAGE is a member of the immunoglobulin receptor family and binds several ligands such as AGEs, HMGB-1, S100 proteins or amyloid beta peptide. Binding of agonists like the AGEs to RAGE results in activation of NADPH-oxidases and other less well described pathways that lead to increased production of ROS. The major downstream target of RAGE is the proinflammatory NFκB-pathway, which in turn leads to elevated RAGE expression and perpetuation of the cellular inflammatory state (Bierhaus et al. 2006). This inflammatory state is characterized by the production of inflammatory cytokines such as IL-6 and TNFα.

AGEs in Aging and Diseases

The presence of AGEs is correlated with age and several important diseases. Only few studies are enumerated as examples. As diabetes is often accompanied with hyperglycemia and oxidative stress, an accelerated rate of AGE-formation is observed. This glycation reaction contributes to morbidity of diabetes, end stage renal and heart diseases, arthritis and cataracts. Immunolocalization of carboxymethyllysin (an AGE compound) in fetal, juvenile, and adult normal tissues (skin, lung, heart, kidney, intestine, intervertebral disc, artery) revealed a clearly age-dependent staining pattern. Only tissues from adults showed significant positive staining and the process was accelerated by diabetes (Ceol et al. 1996). High levels of glycated compounds were found in cerebral, cardiac and renal tissues of patients dead from coronary heart disease, complicated and not complicated from diabetes mellitus and essential hypertension compared to victims dead from traumas. Furthermore, in experimental studies on rabbits and rats, it was shown that the content of glycoconjugates in tissues did not increase in short-term (exogenous or stress-associated) hyperglycemia, while accumulation of glycoconjugates in pericapillary, pericytic, and cellular structures in long-term hyperglycemia caused deformation of tissue ultrastructure (Shkolovoj and Shkolovoj 2011). In a recently published study, Sakul et al. showed that AGEs were significantly elevated in brain, ventricle and kidney, but not in lens and liver of aged rats when compared with young rats and diabetes propagated aging-induced increase in AGEs in brain, ventricle and kidney, and raised significantly lens and liver AGEs levels in aged rats (Sakul et al. 2013).

AGEs are also involved in the pathophysiology of neurodegenerative diseases such as Alzheimer's disease, Parkinson's disease and ALS. The connection between the accumulation of AGEs and neurodegenerative disease, such as cognitive impairment is actively investigated. Different studies demonstrated that AGEs are directly neurotoxic to cultured neurons and methylglyoxal and glyoxal as AGEs inducers increase the aggregation and cytotoxicity of intracellular amyloid-beta carboxy-terminal fragments.

Diabetic nephropathy is a major cause of end-stage renal disease. Pathophysiologically, it is characterized by thickened glomerular basement membrane and increased mesangium (Dronavalli et al. 2008). This is accompanied by increased AGE deposition in these tissues. In contrast, RAGE-deficient animals showed amelioration of diabetic nephropathy (Myint et al. 2006).

The eye is influenced by diabetes at several levels. Firstly, the predominant protein of the lens, crystallines can be modified by AGE-moieties resulting in lens aging and cataract formation (Bhattacharyya et al. 2007). Retinopathy is another form of microvascular disease that correlates with the deposition of AGE-modified proteins in microvasculatory wall (Barile and Schmidt 2007, Murata et al. 1997).

Diabetes is also associated with a significant acceleration of atherosclerosis (Mazzone 2007, Mazzone et al. 2008). That this is at least partially due to glycation was already demonstrated by an early study by Brownlee (Brownlee et al. 1986). It was shown that AGE modification of the crosslinked collagen fraction from vessel grafts correlated with vessel stiffness and function (Hofmann et al. 2013).

Activation of RAGE results in the activation of the transcription factor NFkB, which is a major regulator of inflammation and the immune response. Inflammation is seen to contribute to insulin resistance in diabetes type 2. TNF-α impairs insulin action in many model systems and corresponding to this it was shown that TNF-α knock out mice have improved insulin sensitivity (Wellen and Hotamisligil 2005). Given the fact, that AGE-formation and RAGE activation is elevated in diabetes, a vicious circle can be postulated, that would lead to further increased insulin resistance.

Many studies have shown a link between diabetes and cancer. In these studies, associations between diabetes and for example non-Hodgkin's lymphoma (Chao and Page 2008), liver (Chen et al. 2008) and colorectal (Hart et al. 2008) cancers have been observed. In contrast, for other cancers like lung cancer, no effect or even a reduction of risk was observed (Bartling et al. 2011).

Conclusion

Pathophysiological studies indicated evidence for a link between accumulation of AGEs and degenerative diseases like heart failure. The data suggests that glycation of proteins and binding of such modified proteins to the respective receptors are related to the development and progression of say atherosclerosis and diastolic heart failure. Targeting AGEs therapeutically will hopefully represent a new treatment strategy for patients with degenerative diseases.

Keywords: Advanced glycation endproducts, maillard reaction, nutrition, receptor for AGEs, aging, degenerative diseases

References

Barile, G. R. and A. M. Schmidt. 2007. RAGE and its ligands in retinal disease. Curr Mol Med 7: 758–765.

Bartling, B., M. Desole, R. E. Silber and A. Simm. 2007. Dicarbonyl-mediated protein modifications affect matrix metalloproteinase (MMP) activity. Z Gerontol Geriatr 40: 357–361.

Bartling, B., H. S. Hofmann, A. Sohst, Y. Hatzky, V. Somoza, R. E. Silber et al. 2011. Prognostic potential and tumor growth-inhibiting effect of plasma advanced glycation end products in non-small cell lung carcinoma. Mol Med 17: 980–989.

Bhattacharyya, J., E. V. Shipova, P. Santhoshkumar, K. K. Sharma and B. J. Ortwerth. 2007. Effect of a single AGE modification on the structure and chaperone activity of human alphaB-crystallin. Biochemistry 46: 14682–14692.

Bierhaus, A., P. M. Humpert and P. P. Nawroth. 2006. Linking stress to inflammation. Anesthesiol Clin 24: 325–340.

Brownlee, M., H. Vlassara, A. Kooney, P. Ulrich and A. Cerami. 1986. Aminoguanidine prevents diabetes-induced arterial wall protein cross-linking. Science 232: 1629–1632.

Bulteau, A. L., P. Verbeke, I. Petropoulos, A. F. Chaffotte and B. Friguet. 2001. Proteasome inhibition in glyoxal-treated fibroblasts and resistance of glycated glucose-6-phosphate dehydrogenase to 20 S proteasome degradation *in vitro*. J Biol Chem 276: 45662–45668.

Bunn, H. F. and P. J. Higgins. 1981. Reaction of monosaccharides with proteins: possible evolutionary significance. Science 213: 222–224.

Ceol, M., A. Nerlich, B. Baggio, F. Anglani, U. Sauer, E. Schleicher et al. 1996. Increased glomerular alpha 1 (IV) collagen expression and deposition in long-term diabetic rats is prevented by chronic glycosaminoglycan treatment. Lab Invest 74: 484–495.

Chao, C. and J. H. Page. 2008. Type 2 diabetes mellitus and risk of non-Hodgkin lymphoma: a systematic review and meta-analysis. Am J Epidemiol 168: 471–480.

Chen, C. L., H. I. Yang, W. S. Yang, C. J. Liu, P. J. Chen, S. L. You et al. 2008. Metabolic factors and risk of hepatocellular carcinoma by chronic hepatitis B/C infection: a follow-up study in Taiwan. Gastroenterology 135: 111–121.

Dronavalli, S., I. Duka and G. L. Bakris. 2008. The pathogenesis of diabetic nephropathy. Nat Clin Pract Endocrinol Metab 4: 444–452.

Franks, F. 1987. Physical chemistry of small carbohydrates—equilibrium solution properties. Pure and Applied Chemistry 59: 1189–1202.

Friguet, B., E. R. Stadtman and L. I. Szweda. 1994. Modification of glucose-6-phosphate dehydrogenase by 4-hydroxy-2-nonenal. Formation of cross-linked protein that inhibits the multicatalytic protease. J Biol Chem 269: 21639–21643.

Garner, M. H., A. Bahador and G. Sachs. 1990. Nonenzymatic glycation of Na,K-ATPase. Effects on ATP hydrolysis and K+ occlusion. J Biol Chem 265: 15058–15066.

Gomes, R. A., M. Sousa Silva, H. Vicente Miranda, A. E. Ferreira, C. A. Cordeiro and A. P. Freire. 2005. Protein glycation in Saccharomyces cerevisiae. Argpyrimidine formation and methylglyoxal catabolism. Febs J 272: 4521–4531.

Gomes, R. A., H. V. Miranda, M. S. Silva, G. Graca, A. V. Coelho, A. E. Ferreira et al. 2006. Yeast protein glycation *in vivo* by methylglyoxal. Molecular modification of glycolytic enzymes and heat shock proteins. Febs J 273: 5273–5287.

Hart, A. R., H. Kennedy and I. Harvey. 2008. Pancreatic cancer: a review of the evidence on causation. Clin Gastroenterol Hepatol 6: 275–282.

Hernebring, M., G. Brolen, H. Aguilaniu, H. Semb and T. Nystrom. 2006. Elimination of damaged proteins during differentiation of embryonic stem cells. Proc Natl Acad Sci U S A 103: 7700–7705.

Hodge, J. E. 1953. Dehydrated foods, chemistry of browning reactions in model systems. J Agric Food Chem 1: 928–943.

Hofmann, B., A. C. Adam, K. Jacobs, M. Riemer, C. Erbs, H. Bushnaq et al. 2013. Advanced glycation end product associated skin autofluorescence: a mirror of vascular function? Exp Gerontol 48: 38–44.

Kuniyasu, A., N. Ohgami, S. Hayashi, A. Miyazaki, S. Horiuchi and H. Nakayama. 2003. CD36-mediated endocytic uptake of advanced glycation end products (AGE) in mouse 3T3-L1 and human subcutaneous adipocytes. FEBS Lett 537: 85–90.

Li, Y. M., T. Mitsuhashi, D. Wojciechowicz, N. Shimizu, J. Li, A. Stitt et al. 1996. Molecular identity and cellular distribution of advanced glycation endproduct receptors: relationship of p60 to OST-48 and p90 to 80K-H membrane proteins. Proc Natl Acad Sci U S A 93: 11047–11052.

Lis, H. and N. Sharon. 1993. Protein glycosylation. Structural and functional aspects. Eur J Biochem 218: 1–27.

Lo, T. W., M. E. Westwood, A. C. McLellan, T. Selwood and P. J. Thornalley. 1994. Binding and modification of proteins by methylglyoxal under physiological conditions. A kinetic and mechanistic study with N alpha-acetylarginine, N alpha-acetylcysteine, and N alpha-acetyllysine, and bovine serum albumin. J Biol Chem 269: 32299–32305.

Maillard, L. C. 1912. Action des acides amines sur les sucres: formation des melanoidines par voie methodique. CRAcadSciSer 2 154: 66–68.

Mazzone, T. 2007. Adipose tissue and the vessel wall. Curr Drug Targets 8: 1190–1195.

Mazzone, T., A. Chait and J. Plutzky. 2008. Cardiovascular disease risk in type 2 diabetes mellitus: insights from mechanistic studies. Lancet 371: 1800–1809.

Monnier, V. M. 1990. Nonenzymatic glycosylation, the Maillard reaction and the aging process. J Gerontol 45: B105–111.

Munch, G., D. Schicktanz, A. Behme, M. Gerlach, P. Riederer, D. Palm et al. 1999. Amino acid specificity of glycation and protein-AGE crosslinking reactivities determined with a dipeptide SPOT library. Nat Biotechnol 17: 1006–1010.

Murata, T., R. Nagai, T. Ishibashi, H. Inomuta, K. Ikeda and S. Horiuchi. 1997. The relationship between accumulation of advanced glycation end products and expression of vascular endothelial growth factor in human diabetic retinas. Diabetologia 40: 764–769.

Myint, K. M., Y. Yamamoto, T. Doi, I. Kato, A. Harashima, H. Yonekura et al. 2006. RAGE control of diabetic nephropathy in a mouse model: effects of RAGE gene disruption and administration of low-molecular weight heparin. Diabetes 55: 2510–2522.

Neuberg, C. 1913. The destruction of lactic aldehyde and methylglyoxal by animal organs. Biochem Z 49: 502–506.

Ohgami, N., R. Nagai, M. Ikemoto, H. Arai, A. Kuniyasu, S. Horiuchi et al. 2001. CD36, a member of class B scavenger receptor family, is a receptor for advanced glycation end products. Ann N Y Acad Sci 947: 350–355.

Ohgami, N., R. Nagai, M. Ikemoto, H. Arai, A. Miyazaki, H. Hakamata et al. 2002. CD36, serves as a receptor for advanced glycation endproducts (AGE). J Diabetes Complications 16: 56–59.

Ott, C., K. Jacobs, E. Haucke, A. Navarrete Santos, T. Grune and A. Simm. 2014. Role of advanced glycation end products in cellular signaling. Redox Biol 2: 411–429.

Rabbani, N. and P. J. Thornalley. 2012. Glycation research in amino acids: a place to call home. Amino Acids 42: 1087–1096.

Rahbar, S. 2005. The discovery of glycated hemoglobin: a major event in the study of nonenzymatic chemistry in biological systems. Ann N Y Acad Sci 1043: 9–19.

Sakul, A., A. Cumaoglu, E. Aydin, N. Ari, N. Dilsiz and C. Karasu. 2013. Age- and diabetes-induced regulation of oxidative protein modification in rat brain and peripheral tissues: consequences of treatment with antioxidant pyridoindole. Exp Gerontol 48: 476–484.

Semchyshyn, H. M., L. M. Lozinska, J. Miedzobrodzki and V. I. Lushchak. 2011. Fructose and glucose differentially affect aging and carbonyl/oxidative stress parameters in Saccharomyces cerevisiae cells. Carbohydr Res 346: 933–938.

Shkolovoj, V. V. and S. V. Shkolovoj. 2011. Excessive glycation of pericapillary pericytic matrix components is an essential mechanism of arterial hypertension development in hyperglycemia. Bull Exp Biol Med 150: 572–575.

Stitt, A. W. 2010. AGEs and diabetic retinopathy. Invest Ophthalmol Vis Sci 51: 4867–4874.

Syrovy, I. 1994. Glycation of albumin: reaction with glucose, fructose, galactose, ribose or glyceraldehyde measured using four methods. J Biochem Biophys Methods 28: 115–121.

Szent-Gyorgyi, A. 1977. The living state and cancer. Proc Natl Acad Sci U S A 74: 2844–2847.

Takata, K., S. Horiuchi, N. Araki, M. Shiga, M. Saitoh and Y. Morino. 1988. Endocytic uptake of nonenzymatically glycosylated proteins is mediated by a scavenger receptor for aldehyde-modified proteins. J Biol Chem 263: 14819–14825.

Takata, K., S. Horiuchi, N. Araki, M. Shiga, M. Saitoh and Y. Morino. 1989. Scavenger receptor of human monocytic leukemia cell line (THP-1) and murine macrophages for nonenzymatically glycosylated proteins. Biochim Biophys Acta 986: 18–26.

Thornalley, P. J. 2007. Endogenous alpha-oxoaldehydes and formation of protein and nucleotide advanced glycation endproducts in tissue damage. Novartis Found Symp 285: 229–243; discussion 243-226.

Vlassara, H., Y. M. Li, F. Imani, D. Wojciechowicz, Z. Yang, F. T. Liu et al. 1995. Identification of galectin-3 as a high-affinity binding protein for advanced glycation end products (AGE): a new member of the AGE-receptor complex. Mol Med 1: 634–646.

Wellen, K. E. and G. S. Hotamisligil. 2005. Inflammation, stress, and diabetes. J Clin Invest 115: 1111–1119.

Yao, D., T. Taguchi, T. Matsumura, R. Pestell, D. Edelstein, I. Giardino et al. 2006. Methylglyoxal modification of mSin3A links glycolysis to angiopoietin-2 transcription. Cell 124: 275–286.

Yao, D., T. Taguchi, T. Matsumura, R. Pestell, D. Edelstein, I. Giardino et al. 2007a. Methylglyoxal modification of mSin3A links glycolysis to angiopoietin-2 transcription. Cell 128: 625.

Yao, D., T. Taguchi, T. Matsumura, R. Pestell, D. Edelstein, I. Giardino et al. 2007b. High glucose increases angiopoietin-2 transcription in microvascular endothelial cells through methylglyoxal modification of mSin3A. J Biol Chem 282: 31038–31045.

7

Advanced Glycation and Aging

Alejandro Gugliucci

INTRODUCTION

Aging may be defined as the time-dependent functional decline that affects most living organisms. Almost four decades ago, Monnier and Cerami offered a Maillard (the browning or Maillard reaction is discussed in the Introduction) theory of aging (Bjorksten 1977, Monnier et al. 1981, Monnier and Cerami 1983, Kohn et al. 1984, Monnier et al. 1986, Monnier et al. 1988, Hayase et al. 1989, Monnier 1989)—elaborating on an even earlier cross-linking theory of aging by Bjorksten (Bjorksten 1968, 1977), positing that chronic formation and accumulation of AGE was a determining factor of the rate of aging. Initial evidence for this hypothesis was the measurable increase in browning and fluorescence of collagens and lens crystallins with age, the age-dependent accumulation of AGE in them as well as the age-dependent increase in cross-linking and insolubility of these proteins (Monnier et al. 1981, Monnier and Cerami 1983, Kohn et al. 1984, Monnier et al. 1988). As research on the subject and analyses methodologies were developed, it soon became apparent that a causative role for glycation in aging was difficult to justify. Indeed, AGE are present in merely trace concentrations in proteins, a quantitative argument that they have a substantial effect on protein structure and function and on the viability of mammalian species is hard to maintain (Lee and Cerami 1992, Vlassara 1996, Baynes 2001, 2002). Moreover, AGE concentrations are lower in collagens from short-lived versus long-lived animals, when they are at the same fraction of their maximum life span (Baynes 2002). This suggests that the accumulation of AGE is not an important determinant of the rate of aging. In this regard, the role of AGE in the aging process may be considered as correlative rather than causative (Baynes 2001, 2002).

Professor and Associate Dean, Touro University College of Osteopathic Medicine 1310 Club Drive, Vallejo, CA 94592, USA.
Email: alejandro.gugliucci@tu.edu

However, after the discovery of RAGE and other AGE receptors, a negative impact in the acceleration of the aging process heavily determined by other factors may be envisaged (Ramasamy et al. 2005, Fleming et al. 2011, Falcone et al. 2013). Moreover, even when glycation of proteins surely plays a correlative role, much remains to be investigated on glycation of DNA and RNA.

A series of important questions have emerged concerning the physiological sources of aging-causing damage, the compensatory responses to restore homeostasis, the interconnection between the diverse types of damages and compensatory responses, as well as the potential of exogenous interventions to delay aging.

Recently Lopez-Otin et al. proposed nine candidate hallmarks (Lopez-Otin et al. 2013) that contribute to the aging process and together determine the aging phenotype:

1. Genomic instability
2. Telomere attrition
3. Epigenetic alterations
4. Loss of proteostasis
5. De-regulated nutrient-sensing
6. Mitochondrial dysfunction
7. Cellular senescence
8. Stem cell exhaustion
9. Altered intercellular communication

Many excellent reviews on the role of AGE in aging exist and the reader is referred to them. None has summarized the causative or correlative evidence of the role of AGE in aging in light of these hallmarks and this is precisely the purpose of this present review.

1. Genomic instability

The integrity of DNA is endlessly confronted by exogenous agents and by endogenous aggressions such as DNA replication errors, hydrolytic reactions and reactive oxygen species attack (Baynes 2002). The genetic lesions include point mutations, translocations, chromosomal gains and losses, telomere shortening and gene disruption caused by the integration of viruses or transposons (Lopez-Otin et al. 2013). Causal evidence linking the lifelong increase in genomic damage and aging came from studies in mice and humans indicating that altered DNA repair mechanisms cause augmented aging in mice and trigger numerous human progeroid syndromes (Monnier et al. 1991, Fukada et al. 2014). Overall there is broad evidence that genomic damage occurs in aging and that its artificial induction can aggravate accelerated aging. Can glycation of genetic material participate in aging? As discussed elsewhere in this book, glycation reactions of proteins are the attack of electrophiles on the nucleophilic groups in proteins and also in DNA (Singh et al. 2001, Ulrich and Cerami 2001, Suji and Sivakami 2004, Nass et al. 2007). Even when nuclear DNA is shielded by histones, spermine and spermidine as quenchers of electrophiles (Gugliucci 2005, Gugliucci and Menini 2003), there is evidence that some electrophiles hit their target (Gugliucci 1994, Gugliucci and Menini 2003), not only carcinogenic alkylating agents but also carbonyls. We have shown both *in vitro* and *in vivo* that glycation intermediates have access to the nucleosome and histones are glycated (Gugliucci 1994, Gugliucci and Bendayan 1995, Gugliucci and Menini 2003, Gugliucci 2005, Gugliucci et al. 2009a). AGE or ALE have also been detected in mitochondria (Baynes 2001, 2002). Mitochondrial DNA, which is devoid of a histone shield, may be even more vulnerable to glycation damage, consistent with the tenets of mitochondrial theories of aging (Kowald 2001, Ziegler et al. 2015). There is evidence that the AGE/ALE precursor, MGO, can form adducts with DNA during glycation of DNA by glucose or MG as well as in cells exposed to MG at nontoxic levels (Ahmad et al. 2014, Rabbani et al. 2014, Waris et al. 2015). These stresses usually produce single-strand breaks in DNA. In turn, this produces increase in the synthesis and turnover of poly-ADP-ribose and in the intracellular pool of ADP-ribose monomer (Ahmad et al. 2014, Rabbani et al. 2014, Waris et al. 2015). ADP-ribose is a potent protein glycating agent and could produce AGE in DNA as a reaction to chemical and physical stresses. However, *in vivo* studies in this area are fraught with problems because damage to DNA does not accumulate, as it is repaired by very efficient but not infallible mechanisms (Baynes 2002). There are no published reports on the detection of glycation-modified nucleobases in tissue or body fluids in aging but this is an emerging area under scrutiny for diabetes (Waris et al. 2015, Rabbani et al. 2014). Glycation reactions would lead to a "silent" but gradual loss in the integrity of the genome, resulting from accumulated mutations due to minor errors in repair mechanisms when glycated bases are eliminated (Baynes 2002, 2001). These would be expressed as altered sequences, rather than altered composition of DNA, that would prevent the detection of the original causative damage. According to this hypothesis, chronic changes to the structure of DNA produces cell loss by apoptosis and necrosis in aging and chronic disease and may have a role in the increasing rate of cancer in the elderly as well. Supporting a putative role of the glycation reaction in aging, especially at the DNA level (most damaged proteins are just easily turned over and damage does not accumulate), long-lived species have evolved several mechanisms for protection of DNA against the glycation reaction: histone and polyamine shields and its localization in the nucleus. Glucose is indeed the least reactive hexose (Higgins and Bunn 1981, Bunn and Higgins 1981). For over 3 decades, DNA has been a less attractive subject for study than proteins because glycation adducts do not increase in DNA with age or in disease. However, with the availability of new, more sensitive methodology, chemical proof for the presence of modified nucleotides formed during the glycation reaction is becoming available (Waris et al. 2015, Syslova et al. 2014, Radjei et al. 2014, Rabbani et al. 2014). Studies on the glycation reaction as effector of DNA damage and the effects of AGE or ALE inhibitors on the modification of DNA may herald important data for evaluating the role of the Maillard reaction in aging.

2. Telomere attrition

A corollary of the previous hallmark is the specific damage to DNA at special areas in the chromosomes with substandard capacity for repair. Indeed, replicative DNA polymerases cannot replicate completely the terminal ends of linear DNA. This is achieved by the specialized DNA polymerase known as telomerase. Nonetheless, most mammalian somatic cells do not express telomerase which harbors cumulative loss of telomere-protective sequences from chromosome ends. Telomere collapse illuminates the imperfect proliferative capacity of some types of *in vitro* cultured cells, the classic so-called senescence or Hayflick limit. Significantly, telomere shortening is also detected during normal aging both in human and mice (Jones 2015, Holliday 2014, Lopez-Otin et al. 2013). In humans, recent meta-analyses have shown a robust relation between short telomeres and mortality risk, particularly at younger ages. Normal aging is accompanied by telomere wear and tear in mammals (Jones 2015, Holliday 2014, Lopez-Otin et al. 2013). Furthermore, pathological telomere dysfunction hastens aging in mice and humans, whereas experimental stimulation of telomerase can slow aging in mice, therefore satisfying all the criteria for a hallmark of aging (Lopez-Otin et al. 2013). Other than the putative damage to telomere DNA by the mechanisms proposed above, there is no evidence so far of a specific involvement of Maillard reaction damage to telomeres.

3. Epigenetic alterations

There are several lines of evidence indicating that aging is accompanied by epigenetic changes, and that epigenetic changes can provoke progeroid syndromes. Additionally, sirtuin 6 is an epigenetically relevant enzyme whose loss-of-function reduces longevity and whose gain-of-function extends longevity in mice. It appears that understanding and manipulating the epigenome may result in improvement of age-related pathologies and extending healthy lifespan. Epigenetic changes include modifications in DNA methylation patterns, post-translational modification of histones and chromatin remodeling. The evidence of the role of glycation in epigenetics is indirect and based on cell culture or animal studies. Indeed, histones are target for glycation based on their chemistry, they are basic proteins very rich in lysine and arginine. Amadori glycation of individual histones was demonstrated very early *in vitro* (Lakatos and Jobst 1991, De Bellis and Horowitz 1987). Thereafter, the author's group was the first to show that intact nucleosomes were easily glycated *in vitro* and accumulated AGE (Gugliucci 1994). Others showed ADP ribosylation of histones (Cervantes-Laurean et al. 1993) and we later on demonstrated dramatically increased AGE on histones from diabetic rats (Gugliucci and Bendayan 1995). Histones from the liver of diabetic rats showed AGE levels three-fold higher than those of their age-matched controls. Histone AGE increased with the duration of diabetes and tended to increase with the age as well, even when this was a short-term study (Gugliucci and Bendayan 1995). Histones are then an easy target for glycation as they protect DNA from it. The author's group suggested a possible role for intracellular glycation in the increased theratogeny associated with diabetes mellitus due to this epigenetic change. Regulatory survival information may be encoded epigenetically by levels of histone lysine acetylation/acylation and lysine is a key target for glycation. Is that epigenetic change repaired? Does it interfere with the critical regulation pathways enunciated above? Is the continuous onslaught of electrophils on histones a player in aging through disrupted epigenetic changes? We believe these questions deserve exploration in future studies.

4. Loss of proteostasis

Proteostasis (homeostasis of the proteome) encompasses mechanisms for the stabilization of suitably folded proteins, namely the heat-shock family of proteins, together with mechanisms for the degradation of proteins: proteasome (ubiquitin pathway) or the lysosome (Labbadia and Morimoto 2015). In view of that, various studies have demonstrated that proteostasis is altered with aging (Labbadia and Morimoto 2015, 2014, Jensen and Jasper 2014). Several animal models support a causative impact of chaperone decline on longevity. Overall, there is indication that aging is concomitant with dysfunctional proteostasis and experimental disruption of proteostasis can hasten age-associated pathologies (Vilchez et al. 2014, Labbadia and Morimoto 2014, Jensen and Jasper 2014). AGE may play a role in perturbing the proteostasis: under chronic glycative stress and/or aging these abilities are inadequate to maintain the proteome and are actually diminished in absolute terms. Dicarbonyls or other sugar metabolites promote protein modification, aggregation and crosslinking (Gomes et al. 2006). Accordingly, degrons (amino acid residues that mark substrates for degradation) may be blocked (Uchiki et al. 2012). Glycated ubiquitin may be incorporated into conjugates, rendering them less susceptible to degradation (Takizawa et al. 1993). Although there is no human correlate to these results, a role for AGE in the aging process via impairment of proteostasis is worthy of further analysis (Uchiki et al. 2012, Queisser et al. 2010).

5. Deregulated nutrient-sensing

A deregulated nutrient-sensing is a key hallmark of aging. Strong support for this contention is provided by the fact that dietary restriction increases lifespan or healthspan in all eukaryote species that have been examined, from unicellular to non-human primates (Lopez-Otin et al. 2013). Overall, available evidence strongly supports the contention that anabolic signaling accelerates aging and decreased nutrient signaling extends longevity (Michan 2014, Kitada et al. 2013, McCarter et al. 2007). The somatotrophic axis in mammals includes GH and its mediator, the IGF-1. IGF-1 and insulin signaling may be referred to as the pathway (Michan 2014, McCarty 2014, Parr 1997). The IIS pathway is the most conserved aging-controlling pathway in evolution. Numerous genetic manipulations that decrease signaling intensity of the IIS pathway reliably extend the lifespan of worms, flies and mice (Michan 2014, McCarty 2014). Conversely, GH and IGF-1 levels decline during normal aging (Lopez-Otin et al. 2013). Accordingly, a reduced IIS is a shared feature of both physiological and accelerated aging, while a constitutively diminished IIS extends longevity. There is a network of other pathways that function in nutrient sensing. One of them is AMPK, which monitors low energy levels by detecting buildup of AMP. AMPK activation has manifold effects on metabolism. AMPK activation may facilitate lifespan-extension by metformin in worms (Burkewitz et al. 2014). Is there a link between AGE and aging that implicates AMPK? The exquisite allosteric sensing of AMP is achieved by a domain with three arginine residues, which makes it very vulnerable to glycation, especially by MG (Chung et al. 2015). MG accumulates in hyperglycemia, insulin resistance, diabetes and when there is excess flux of reactive oxygen species coming from the mitochondria and as shown elsewhere, may be linked to aging. We have hypothesized that excess MG in the above-mentioned conditions blocks the sensing of AMP by AMPK, thereby favoring hallmarks of insulin resistance, diabetes and aging (Gugliucci 2009a). Metformin quenches MG very effectively (Chung et al. 2015). Hence, our hypothesis may explain, for instance, part of the action of metformin, which is a potent anti-glycation agent (Beisswenger and Ruggiero-Lopez 2003, Ruggiero-Lopez et al. 1999). Although not proven experimentally, publications appeared since converge to indirectly support such a mechanism (Chung et al. 2015, Burkewitz et al. 2014). Unraveling its complete potential will necessitate discriminating targeting toward substrates involved in longevity assurance and not others that may be deleterious.

6. Mitochondrial dysfunction

Mitochondrial function has a deep influence on the aging process. Mitochondrial dysfunction can hasten aging in mammals (Ziegler et al. 2015, Liu et al. 2014, Jensen and Jasper 2014, Kowald 2001). Suggestive evidence indicates that improving mitochondrial function may extend lifespan in mammals (Ziegler et al. 2015). The 50- year-old free radical theory of aging, contested in recent years, posits that the increasing mitochondrial dysfunction that ensues with aging augments ROS, which then provokes additional mitochondrial decline and cellular damage (Harman 2003, 1965, 1960). In the last eight years, a deep re-analyses of the free radical theory of aging has occurred since contrary results have emerged for the case of age-related macular degeneration (Liu et al. 2014, Dai et al. 2014, Barja 2014). Increased ROS may prolong lifespan in yeast and C. elegans. Data from mice have shown that increased mitochondrial ROS and oxidative damage do not accelerate aging, intensification of antioxidant defenses do not extend longevity and weakening mitochondrial function (without changing ROS) accelerates aging (Lapointe and Hekimi 2010, Afanas'ev 2010, Perez et al. 2009). Conversely ROS trigger proliferative and survival signals. A synthesis of both lines of evidence would indicate that ROS are a stress-elicited survival signal aimed at compensating for the progressive decline associated with aging. While deep mitochondrial dysfunction is deleterious, minor respiratory insufficiencies may increase lifespan, possibly due to the hormetic response (Lapointe and Hekimi 2010, Afanas'ev 2010, Perez et al. 2009). The link between glycation and mitochondrial dysfunction is complex. On the one hand, according to Brownlee, excessive ROS-induced DNA damage activates PARP and modifies GAPDH leading to upstream accumulation of AGE-generating molecules, protein kinase C activation, hexosamine pathway and other links to diabetic complications (Brownlee 2005, Nishikawa et al. 2000). Some of these are also present during aging. If this is proven to be true, these mechanisms should go beyond the beneficial hormetic responses described above. On the other hand, according to what was discussed above under genomic instability glycation should produce more profound damage in mitochondrial DNA, which has an inefficient repair mechanism. Experimental evidence for this contention, however, is still lacking.

7. Cellular senescence

Cellular senescence is the stable arrest of the cell cycle coupled to stereotyped phenotypic changes (Holliday 2014, Lopez-Otin et al. 2013). Initially described by Hayflick in cells serially passaged in culture, we now know it is caused by telomere shortening. There are other aging-associated stimuli that activate senescence independently of this telomeric route. Cellular senescence may be a helpful compensatory answer to injury that comes to be harmful and hastens aging when tissues lose their renewing capacity (Holliday 2014, Lopez-Otin et al. 2013). Glycation may contribute to senescence by acting on DNA as proposed previously in the section of genomic instability although no direct evidence has been found.

8. Stem cell exhaustion

Stem cell exhaustion is a final common integrative result of various types of aging-associated damages and probably represents one of the crucial effectors of tissue and whole body aging (Fukada et al. 2014, Lopez-Otin et al. 2013). Recent encouraging studies submit that stem cell rejuvenation could reverse the aging phenotype. The deterioration in the regenerative capability of tissues is one of the most noticeable features of aging. Telomere shortening is also a significant cause of stem cell waning with aging. No specific information for direct participation of AGE in stem cell decline exists, other than their influence in other mechanisms, as summarized above that may indirectly contribute to this common pathway.

9. Altered intercellular communication

Aging is not a cell biology phenomenon, the whole organism ages and this encompasses modifications at the level of intercellular communication, endocrine, neuroendocrine or neuronal. Consequently, neurohormonal responses have a tendency to be deregulated in aging as inflammatory reactions increase, immunosurveillance against pathogens and neoplastic cells declines and the peri- and extracellular milieu changes, which affects the mechanical and functional properties of all tissues (Lopez-Otin et al. 2013). There is a characteristic aging-associated modification in intercellular communication known as 'inflammaging': a pro-inflammatory phenotype (Lopez-Otin et al. 2013). Hyper-activation of the NF-κB pro-inflammatory pathway is one of the transcriptional signatures of aging. AGE may be an integral part of these responses via their recognition by the RAGE, which precisely elicits a pro-inflammatory phenotype where the NF-κB pathway is central (Schmidt et al. 1996). Besides recognizing AGEs, RAGE also binds a large number of different ligands and can exert pro-inflammatory activities among many others (Ramasamy et al. 2005, Schmidt et al. 1996). Although RAGE is expressed at basal levels in healthy tissues, its expression is markedly augmented by a classical amplification loop through a sustained activation of NF-κB and triggered by the increased levels of ligands found in aging and pathologic states. Expression of RAGE is determined by the cell type and the developmental stage (Sorci et al. 2013). Throughout embryonic development, RAGE is constitutively expressed and subsequently down-regulated in adult life, but increases during aging (Sorci et al. 2013). After a period of favor for the Maillard theory of aging three decades ago, research failed to show a distinct relationship between the AGE content of long-lived proteins and an organism's lifespan. Although these observations challenge AGE as a causative factor, accumulation of AGE, either by endogenous formation or from increased dietary intake, can speed up the multisystem functional deterioration that occurs in aging and was described earlier. In this regard, the involvement of RAGE may act as an accelerant and help perpetuate the aging process and might shorten lifespan. AGEs accumulate during aging in collagens and other proteins of the extracellular matrix (Sveen et al. 2015, Monnier et al. 1988). The most direct mechanism by which RAGE could affect aging is precisely via their interaction with these AGEs. A case could be made for a cycle of events in which RAGE is upregulated and later activated with time and as the individual ages and simultaneously accumulates more RAGE ligands, leading to a continued pro-inflammatory phenotype, a distinctive pathology associated with age-related vascular, musculoskeletal and neurodegenerative diseases. The interaction of AGE with the RAGE is likely to accelerate and perpetuate the aging process through the maintenance of inflammatory reactions. Moreover RAGE down-regulates glyoxylase I, the key detoxification mechanism that eliminates MG as the main AGE precursor (Ramasamy et al. 2012). The AGE/RAGE axis may consequently offer an innovative therapeutic approach in the treatment of age-related diseases and achieving "healthy aging".

Conclusions

Accrual of AGE in proteins correlates with age, but does not seem to determine life span. One good example is studies on the Fukomys mole-rats, which are mammals whose life span is powerfully influenced by reproductive status: breeders far outlive nonbreeders. Yet, surprisingly, total AGE, glucosepane and CML were significantly higher in breeders versus nonbreeders (Dammann et al. 2012). This is not to say that AGE, ALE and other adducts are inconsequential in aging and life span. The current evidence suggests that AGE accumulation is correlative but not causative of aging. Besides, evidence seems to indicate that AGE may participate in aging as an aggravating factor. In humans, the steady buildup of these compounds may alter the structure and function of proteins and affect several of the hallmarks of aging described above and may also contribute to the progress of pathology in age-related diseases, such as diabetes and atherosclerosis as well to oxidative stress and inflammation associated with neurodegenerative diseases of aging.

Figure 1 summarizes the contribution of AGE and RAGE to the nine candidate hallmarks that characterize the aging process and together determine the aging phenotype.

The current evidence suggests that AGE accumulation is correlative but not causative of aging and is therefore depicted in broken arrows. In humans, the steady buildup of these compounds may alter the structure and function of proteins and affect several of the hallmarks of aging described in the text and may also contribute to the progress of pathology in age-related diseases.

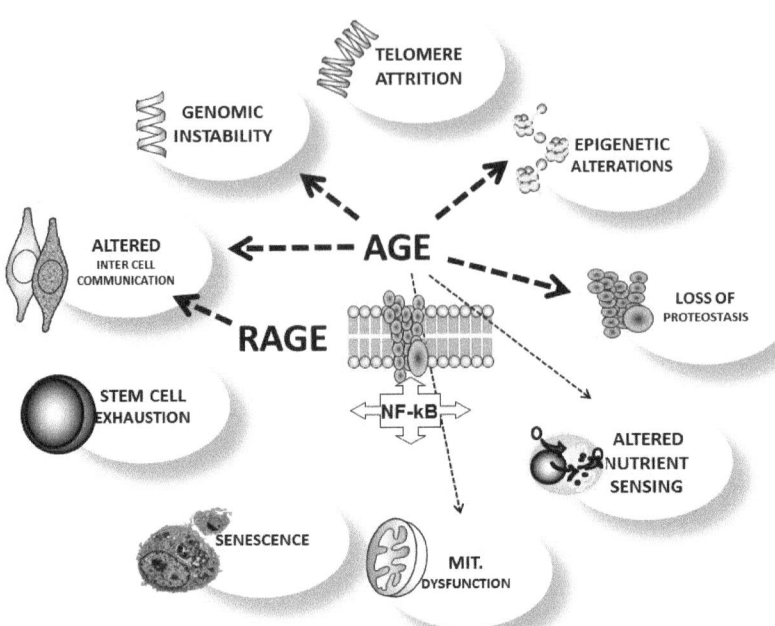

Figure 1. Contributions of AGE and RAGE to the nine candidate hallmarks that characterize the aging process and together determine the aging phenotype.

Keywords: Aging, advanced glycation, adduct, receptor for advanced glycation endproducts (RAGE), reactive oxygen species, Hayflick limit, senescence, inflammaging, telomerase, hormesis, proteostasis

References

Afanas'ev, I. 2010. Signaling and damaging functions of free radicals in aging-free radical theory, hormesis, and TOR. Aging Dis 1(2): 75–88.

Ahmad, S., U. Moinuddin, S. Shahab, M. Habib, K. Salman Khan, A. Alam et al. 2014. Glycoxidative damage to human DNA: Neo-antigenic epitopes on DNA molecule could be a possible reason for autoimmune response in type 1 diabetes. Glycobiology 24(3): 281–91.

Barja, G. 2014. The mitochondrial free radical theory of aging. Prog Mol Biol Transl Sci 127: 1–27.

Baynes, J. W. 2001. The role of AGEs in aging: causation or correlation. Exp Gerontol 36(9): 1527–37.

Baynes, J. W. 2002. The Maillard hypothesis on aging: time to focus on DNA. Ann N Y Acad Sci 959: 360–7.

Beisswenger, P. and D. Ruggiero-Lopez. 2003. Metformin inhibition of glycation processes. Diabetes Metab 29(4 Pt 2): 6S95–103.

Bjorksten, J. 1968. The crosslinkage theory of aging. J Am Geriatr Soc 16(4): 408–27.

Bjorksten, J. 1977. Some therapeutic implications of the crosslinkage theory of aging. Adv Exp Med Biol 86B: 579–602.

Brownlee, M. 2005. The pathobiology of diabetic complications: a unifying mechanism. Diabetes 54(6): 1615–25.

Bunn, H. F. and P. J. Higgins. 1981. Reaction of monosaccharides with proteins: possible evolutionary significance. Science 213(4504): 222–4.

Burkewitz, K., Y. Zhang and W. B. Mair. 2014. AMPK at the nexus of energetics and aging. Cell Metab 20(1): 10–25.

Cervantes-Laurean, D., D. E. Minter, E. L. Jacobson and M. K. Jacobson. 1993. Protein glycation by ADP-ribose: studies of model conjugates. Biochemistry 32(6): 1528–34.

Chung, M. M., Y. L. Chen, D. Pei, Y. C. Cheng, B. Sun, C. J. Nicol et al. 2015. The neuroprotective role of metformin in advanced glycation end product treated human neural stem cells is AMPK-dependent. Biochim Biophys Acta 1852(5): 720–31.

Dai, D. F., Y. A. Chiao, D. J. Marcinek, H. H. Szeto and P. S. Rabinovitch. 2014. Mitochondrial oxidative stress in aging and healthspan. Longev Healthspan 3: 6.

Dammann, P., D. R. Sell, S. Begall, C. Strauch and V. M. Monnier. 2012. Advanced glycation end-products as markers of aging and longevity in the long-lived Ansell's mole-rat (Fukomys anselli). J Gerontol A Biol Sci Med Sci 67(6): 573–83.

De Bellis, D. and M. I. Horowitz. 1987. *In vitro* studies of histone glycation. Biochim Biophys Acta 926(3): 365–8.

Falcone, C., S. Bozzini, A. Colonna, B. Matrone, E. M. Paganini, R. Falcone et al. 2013. Possible role of -374T/A polymorphism of RAGE gene in longevity. Int J Mol Sci 14(11): 23203–11.

Fleming, T. H., P. M. Humpert, P. P. Nawroth and A. Bierhaus. 2011. Reactive metabolites and AGE/RAGE-mediated cellular dysfunction affect the aging process: a mini-review. Gerontology 57(5): 435–43.

Fukada, S., Y. Ma and A. Uezumi. 2014. Adult stem cell and mesenchymal progenitor theories of aging. Front Cell Dev Biol 2: 10.

Gomes, R. A., H. Vicente Miranda, M. S. Silva, G. Graca, A. V. Coelho, A. E. Ferreira et al. 2006. Yeast protein glycation *in vivo* by methylglyoxal. Molecular modification of glycolytic enzymes and heat shock proteins. FEBS J 273(23): 5273–87.

Gugliucci, A. 1994. Advanced glycation of rat liver histone octamers: an *in vitro* study. Biochem Biophys Res Commun 203(1): 588–93.

Gugliucci, A. and M. Bendayan. 1995. Histones from diabetic rats contain increased levels of advanced glycation end products. Biochem Biophys Res Commun 212(1): 56–62.

Gugliucci, A. and T. Menini. 2003. The polyamines spermine and spermidine protect proteins from structural and functional damage by AGE precursors: a new role for old molecules? Life Sci 72(23): 2603–16.

Gugliucci, A. 2005. Alternative antiglycation mechanisms: are spermine and fructosamine-3-kinase part of a carbonyl damage control pathway? Med Hypotheses 64(4): 770–7.

Gugliucci, A. 2009a. "Blinding" of AMP-dependent kinase by methylglyoxal: a mechanism that allows perpetuation of hepatic insulin resistance? Med Hypotheses 73(6): 921–4.

Gugliucci, A., D. H. Bastos, J. Schulze and M. F. Souza. 2009b. Caffeic and chlorogenic acids in Ilex paraguariensis extracts are the main inhibitors of AGE generation by methylglyoxal in model proteins. Fitoterapia 80(6): 339–44. doi: 10.1016/j.fitote.2009.04.007.

Harman, D. 1960. The free radical theory of aging: the effect of age on serum mercaptan levels. J Gerontol 15: 38–40.

Harman, D. 1965. The free radical theory of aging: effect of age on serum copper levels. J Gerontol 20: 151–3.

Harman, D. 2003. The free radical theory of aging. Antioxid Redox Signal 5(5): 557–61.

Hayase, F., R. H. Nagaraj, S. Miyata, F. G. Njoroge and V. M. Monnier. 1989. Aging of proteins: immunological detection of a glucose-derived pyrrole formed during maillard reaction *in vivo*. J Biol Chem 264(7): 3758–64.

Higgins, P. J. and H. F. Bunn. 1981. Kinetic analysis of the nonenzymatic glycosylation of hemoglobin. J Biol Chem 256(10): 5204–8.

Holliday, R. 2014. The commitment of human cells to senescence. Interdiscip Top Gerontol 39: 1–7.

Jensen, M. B. and H. Jasper. 2014. Mitochondrial proteostasis in the control of aging and longevity. Cell Metab 20(2): 214–25.

Jones, D. P. 2015. Redox theory of aging. Redox Biol 5: 71–79.

Kitada, M., S. Kume, A. Takeda-Watanabe, S. Tsuda, K. Kanasaki and D. Koya. 2013. Calorie restriction in overweight males ameliorates obesity-related metabolic alterations and cellular adaptations through anti-aging effects, possibly including AMPK and SIRT1 activation. Biochim Biophys Acta 1830(10): 4820–7.

Kohn, R. R., A. Cerami and V. M. Monnier. 1984. Collagen aging *in vitro* by nonenzymatic glycosylation and browning. Diabetes 33(1): 57–9.

Kowald, A. 2001. The mitochondrial theory of aging. Biol Signals Recept 10(3-4): 162–75.

Labbadia, J. and R. I. Morimoto. 2014. Proteostasis and longevity: when does aging really begin? F1000Prime Rep 6: 7.

Labbadia, J. and R. I. Morimoto. 2015. The biology of proteostasis in aging and disease. Annu Rev Biochem 84: 435–64. doi: 10.1146/annurev-biochem-060614-033955.

Lakatos, A. and K. Jobst. 1991. Kinetics of histone protein glycation. Acta Biochim Biophys Hung 26(1-4): 33–8.

Lapointe, J. and S. Hekimi. 2010. When a theory of aging ages badly. Cell Mol Life Sci 67(1): 1–8.

Lee, A. T. and A. Cerami. 1992. Role of glycation in aging. Ann N Y Acad Sci 663: 63–70.

Liu, Y., J. Long and J. Liu. 2014. Mitochondrial free radical theory of aging: who moved my premise? Geriatr Gerontol Int 14(4): 740–9.

Lopez-Otin, C., M. A. Blasco, L. Partridge, M. Serrano and G. Kroemer. 2013. The hallmarks of aging. Cell 153(6): 1194–217.

McCarter, R., W. Mejia, Y. Ikeno, V. Monnier, K. Kewitt, M. Gibbs et al. 2007. Plasma glucose and the action of calorie restriction on aging. J Gerontol A Biol Sci Med Sci 62(10): 1059–70.

McCarty, M. F. 2014. AMPK activation—protean potential for boosting healthspan. Age (Dordr) 36(2): 641–63.

Michan, S. 2014. Calorie restriction and NAD(+)/sirtuin counteract the hallmarks of aging. Front Biosci (Landmark Ed) 19: 1300–19.

Monnier, V. M., V. J. Stevens and A. Cerami. 1981. Maillard reactions involving proteins and carbohydrates *in vivo*: relevance to diabetes mellitus and aging. Prog Food Nutr Sci 5(1-6): 315–27.

Monnier, V. M. and A. Cerami. 1983. Detection of nonenzymatic browning products in the human lens. Biochim Biophys Acta 760(1): 97–103.

Monnier, V. M., C. A. Elmets, K. E. Frank, V. Vishwanath and T. Yamashita. 1986. Age-related normalization of the browning rate of collagen in diabetic subjects without retinopathy. J Clin Invest 78(3): 832–5.

Monnier, V. M., D. R. Sell, F. W. Abdul-Karim and S. N. Emancipator. 1988. Collagen browning and cross-linking are increased in chronic experimental hyperglycemia. Relevance to diabetes and aging. Diabetes 37(7): 867–72.

Monnier, V. M. 1989. Toward a Maillard reaction theory of aging. Prog Clin Biol Res 304: 1–22.

Monnier, V. M., D. R. Sell, R. H. Nagaraj and S. Miyata. 1991. Mechanisms of protection against damage mediated by the Maillard reaction in aging. Gerontology 37(1-3): 152–65.

Nass, N., B. Bartling, A. Navarrete Santos, R. J. Scheubel, J. Borgermann, R. E. Silber et al. 2007. Advanced glycation end products, diabetes and ageing. Z Gerontol Geriatr 40(5): 349–56.

Nishikawa, T., D. Edelstein and M. Brownlee. 2000. The missing link: a single unifying mechanism for diabetic complications. Kidney Int Suppl 77: S26–30.

Parr, T. 1997. Insulin exposure and aging theory. Gerontology 43(3): 182–200.

Perez, V. I., A. Bokov, H. Van Remmen, J. Mele, Q. Ran, Y. Ikeno et al. 2009. Is the oxidative stress theory of aging dead? Biochim Biophys Acta 1790(10): 1005–14.

Queisser, M. A., D. Yao, S. Geisler, H. P. Hammes, G. Lochnit, E. D. Schleicher et al. 2010. Hyperglycemia impairs proteasome function by methylglyoxal. Diabetes 59(3): 670–8.

Rabbani, N., F. Shaheen, A. Anwar, J. Masania and P. J. Thornalley. 2014. Assay of methylglyoxal-derived protein and nucleotide AGEs. Biochem Soc Trans 42(2): 511–7.

Radjei, S., B. Friguet, C. Nizard and I. Petropoulos. 2014. Prevention of dicarbonyl-mediated advanced glycation by glyoxalases: implication in skin aging. Biochem Soc Trans 42(2): 518–22.

Ramasamy, R., S. J. Vannucci, S. S. Yan, K. Herold, S. F. Yan and A. M. Schmidt. 2005. Advanced glycation end products and RAGE: a common thread in aging, diabetes, neurodegeneration, and inflammation. Glycobiology 15(7): 16R–28R.

Ramasamy, R., S. F. Yan and A. M. Schmidt. 2012. Advanced glycation endproducts: from precursors to RAGE: round and round we go. Amino Acids 42(4): 1151–61.

Ruggiero-Lopez, D., M. Lecomte, G. Moinet, G. Patereau, M. Lagarde and N. Wiernsperger. 1999. Reaction of metformin with dicarbonyl compounds. Possible implication in the inhibition of advanced glycation end product formation. Biochem Pharmacol 58(11): 1765–73.

Schmidt, A. M., O. Hori, R. Cao, S. D. Yan, J. Brett, J. L. Wautier et al. 1996. RAGE: a novel cellular receptor for advanced glycation end products. Diabetes 45 Suppl 3: S77–80.

Singh, R., A. Barden, T. Mori and L. Beilin. 2001. Advanced glycation end-products: a review. Diabetologia 44(2): 129–46.

Sorci, G., F. Riuzzi, I. Giambanco and R. Donato. 2013. RAGE in tissue homeostasis, repair and regeneration. Biochim Biophys Acta 1833(1): 101–9. doi: 10.1016/j.bbamcr.2012.10.021.

Suji, G. and S. Sivakami. 2004. Glucose, glycation and aging. Biogerontology 5(6): 365–73.

Sveen, K. A., K. Dahl-Jorgensen, K. H. Stensaeth, K. Angel, I. Seljeflot, D. R. Sell et al. 2015. Glucosepane and oxidative markers in skin collagen correlate with intima media thickness and arterial stiffness in long-term type 1 diabetes. J Diabetes Complications 29(3): 407–12.

Syslova, K., A. Bohmova, M. Mikoska, M. Kuzma, D. Pelclova and P. Kacer. 2014. Multimarker screening of oxidative stress in aging. Oxid Med Cell Longev 2014: 562860.

Takizawa, N., K. Takada and K. Ohkawa. 1993. Inhibitory effect of nonenzymatic glycation on ubiquitination and ubiquitin-mediated degradation of lysozyme. Biochem Biophys Res Commun 192(2): 700–6.

Uchiki, T., K. A. Weikel, W. Jiao, F. Shang, A. Caceres, D. Pawlak et al. 2012. Glycation-altered proteolysis as a pathobiologic mechanism that links dietary glycemic index, aging, and age-related disease (in nondiabetics). Aging Cell 11(1): 1–13.

Ulrich, P. and A. Cerami. 2001. Protein glycation, diabetes, and aging. Recent Prog Horm Res 56: 1–21.

Vilchez, D., M. S. Simic and A. Dillin. 2014. Proteostasis and aging of stem cells. Trends Cell Biol 24(3): 161–70.

Vlassara, H. 1996. Protein glycation in the kidney: role in diabetes and aging. Kidney Int 49(6): 1795–804.

Waris, S., B. M. Winklhofer-Roob, J. M. Roob, S. Fuchs, H. Sourij, N. Rabbani et al. 2015. Increased DNA dicarbonyl glycation and oxidation markers in patients with type 2 diabetes and link to diabetic nephropathy. J Diabetes Res 2015: 915486.

Ziegler, D. V., C. D. Wiley and M. C. Velarde. 2015. Mitochondrial effectors of cellular senescence: beyond the free radical theory of aging. Aging Cell 14(1): 1–7.

8

AGE-RAGE Axis in the Aging and Diabetic Heart
Prime Target for Adjunctive Therapeutic Intervention

Karen M. O'Shea,[a] Ann Marie Schmidt[b] and
Ravichandran Ramasamy*

INTRODUCTION

Cardiovascular aging in humans is accompanied by fundamental biochemical, metabolic, and molecular changes that ultimately result in basal dysfunction and increased susceptibility to injury, triggered by acute stress, such as ischemia/reperfusion. In fact, the primary factor underlying the enhanced propensity to cardiovascular diseases in humans is advanced age (Lakatta 2003, Lakatta and Levy 2003a,b). As humans age, progressive insulin resistance and hyperglycemia also occur, contributing to the functional decline of the heart. Cardiovascular diseases and complications secondary to myocardial infarction are the primary causes of morbidity and mortality in the elderly (Mozaffarian et al. 2015, Saunderson et al. 2014).

Diabetes is also a major risk factor for the development of cardiovascular disease, such as atherosclerosis, and patients with diabetes have an increased risk of myocardial infarction (Libby 2002). Hyperglycemia and insulin resistance induce changes in cell signaling and metabolism within cardiomyocytes leading to contractile dysfunction, fibrosis and heart failure (Wang et al. 2006). Aged human hearts, from both diabetic and non-diabetic subjects, display increased accumulation of Advanced Glycation Endproducts (AGEs), species formed by non enzymatic glycation/oxidation of proteins/lipids. AGEs have been demonstrated to mediate both microvascular and

New York University, Department of Medicine, 550 1st Avenue New York, NY USA, 10016.
[a] Email: osheak05@nyumc.org
[b] Email: AnnMarie.Schmidt@nyumc.org
* Corresponding author: ramasr02@nyumc.org

macrovascular complications in diabetes, including nephropathy, retinopathy, and neuropathy (Ramasamy et al. 2010).

AGE accumulation occurs when reducing sugars, such as glucose and fructose, interact with the amino group of proteins, lipids and nucleic acids. AGEs are comprised of a heterogeneous class of irreversibly formed molecules that may fluoresce and can directly modify proteins or bind to cell surface receptors (Schmidt et al. 1994). Early investigation of AGEs uncovered their role in mediating the complications of diabetes (Schmidt et al. 1992); however, accumulation of AGEs has been observed in various other pathological conditions, including myocardial ischemic injury, atherosclerosis, heart failure, renal failure and inflammation (Ramasamy et al. 2010). Aging tissues, such as vessels and myocardium, even in the absence of diabetes or inflammation, display increased AGEs (Schleicher et al. 1997, Pamplona et al. 2002). Accumulation of AGEs in the circulation and tissue is determined by a number of factors; primarily, hyperglycemia, renal function, inflammation and oxidative stress (Miyata et al. 1997). In aging, AGEs accumulate due to reduced turnover of circulating and structural proteins. Such proteins become modified even in ambient levels of glucose. An additional mechanism is the decreased expression/activity of glyoxalase I, a major defense mechanism in detoxification/removal of AGE precursors. Tissue levels of anti-AGE defense mechanisms, specifically glyoxalase I, decrease with aging (Amicarelli et al. 1997, Thornalley 2003, Shinohara et al. 1998).

AGEs lead to various complications, in part due to their ability to modify and cross-link proteins. AGEs interact with the proteins of the extracellular matrix in the heart and vasculature, including collagen and elastins, causing myocardial and vascular stiffness and subsequent contractile dysfunction and fibrosis (Funk et al. 2012, Zieman et al. 2005). AGEs may directly impact the structural integrity of the vessel wall and underlying basement membrane. Excessive cross-linking of matrix molecules such as collagen may disrupt matrix-matrix and matrix-cell interactions (Tanaka et al. 1988, Haitoglou et al. 1992, Giardino et al. 1994). AGEs also modify signaling proteins and scavenge nitric oxide, leading to impaired vascular relaxation, oxidative stress and the formation of reactive nitrogen species (Munch et al. 2012, Loske et al. 1998). Reduced arterial elasticity of aging is associated with endothelial dysfunction in the elderly (Tao et al. 2004). Additionally, AGEs can induce deleterious signaling in cells by binding to cell surface receptors, such as the receptor for AGEs (RAGE) (Schmidt et al. 1999).

Under physiological conditions, endogenous AGE formation occurs relatively slowly, over a period of weeks to months, mainly impacting proteins with a long half-life (Singh et al. 2001). Under diabetic conditions, hyperglycemia accelerates the formation of AGEs. However, AGEs have been reported to accumulate from exogenous sources, particularly from food prepared at high temperatures or tobacco smoke (Nicholl and Bucala 1998, O'Brien and Morrissey 1989). Therefore, the total level of AGEs in the body is a balance of exogenous AGE consumption, endogenous AGE production and clearance or detoxification by enzyme systems (Koschinsky et al. 1997). The majority of AGEs in the circulation are bound to albumin or other carrier proteins (Miyata et al. 1996).

In addition to endogenous AGE formation, it has been proposed that diets containing AGEs may contribute to deterioration of function associated with aging and diabetes (Goldberg et al. 2004, Uribarri et al. 2005). Diabetic patients consuming a diet low in AGEs displayed decreased inflammatory mediators (Vlassara et al. 2002). In addition, limiting glycotoxins have been shown to reduce AGE accumulation in patients with renal failure (Uribarri et al. 2003). Chronic administration of methylglyoxal for three months to normal non-diabetic rats has been shown to mimic the effects observed in type 2 diabetic non-obese Goto-Kakizaki rats, such as oxidative stress, AGE accumulation, RAGE activation and apoptotic signaling (Crisostomo et al. 2013). Taken together, these observations indicate that reducing dietary intake of AGEs may have beneficial effects in terms of inflammation and aging.

The AGEing hypothesis of aging has been proposed primarily based on the evidence from several studies demonstrating AGE accumulation in aging tissues. Accumulation of CML adducts, CEL lysine adducts, methylglyoxal, pentosidine and others has been observed as a function of age in various tissues. Furthermore, modification of DNA by AGEs can lead to changes in regulatory and epigenetic components of the aging process (Baynes 2002). AGEs form as a consequence of natural aging and due to exposure of long-lived proteins to even euglycemic levels of glucose. Aging-associated insulin resistance also promotes the formation of AGEs in aging tissues. Aging is also associated with decreased defenses against AGE generation, such as glyoxalase I, which is a cytosolic enzyme that reduces production of the AGE precursor, methylglyoxal (Thornalley 2003). In aging, mechanisms that promote injury may be heightened while anti-injury defense mechanisms decline.

Although not fully independent of aging-associated diseases, innate dysfunction in the aging cardiovascular system in the vasculature and myocardium primes the aged organism for heightened vulnerability to superimposed stresses (Lakatta 2003, Lakatta and Levy 2003a, b). This chapter focuses on the effects of AGEs in the heart in the context of aging and cardiovascular disease. While much of the previous investigation into AGEs in the heart revolves around hyperglycemia and diabetic cardiomyopathy, AGEs accumulate even in conditions of ambient glucose, such as increased age and ischemia/reperfusion injury. In order to understand the role of AGEs in aging in the heart, the effects of AGEs in various disease models must be investigated. AGEs and the systems that generate them have been studied in the context of ischemia/reperfusion injury and heart failure, involving mitochondrial function, contractile function and fibrosis. The enhanced susceptibility of aged human subjects to I/R has been modeled in animal systems. Although multiple etiologies underlie these findings, the role of AGE-generating systems and the multi-ligand receptor, RAGE, will be examined in this chapter to understand how the effects of AGEs are mediated in aging and cardiac stress.

Advanced Glycation Endproduct-generating Systems

The formation of AGEs occurs as a result of glycation and oxidation by the carbonyl group of reducing sugars, such as glucose or fructose, with the amino group of proteins, nucleic acids and lipids. One way that this occurs is through a process called the Maillard reaction, when glucose non-enzymatically attaches to free amino acids on proteins to form a Schiff base, which can spontaneously rearrange to generate Amadori products (Tanaka et al. 1988, John and Lamb 1993). This initial step is reversible, but further oxidation leads to the production of irreversible AGEs, including Nε-carboxymethyllysine (CML), Nε-carboxyethyl-lysine, pyrilline, pentosidine and argpyramidine (Ahmed et al. 1986). AGEs derived from CML are the most prevalent *in vivo* (Dyer et al. 1993). AGEs can also be formed by oxidation of glucose and the peroxidation of lipids into dicarbonyl derivatives, called α-dicarbonyls or oxoaldehydes, which are highly reactive and accumulate as a result of aging, hyperglycemia and increased oxidative stress (Suzuki et al. 1999). Dicarbonyl stress involves the accumulation of intermediates, such as 3-deoxyglucosone, glyoxal and methylglyoxal, that can react with amino, sulfhydryl, and guanidine functional groups to brown, denature or cross-link proteins (Lo et al. 1994). Dicarbonyl stress contributes to the damage induced by chronic inflammation in the heart and vasculature in diabetes (Vulesevic et al. 2014).

Methylglyoxal, in particular, is one the most abundant precursors for AGEs and has been implicated in mediating the complications of diabetes (Shamsi et al. 1998). Diabetic patients with optimal glycemic control experience the development of complications due to elevated methylglyoxal (Turk 2010). Methylglyoxal is a 2-oxoaldehyde that can be derived from various metabolic pathways involving glycolytic intermediates, polyol pathway intermediates and ketone metabolism (Thornalley 1993). The formation of methylglyoxal from ketone metabolism may account for the especially high levels of methylglyoxal observed in patients with Type 1 diabetes (McLellan et al. 1994).

Glucose metabolism via the polyol pathway plays an important role in the generation of reactive carbonyl intermediates and AGEs. Aldose reductase, encoded by the human gene AKR1B1, is a cytoplasmic enzyme that reduces aldehydes. As the rate-limiting enzyme that catalyzes the first step of the polyol pathway, aldose reductase converts glucose to sorbitol in a NADPH-dependent manner. Sorbitol is then converted to fructose by sorbitol dehydrogenase at the cost of depletion of NAD^+. Sorbitol and fructose accumulate in diabetic hearts when flux through the polyol pathway is increased due to high levels of glucose, leading to the generation of AGEs (Chung and Chung 2003). Fructose can be phosphorylated by fructose-3-phosphokinase into fructose-3-phosphate, which itself can modify and directly interact with proteins, lipids, and nucleic acids or be converted into AGE precursor3-deoxyglucosone. Additionally, the depletion of NAD^+ and increased $NADH/NAD^+$ that occurs as a result of increased flux through the polyol pathway inhibits NAD^+-dependent enzymes, including GAPDH. This results in accumulation of glyceraldehyde-3-phosphate and dihydroxyacetone phosphate, both of which are precursors for methylglyoxal (Rabbani and Thornalley 2015).

Although *in vitro* studies have suggested that methylglyoxal is a substrate for aldose reductase, it is primarily metabolized through the glutathione-dependent glyoxalase system. The glyoxalase system is present in the cytosol of all cells and has been identified in all human tissues. In this system, tissue methylglyoxal is maintained at a relatively low level under physiological conditions. The glyoxalase system is present in the cytosol of all cells and has been identified in all human tissues. Methylglyoxal is ultimately catabolized to D-lactate as a result of the activity of Glo-1, reduced GSH and Glo-2 (McLellan et al. 1993).

Glutathione depletion occurs as result of hyperglycemia and oxidative stress, both of which occur in diabetic cardiomyopathy and aging. This is one of the mechanisms of reduced detoxification and subsequent accumulation of methylglyoxal in diabetes and aging. Glo-1 activity is reduced in the diabetic heart, despite no change in mRNA or protein expression (Brouwers et al. 2013). Transgenic models of Glo-1 overexpression have indicated that reducing AGEs decreases oxidative stress, inflammation and fibrosis in the diabetic heart. Genetic overexpression of Glo-1 in rats with 24 weeks of streptozotocin-induced diabetes attenuated the diabetes-induced increase in methylglyoxal and 3-deoxyglucasone in the heart, both of which have been linked to promoting oxidative stress and fibrosis. Glo-1 overexpression also reduces the diabetes-induced increase in expression of glutathione peroxidase-1, indicating an attenuation of oxidative stress. Furthermore, overexpression of Glo-1 increased the diabetes-induced reduction in genes responsible for fibrosis, including collagen and MMP-2 (Brouwers et al. 2013). There is a clear need for further investigation into the role of Glo-1 in detoxifying AGEs in the aging heart.

AGEs and their Receptor: RAGE

AGEs may also exert their pathogenic effects by engagement of cell surface binding sites/receptors. Although AGEs can bind to various receptors, such as AGE-R1 (oligosaccharyl transferase-48), AGE-R2 (80K-H phosphoprotein), AGE-R3 (galectin-3), CD36 and the class A macrophage scavenger receptor types I and II, they do not activate intracellular signals after binding of AGEs (Wendt et al. 2002, Vlassara et al. 1995, el Khoury et al. 1994). In contrast, the interaction of AGEs with the receptor for AGEs (RAGE) has been best described and induces intracellular signaling cascades. RAGE is a member of the immunoglobulin super-family of receptors (Schmidt et al. 1992). RAGE was discovered in 1992 for its ability to recognize and bind AGEs, but since then a variety of other classes of ligands have been identified for RAGE, such as members of the S100/calgranulin family (Hofmann et al. 1999), high mobility group box 1 (Taguchi et al. 2000), amyloid beta-peptide and beta-sheet fibrils (Yan et al. 1996) and lysophosphatidic acid (Rai et al. 2012). The result of the role of RAGE as a multi-ligand receptor is activation of signaling cascades, generation of cytokines and proinflammatory adhesion molecules and cellular migration (Taguchi et al. 2000, Sakaguchi et al. 2003).

At least two classes of AGEs, which are increased in aging, bind RAGE. First, CML-adduct AGEs bind RAGE in multiple cell types, activate signal transduction and modulate gene expression (Kislinger et al. 1999, Boulanger et al. 2004). As mentioned in the previous section, CML-AGEs accumulate in aging and may be formed by multiple mechanisms, involving the precursor 3-DG or glyoxal intermediates (Niwa 1999, Niwa et al. 1998). Direct phosphorylation of fructose by fructose-3 phosphokinase also results in a potent glycating agent, fructose-3-phosphate. Plasma levels of 3-DG increase, along with increased AR, in erythrocytes, in renal failure (Hasuike et al. 2002). Administration of epalrestat, an inhibitor of aldose reductase, reduced the levels of CML adducts and their precursors in erythrocytes in diabetic patients (Hamada et al. 2000). Second, hydroimidazolones are primarily derived from MH precursors and also bind RAGE (Thornalley 1998).

RAGE expression is present in various cell types, including cardiomyocytes, vascular cells, inflammatory cells and smooth muscle cells (Brett et al. 1993). Under physiological conditions, RAGE expression is relatively low; however, under pathological conditions involving the accumulation of AGEs and inflammation, such as cardiovascular disease, diabetes and aging, RAGE expression increases (Yan et al. 2010a). In human atrial tissue retrieved at the time of surgery, in aged (mean age, 76 years) vs. young (mean age, 2 years) hearts, an increase in RAGE antigen was observed (Casselmann et al. 2004, Simm et al. 2004). Thus, in natural aging, without overt aging-associated disease, AGE-RAGE is upregulated, suggesting that RAGE protein expression is increased with aging in the heart and is independent of established aging-related diseases.

The precise consequences of RAGE signaling vary based on cell type, as well as duration of RAGE stimulation. RAGE transduces extracellular signals through ligand binding to its 332-amino acid extracellular component, consisting of 2 "C"-type domains preceded by 1 "V"-type immunoglobulin-like domain (Yan et al. 2003). RAGE also contains a transmembrane domain and a highly charged 43-amino acid cytosolic tail, which is responsible for mediating intracellular signaling (Schmidt et al. 2001). While the precise mechanisms of intracellular RAGE signaling have not been fully elucidated, the cytosolic tail interacts with the formin protein, Diaph1 (also known as mDia1 or Drf1) (Hudson et al. 2008). The cytosolic domain of RAGE is critical for RAGE-dependent signaling and modulation of gene expression. When the cytosolic domain is deleted, this imparts a "dominant negative" effect; although cells may bind ligand, ligands are not able to activate RAGE signaling (Bucciarelli et al. 2008, Harja et al. 2008).

Activation of RAGE contributes to the cellular response to stress in both acute and chronic states, leading to inflammation, oxidative stress and fibrosis (Ramasamy et al. 2011). This occurs through ligand engagement of RAGE and subsequent activation of multiple signaling pathways, including p21ras, ERK1/2 (p44/p42) MAP kinases, p38 and SAPK/JNK MAP kinases, Rho GTPases, phosphoinositol-3 kinase and the JAK/STAT pathway and downstream consequences such as activation of NF-kB, Egr-1 and CREB (Huang et al. 2001, Yeh et al. 2001, Huttunen et al. 2002). At least in part, ligand-engagement of RAGE triggers generation of ROS, key species linked to triggering of cell signaling. Studies have shown that both NADPH oxidase and mitochondrial sources of ROS resulted from AGE-RAGE interaction (Wautier et al. 2001, Basta et al. 2005).

In addition to the transmembrane receptor, RAGE, there are two forms of soluble RAGE, or sRAGE in the circulation. One is the extracellular form of RAGE cleaved by matrix metalloproteinasesor A-Distintegrin and Metalloprotease (ADAM)-10 from the cell surface, and the second is called endogenous secretory RAGE (esRAGE). esRAGE results from a splice variant lacking the transmembrane and cytoplasmic domains and comprises about 20% of total sRAGE (Yan et al. 2010b). Pharmacological sRAGE treatment of animals in experimental conditions, such as cardiovascular disease, diabetes and chronic inflammation, has been shown to be effective in reducing markers of injury. The mechanism of protection conferred by sRAGE treatment is thought to be sequestration of RAGE ligands, thereby blocking their binding to RAGE and other cell surface receptors (Schmidt 2015). Furthermore, high levels of sRAGE were associated with extreme and "healthy" longevity in aging (Geroldi et al. 2006). Healthy centenarians displayed the highest levels of sRAGE when compared to young subjects with acute MI (the lowest levels of sRAGE) and when compared to young "healthy" subjects (levels were midway between the other two groups). These considerations suggest that higher RAGE ligand binding potential (i.e., higher sRAGE) may be protective in disease and this is now suggested in extreme longevity. These findings link RAGE to human aging and suggest that sRAGE may be a biomarker to predict relative "health" in aging.

The role of the AGE/RAGE axis in ischemia/reperfusion injury

Ischemic heart disease is one of the most prevalent causes of mortality in the United States and around the world (Roger et al. 2012). Ischemia causes severe myocardial dysfunction and tissue damage, which is worsened by subsequent reperfusion, a phenomenon known as ischemia/reperfusion injury. Reperfusion is a mainstay of treatment for patients presenting with myocardial infarction, but methods of reducing injury following reperfusion are necessary to improve outcomes. Ischemia/reperfusion is a complex clinical condition with several underlying mechanisms of injury, including oxidative stress, ionic imbalances, cell death, inflammation and metabolic changes (Hausenloy and Yellon 2013). One of the main contributors to cell death and irreversible damage after ischemia/reperfusion injury is mPTP opening. The mPTP is a nonspecific pore that opens during reperfusion in response to elevated cytosolic calcium, ROS and inorganic phosphates, resulting in dissolution of the mitochondrial membrane potential, electron transport chain dysfunction and release of cytochrome c and other pro-apoptotic proteins into the cytosol (Borutaite et al. 2001).

The aging heart is susceptible to glycation processes, which can lead to ventricular dysfunction, elevated ventricular and vascular stiffness and impaired regulation of vascular tone (Lakatta and Levy 2003a). Evidence of the role of AGEs in mediating ischemia/reperfusion injury can be found in experimental models. Methylglyoxal accumulates after ischemia/reperfusion, blocking the necessary activation of cell survival pathways. Cardiomyocytes incubated with methylglyoxal and exposed to experimental ischemia/reperfusion injury demonstrate glycation of anti-apoptotic thioredoxin, indicating one mechanism for the apoptotic effect of methylglyoxal (Wang et al. 2010).

One of the hallmarks of injury is altered myocardial metabolism, including increased flux via the AGE precursor-generating polyol pathway. Extensive research from our laboratory has established that flux via aldose reductase drives ischemic injury in the heart. The polyol pathway depletes NAD^+ as a result of sorbitol dehydrogenase activity disrupting the activity of other dehydrogenases that require NAD^+. Genetic deletion or pharmacological inhibition of aldose reductase has shown that aldose reductase mediates ischemic injury. Rat hearts perfused with the aldose reductase inhibitor, zopolrestat, and then subjected to ischemia/reperfusion exhibited reduced ischemic injury, a lower $NADH/NAD^+$, increased tissue content of ATP, and improved cardiac function (Hwang et al. 2002, Ramasamy et al. 1997). Rabbits subjected to *in vivo* ischemia/reperfusion by temporary occlusion of a prominent left coronary artery also benefited from aldose reductase inhibition as indicated by reduced infarct size (Tracey et al. 2000). These changes appeared to be due to JAK and subsequent STAT5 activation in a PKC-dependent manner (Hwang et al. 2005). Mice have relatively low expression of aldose reductase compared to humans and rats; hence, mice overexpressing human ARTg have been generated. Hearts isolated from ARTg and subjected to ischemia/reperfusion mice display worsened contractile function, increased infarct size and increased creatine kinase release (a marker of ischemic injury). These parameters are all improved by aldose reductase inhibition (Hwang et al. 2004). In mice and rats, we demonstrated that flux via the aldose reductase pathway is significantly increased in aging hearts and consequently increases vulnerability of hearts to I/R injury (Ananthakrishnan et al. 2011). Importantly, we showed that increased aldose reductase pathway activity in aging impacts AGE levels and mediates cardiovascular dysfunction, in part, via RAGE (Hallam et al. 2010). Further investigation into the mechanism behind the effects of aldose reductase in the heart revealed a role for aldose reductase in promoting mPTP opening via increased ROS production (Ananthakrishnan et al. 2009). These results indicate that aldose reductase is a novel adjunctive target for the protection of hearts in diabetic and aging patients with evolving myocardial infarction. In fact, treatment with the specific aldose reductase inhibitor, zopolrestat, for one year in human diabetic subjects resulted in increased resting left ventricle ejection fraction, cardiac output, left ventricular stroke volume and exercise ejection fraction (Johnson et al. 2004).

RAGE has been implicated in mediating myocardial dysfunction that occurs due to natural aging. In human subjects undergoing cardiac surgery, an age-dependent increase in atrial RAGE protein expression was observed, with the highest levels observed in the senescent population and the lowest levels observed in the youngest children (Simm et al. 2004). Regression analyses revealed that the level of RAGE expression was associated with reduced ventricular function in these subjects. Although the precise mechanisms regulating this effect have not been described in detail, these results indicate that RAGE is upregulated in the aging human heart.

Genetic deletion of RAGE as well as pharmacological treatment with sRAGE has been shown to improve functional outcomes after ischemia/reperfusion. Mouse models of myocardial infarction, involving ligation of the left anterior descending coronary artery for 30 minutes, followed by reperfusion have provided much insight into the role of RAGE in the heart. Multiple studies have shown that genetic deletion of RAGE is cardio protective. Infarct size is reduced, contractile function is improved, apoptosis is decreased in RAGE-null hearts compared to wild type after *in vivo* ischemia/reperfusion. These changes were associated with decreased JAK and STAT5 phosphorylation in the RAGE-null hearts (Aleshin et al. 2008). Pretreatment of rats with sRAGE before ischemia/reperfusion reduced heart tissue levels of methylglyoxal, decreased RAGE expression, improved functional recovery, and decreased lactate dehydrogenase release (a marker of ischemic injury). Furthermore, sRAGE pretreatment reduced ischemia/reperfusion-induced mitochondrial protein tyrosine nitration and increased inducible nitric oxide synthase, markers of oxidant stress. In the same study, similar results were observed with genetic deletion of RAGE after ischemia/reperfusion injury (Bucciarelli et al. 2006).

In order to determine the effects of RAGE in cardiomyocytes as opposed to all cells types in the heart, including vascular cells, inflammatory cells and fibroblasts, primary adult ventricular myocytes have been used. These cells were isolated from the hearts of wild type and RAGE-null mice and then subjected to *in vitro* hypoxia (0.5% oxygen) for 30 minutes followed by 60 minutes of reoxygenation. Hypoxia/

reoxygenation led to increased AGEs and RAGE expression in wild type cells (Shang et al. 2010). These effects were associated with increased phosphorylation of JNK and reduction of phosphorylation of GSK-3β at serine 9, and were blocked by RAGE deletion or pretreatment with sRAGE. In addition, apoptosis was reduced by RAGE deletion or pretreatment with sRAGE.

AGEs and RAGE in the Failing Heart

Aging is accompanied by progressive systolic and diastolic function, which can ultimately result in heart failure. Heart failure is a complex clinical syndrome defined as any impairment of the ventricle to fill with or eject blood. It is characterized by inflammation, oxidative stress, tissue injury, mitochondrial dysfunction and fibrosis, all culminating in contractile dysfunction. Heart failure is the number one cause for hospitalization in patients over the age of 65. The prognosis for even optimally treated patients remains poor. Heart failure is a progressive syndrome that can develop from a variety of pathologies, involving hemodynamic stress or myocardial injury, most commonly secondary to myocardial infarction. Heart failure is associated with abnormalities of myocardial structure, often derived from cross-linking of extracellular matrix proteins by AGEs. AGEs have been implicated in promoting injury in the heart even in the absence of diabetes. Plasma levels of CML correlate with the severity of heart failure and can be used to track prognosis of patients (Hartog et al. 2007). Serum levels of AGEs correlate with prolonged isovolumic relaxation time, which is a measurement of diastolic function (Avendano et al. 1999). Pentosidine, in particular, has been examined in heart failure patients. Serum pentosidine concentration is an independent risk factor for heart failure, and is increased in patients with cardiac events versus event-free patients. In fact, the highest quartile of serum pentosidine was demonstrated to be associated with the highest risk for cardiac events (Koyama et al. 2007). In our laboratory, we have demonstrated that CML-AGEs accumulate in the diabetic mouse and diabetic rat hearts. Interestingly, we showed that RAGE itself appears to regulate AGE levels in heart tissue. Genetic deletion of RAGE in diabetic mice reduced CML and pentosidine levels in heart tissue subjected to ischemia/reperfusion or normoxic conditions (Bucciarelli et al. 2008). In diabetic transgenic mice expressing DN RAGE in endothelial cells (Tg DN PPET RAGE) and in macrophages (Tg DN MSR RAGE), only CML levels are attenuated in hearts under all perfusion conditions. Thus, these data indicate that suppressive effect of DN RAGE on pentosidine is cell specific, i.e., expression of DN RAGE reduced pentosidine in Tg DN MSR but not in Tg DN PPET mice hearts. Furthermore, data indicate that most of the CML effects appear to be associated with the endothelial cells, whereas pentosidine effects are pronounced in the mononuclear phagocytes (Bucciarelli et al. 2008).

Impaired calcium homeostasis is a hallmark of diabetes and heart failure. The tight regulation of calcium transients during systole and diastole is directly responsible for regulating contraction and relaxation of the contractile apparatus in cardiomyocytes. For a more detailed review of the role of calcium in the heart and the progression to heart failure, the reader is referred to (Gorski et al. 2015). Depolarization of the sarcolemma triggers the L-type caclium channels to allow entry of a small amount of extracellular calicum into the cytosol. This calcium then binds to RyR on the sarcoplasmic reticulum, prompting the release of stored calcium from the sarcoplasmic reticulum into the cytosol, a process termed calcium-induced calcium release. This results in an approximate 10-fold increase in cytosolic calcium, which can then bind to troponin and induce a conformational change that allows for contraction of the contractile apparatus. In order for relaxation to occur, the calcium must be cleared from the cytosol. This primarily occurs through the activity of the SERCA, which pumps calcium back into the sarcoplasmic reticulum at the cost of ATP. Additionally, the sodium-calcium exchanger (NCX) at the sarcolemma extrudes calcium from the cytosol in exchange for sodium entry into the cytosol.

In heart failure, diabetes and aging, the expression and activity of these and other calcium-handling proteins is impaired, resulting in contractile dysfunction. RyR dysfunction has been observed in contractile dysfunction, secondary to myocardial infarction and heart failure. In particular, leakage of calcium from the sarcoplasmic reticulum via RyR during diastole has been observed (Fischer et al. 2013). It has been well-established that SERCA expression and activity is also reduced or impaired in diabetes and heart failure (Hovnanian 2007). SERCA is particularly susceptible to modification by AGEs due to its abundance of arginine and lysine residues at its ATP binding site and phosphorylation site (Horakova et al. 2013). The polyol pathway has been shown to directly impact SERCA activity in diabetic and non-diabetic hearts (Ramasamy et al. 1997), and this can impact SERCA and RyR downstream through increases in oxidative stress (Tang et al. 2010). In hearts subjected to acute hyperglycemia, the NAD/NADH is increased due to elevated flux through the polyol pathway. This results in the activation of the NADH oxidase to produce reactive oxygen species that interact with SERCA and inhibit its ability to bind ATP. In addition, *ex vivo* perfusion of hearts with a high glucose buffer results in nitration of SERCA that can be reversed by addition of an aldose reductase inhibitor to the buffer (Tang et al. 2010). For a more detailed review of the effects of hyperglycemia and oxidative stress on SERCA activity, the reader is referred to (Horakova et al. 2013).

AGEs have been implicated in contributing to the impairment of calcium homeostasis in cardiomyocytes. Cardiomyocytes isolated from neonatal mice displayed prolonged calcium transients after exposure to glyceraldehyde-derived AGE-BSA, an effect that was exacerbated in transgenic mice overexpressing human RAGE. Furthermore, human RAGE overexpression alone resulted in reduced systolic and diastolic calcium levels compared to wild type, indicating that AGE-RAGE signaling plays a role in intracellular calcium regulation (Petrova et al. 2002). In addition, neonatal rat ventricular myocytes treated with AGE-BSA displayed increased frequency of calcium sparks, indicating calcium leakage through RyR (Yan et al. 2014). In adult cardiomyocytes isolated from rats, incubation with methylglyoxal resulted in decreased contractile indices, increased intracellular calcium, delayed calcium clearance from the cytosol, oxidative stress and impaired mitochondrial enzyme activities. Treatment with a PKCβII inhibitor, LY333531, attenuated contractile dysfunction and improved calcium transients, indicating that PKC is a mediator of AGE-induced contractile dysfunction and calcium homeostasis (Zhang et al. 2014).

The heart is comprised of cardiomyocytes, fibroblasts, vascular smooth muscle cells and endothelial cells. Fibroblasts are responsible for extracellular matrix remodeling, which occurs under normal physiological and pathological conditions. Communication between fibroblasts and the extracellular matrix is essential for structural organization in the heart and cell viability. Under pathological conditions, the communication between the fibroblasts and extracellular matrix is altered. The AGE/RAGE signaling axis is a major regulator of this process in the diabetic heart. Profibrotic signaling activated by hyperglycemia is thought to be an important mechanism in the development of heart failure. AGEs modify collagen and block its hydrolytic turnover. The result is an accumulation of extracellular matrix proteins and matrix stiffness, culminating in the development of fibrosis in various tissues, including the heart.

Investigation into the diabetic and aging heart in animals has revealed that AGEs may play a central role in mediating contractile function and fibrosis. Treatment of rats with diabetes induced by streptozotocin with a breaker of AGE-collagen crosslinks, ALT-711, resulted in attenuated RAGE expression, reduced BNP and improved collagen solubility (Candido et al. 2003). Treatment with ALT-711 also reduced collagen and improved cardiac function in aging diabetic dogs (Liu et al. 2003) and reduced arterial stiffness in primates (Vaitkevicius et al. 2001). In addition, rodents treated with aminoguanidine, an inhibitor of AGE formation, displayed reduced aortic wall stiffness in aged rats (Chang et al. 2003, Cantini et al. 2001, Aronson 2003). Support for the AGEs as modulators of arterial compliance in human subjects was demonstrated when administration of ALT-711 to patients (mean age, 67 years) for 56 days resulted in improved arterial compliance, but no change in cardiac output (Kass et al. 2001). This indicates that the relationship between the AGE/RAGE axis and diabetic complications in the heart may be an effective therapy for diabetic and aging patients and merits further investigation.

In addition to cross-linking collagens, AGEs promote fibrosis through interaction with RAGE and activation of signaling cascades that increase expression and deposition of extracellular matrix proteins. Signaling kinases, such as p38, extracellular signal-regulated kinase 1/2 (ERK 1/2), NF-κB, and c-JNK, regulate a variety of intracellular processes, including fibrosis, through activation of transcription factors. In a leptin receptor-deficient (*db/db*) mouse model of type 2 diabetes, a RAGE blocking antibody attenuated left ventricular stiffness, improved contractile function and decreased fibrosis (Nielsen et al. 2009).

The AGE antagonist, pyridoxamine, has been tested in rodent models of heart failure. It has been shown to attenuate arterial stiffening and improve age-associated deterioration of myocardial function (Wang et al. 2014, Wu et al. 2011). Pyridoxamine reduced methylglyoxal accumulation and subsequent impairment of cell survival pathways in rat hearts subjected to ischemia (Almeida et al. 2013). In rats subjected to streptozotocin-induced diabetes followed by infarction by left anterior descending coronary artery ligation, pyridoxamine reduced infarction-induced AGE accumulation and inflammatory mediator release. Further investigation into this effect using *in vitro* studies showed that AGEs stimulate dendritic cell differentiation, which release cytokines in a NF-κB controlled manner and stimulate hypertrophic gene expression in cardiomyocytes (Cao et al. 2015). These results indicate that blockade of AGE formation is an important mechanism for reducing myocardial injury.

sRAGE has been shown to have cardio-protective properties by acting as a decoy for pro-injury RAGE signaling. Endogenous sRAGE is downregulated in the heart following ischemia/reperfusion (Dong et al. 2011). Pretreatment of neonatal cardiomyocytes with exogenous sRAGE decreases hypoxia/reoxygenation induced apoptosis, mPTP opening, caspase activity, and cytochrome c release from the mitochondria (Guo et al. 2012). Paradoxically, however, circulating sRAGE levels are positively correlated with severity of heart failure in patients (Raposeiras-Roubin et al. 2010, Wang et al. 2011). A recent study showed that patients with ST-elevation myocardial infarction, who were treated with remote ischemic conditioning, did not display altered sRAGE concentrations despite an improved myocardial salvage index compared to control patients receiving primary percutaneous intervention (Jensen et al. 2015). These results indicate that sRAGE concentrations do not always reflect functional changes in the heart. A prospective clinical study demonstrated that in patients with acute coronary syndrome, increased sRAGE levels were associated with worsened in-hospital prognosis. In the same study, the investigators reported that fluorescent AGE levels were associated with long-term prognosis (Raposeiras-Roubin et al. 2013). These results show that sRAGE and AGEs may both be important prognostic tools for patients with acute coronary syndrome; however, additional studies are needed to better understand their roles as biomarkers. Furthermore, the exact reason for cardio-protection by sRAGE treatment yet association of endogenous sRAGE levels with worsened function in the heart is not understood.

Mitochondrial dysfunction is another hallmark of aging, heart failure and diabetes. The inability of cardiomyocytes to maintain ATP production through dysfunction of the electron transport chain and generation of ROS directly impairs the ability of the ventricle to continue contractile work. The mitochondria are one of the major sources of increased ROS production observed in the heart. Under physiological conditions, ROS are produced as a result of the 0.1% of total oxygen that leaks from the electron chain. Hyperglycemia causes an increase in intracellular glucose, which promotes the entry of NADH and $FADH_2$ from glucose-derived pyruvate into the electron transport chain. As a result, the inner mitochondrial membrane becomes hyperpolarized, reducing the ability of complex III to transfer elections to ubisemiquinone and producing superoxide (Nishikawa et al. 2000). Mitochondrial DNA is especially susceptible to oxidative damage due to its proximity to the site of origin of ROS production (Savu et al. 2011). While there is not a direct link between AGEs and mitochondrial ROS production, there is evidence that blockage of AGE production is associated with reduced ROS formation by the mitochondria in primary renal cells (Coughlan et al. 2009). Mitochondrial dysfunction can also lead to cell death via apoptosis or necrosis through mPTP opening. When H9C2 rat cardiomyoblasts were treated with CML, mitochondrial respiration was impaired in a ceramide-dependent manner (Nelson et al. 2015). This mechanism merits further investigation to better understand the potential impact of AGEs in heart mitochondria.

Conclusions

It is clear that AGEs play an important role in the heart, regulating the progression of disease states and the aging process (Fig. 1). A fuller, broader profiling of the different species of AGEs is necessary to better understand their role in the heart. The precise role of the various AGE species has not been well-defined; therefore, more in-depth studies defining the exact signaling and functional role of AGEs would be beneficial. Ideally, a better understanding of AGEs at each stage of the life cycle would provide much needed clarification of the role of AGEs in the heart. It is likely that specific AGEs elicit specific responses through RAGE or other cell surface receptors, or through interaction with circulating and structural proteins in the heart and vasculature. Blockade of either the generation of AGEs through aldose reductase inhibition or modulation of Glo-1 activity, or the intracellular effects of AGEs through RAGE blockade is an attractive target for protection of the aging, diabetic, and ischemic myocardium.

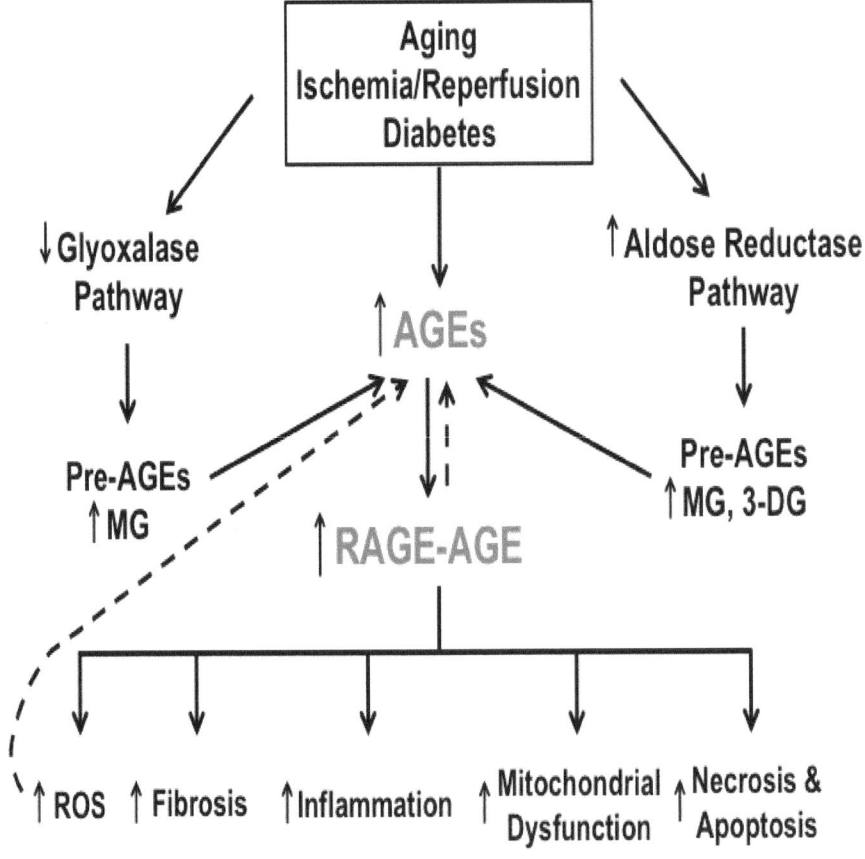

Figure 1. Aging, ischemia/reperfusion injury, and diabetes activate aldose reductase and downregulate the glyoxalase pathway, resulting in the accumulation of AGEs. Signaling via the AGE-RAGE axis leads to increased reactive oxygen species, fibrosis, mitochondrial dysfunction, inflammation, necrosis and apoptosis, all of which result in dysfunction in the heart. Generation of AGEs is also facilitated by sustained by generation of reactive oxygen species. AGE: advanced glycation endproducts, MG: methylglyoxal, 3-DG: 3-deoxyglucosone, RAGE: receptor for AGEs, ROS: reactive oxygen species. Hatched line indicates possible, yet untested mechanisms by which AGEs generation are likely to be amplified in aging, ischemia/reperfusion injury, and diabetes.

Keywords: Advanced glycation endproducts, RAGE, aldose reductase, glyoxylase, cardiac aging, diabetes, heart failure, ischemia-reperfusion injury, AGE-RAGE mechanisms

References

Ahmed, M. U., S. R. Thorpe and J. W. Baynes. 1986. Identification of N epsilon-carboxymethyllysine as a degradation product of fructoselysine in glycated protein. J Biol Chem 261(11): 4889–94.

Aleshin, A., R. Ananthakrishnan, Q. Li, R. Rosario, Y. Lu, W. Qu et al. 2008. RAGE modulates myocardial injury consequent to LAD infarction via impact on JNK and STAT signaling in a murine model. Am J Physiol Heart Circ Physiol 294(4): H1823–32. doi: 10.1152/ajpheart.01210.2007.

Almeida, F., D. Santos-Silva, T. Rodrigues, P. Matafome, J. Crisostomo, C. Sena et al. 2013. Pyridoxamine reverts methylglyoxal-induced impairment of survival pathways during heart ischemia. Cardiovasc Ther 31(6): e79–85. doi: 10.1111/1755-5922.12039.

Amicarelli, F., C. Di Ilio, L. Masciocco, A. Bonfigli, O. Zarivi, M. R. D'Andrea et al. 1997. Aging and detoxifying enzymes responses to hypoxic or hyperoxic treatment. Mech Ageing Dev 97(3): 215–26.

Ananthakrishnan, R., M. Kaneko, Y. C. Hwang, N. Quadri, T. Gomez, Q. Li et al. 2009. Aldose reductase mediates myocardial ischemia-reperfusion injury in part by opening mitochondrial permeability transition pore. Am J Physiol Heart Circ Physiol 296(2): H333–41. doi: 10.1152/ajpheart.01012.2008.

Ananthakrishnan, R., Q. Li, T. Gomes, A. M. Schmidt and R. Ramasamy. 2011. Aldose reductase pathway contributes to vulnerability of aging myocardium to ischemic injury. Exp Gerontol 46(9): 762–7. doi: 10.1016/j.exger.2011.05.001.

Aronson, D. 2003. Cross-linking of glycated collagen in the pathogenesis of arterial and myocardial stiffening of aging and diabetes. J Hypertens 21(1): 3–12. doi: 10.1097/01.hjh.0000042892.24999.92.

Avendano, G. F., R. K. Agarwal, R. I. Bashey, M. M. Lyons, B. J. Soni, G. N. Jyothirmayi et al. 1999. Effects of glucose intolerance on myocardial function and collagen-linked glycation. Diabetes 48(7): 1443–7.

Basta, G., G. Lazzerini, S. Del Turco, G. M. Ratto, A. M. Schmidt and R. De Caterina. 2005. At least 2 distinct pathways generating reactive oxygen species mediate vascular cell adhesion molecule-1 induction by advanced glycation end products. Arterioscler Thromb Vasc Biol 25(7): 1401–7. doi: 10.1161/01.ATV.0000167522.48370.5e.

Baynes, J. W. 2002. The Maillard hypothesis on aging: time to focus on DNA. Ann N Y Acad Sci 959: 360–7.

Borutaite, V., A. Budriunaite, R. Morkuniene and G. C. Brown. 2001. Release of mitochondrial cytochrome c and activation of cytosolic caspases induced by myocardial ischaemia. Biochim Biophys Acta 1537(2): 101–9.

Boulanger, E., M. P. Wautier, P. Gane, C. Mariette, O. Devuyst and J. L. Wautier. 2004. The triggering of human peritoneal mesothelial cell apoptosis and oncosis by glucose and glycoxydation products. Nephrol Dial Transplant 19(9): 2208–16. doi: 10.1093/ndt/gfh277.

Brett, J., A. M. Schmidt, S. D. Yan, Y. S. Zou, E. Weidman, D. Pinsky et al. 1993. Survey of the distribution of a newly characterized receptor for advanced glycation end products in tissues. Am J Pathol 143(6): 1699–712.

Brouwers, O., J. M. de Vos-Houben, P. M. Niessen, T. Miyata, F. van Nieuwenhoven, B. J. Janssen et al. 2013. Mild oxidative damage in the diabetic rat heart is attenuated by glyoxalase-1 overexpression. Int J Mol Sci 14(8): 15724–39. doi: 10.3390/ijms140815724.

Bucciarelli, L. G., M. Kaneko, R. Ananthakrishnan, E. Harja, L. K. Lee, Y. C. Hwang et al. 2006. Receptor for advanced-glycation end products: key modulator of myocardial ischemic injury. Circulation 113(9): 1226–34. doi: 10.1161/CIRCULATIONAHA.105.575993.

Bucciarelli, L. G., R. Ananthakrishnan, Y. C. Hwang, M. Kaneko, F. Song, D. R. Sell et al. 2008. RAGE and modulation of ischemic injury in the diabetic myocardium. Diabetes 57(7): 1941–51. doi: 10.2337/db07-0326.

Candido, R., J. M. Forbes, M. C. Thomas, V. Thallas, R. G. Dean, W. C. Burns et al. 2003. A breaker of advanced glycation end products attenuates diabetes-induced myocardial structural changes. Circ Res 92(7): 785–92. doi: 10.1161/01.RES.0000065620.39919.20.

Cantini, C., P. Kieffer, B. Corman, P. Liminana, J. Atkinson and I. Lartaud-Idjouadiene. 2001. Aminoguanidine and aortic wall mechanics, structure, and composition in aged rats. Hypertension 38(4): 943–8.

Cao, W., J. Chen, Y. Chen, S. Chen, X. Chen, H. Huang et al. 2015. Advanced glycation end products induced immune maturation of dendritic cells controls heart failure through NF-kappaB signaling pathway. Arch Biochem Biophys 580: 112–20. doi: 10.1016/j.abb.2015.07.003.

Casselmann, C., A. Reimann, I. Friedrich, A. Schubert, R. E. Silber and A. Simm. 2004. Age-dependent expression of advanced glycation end product receptor genes in the human heart. Gerontology 50(3): 127–34. doi: 10.1159/000076770.

Chang, K. C., K. L. Hsu, Y. I. Peng, F. C. Lee and Y. Z. Tseng. 2003. Aminoguanidine prevents age-related aortic stiffening in Fisher 344 rats: aortic impedance analysis. Br J Pharmacol 140(1): 107–14. doi: 10.1038/sj.bjp.0705410.

Chung, S. S. and S. K. Chung. 2003. Genetic analysis of aldose reductase in diabetic complications. Curr Med Chem 10(15): 1375–87.

Coughlan, M. T., D. R. Thorburn, S. A. Penfold, A. Laskowski, B. E. Harcourt, K. C. Sourris et al. 2009. RAGE-induced cytosolic ROS promote mitochondrial superoxide generation in diabetes. J Am Soc Nephrol 20(4): 742–52. doi: 10.1681/ASN.2008050514.

Crisostomo, J., P. Matafome, D. Santos-Silva, L. Rodrigues, C. M. Sena, P. Pereira et al. 2013. Methylglyoxal chronic administration promotes diabetes-like cardiac ischaemia disease in Wistar normal rats. Nutr Metab Cardiovasc Dis 23(12): 1223–30. doi: 10.1016/j.numecd.2013.01.005.

Dong, X. N., A. Qin, J. Xu and X. Wang. 2011. *In situ* accumulation of advanced glycation endproducts (AGEs) in bone matrix and its correlation with osteoclastic bone resorption. Bone 49(2): 174–83. doi: 10.1016/j.bone.2011.04.009.

Dyer, D. G., J. A. Dunn, S. R. Thorpe, K. E. Bailie, T. J. Lyons, D. R. McCance et al. 1993. Accumulation of Maillard reaction products in skin collagen in diabetes and aging. J Clin Invest 91(6): 2463–9. doi: 10.1172/JCI116481.

el Khoury, J., C. A. Thomas, J. D. Loike, S. E. Hickman, L. Cao and S. C. Silverstein. 1994. Macrophages adhere to glucose-modified basement membrane collagen IV via their scavenger receptors. J Biol Chem 269(14): 10197–200.

Fischer, T. H., L. S. Maier and S. Sossalla. 2013. The ryanodine receptor leak: how a tattered receptor plunges the failing heart into crisis. Heart Fail Rev 18(4): 475–83. doi: 10.1007/s10741-012-9339-6.

Funk, S. D., A. Yurdagul, Jr. and A. W. Orr. 2012. Hyperglycemia and endothelial dysfunction in atherosclerosis: lessons from type 1 diabetes. Int J Vasc Med 2012: 569654. doi: 10.1155/2012/569654.

Geroldi, D., C. Falcone, P. Minoretti, E. Emanuele, M. Arra and A. D'Angelo. 2006. High levels of soluble receptor for advanced glycation end products may be a marker of extreme longevity in humans. J Am Geriatr Soc 54(7): 1149–50. doi: 10.1111/j.1532-5415.2006.00776.x.

Giardino, I., D. Edelstein and M. Brownlee. 1994. Nonenzymatic glycosylation *in vitro* and in bovine endothelial cells alters basic fibroblast growth factor activity. A model for intracellular glycosylation in diabetes. J Clin Invest 94(1): 110–7. doi: 10.1172/JCI117296.

Goldberg, T., W. Cai, M. Peppa, V. Dardaine, B. S. Baliga, J. Uribarri et al. 2004. Advanced glycoxidation end products in commonly consumed foods. J Am Diet Assoc 104(8): 1287–91. doi: 10.1016/j.jada.2004.05.214.

Gorski, P. A., D. K. Ceholski and R. J. Hajjar. 2015. Altered myocardial calcium cycling and energetics in heart failure-a rational approach for disease treatment. Cell Metab 21(2): 183–94. doi: 10.1016/j.cmet.2015.01.005.

Guo, C., X. Zeng, J. Song, M. Zhang, H. Wang, X. Xu et al. 2012. A soluble receptor for advanced glycation end-products inhibits hypoxia/reoxygenation-induced apoptosis in rat cardiomyocytes via the mitochondrial pathway. Int J Mol Sci 13(9): 11923–40. doi: 10.3390/ijms130911923.

Haitoglou, C. S., E. C. Tsilibary, M. Brownlee and A. S. Charonis. 1992. Altered cellular interactions between endothelial cells and nonenzymatically glucosylated laminin/type IV collagen. J Biol Chem 267(18): 12404–7.

Hallam, K. M., Q. Li, R. Ananthakrishnan, A. Kalea, Y. S. Zou, S. Vedantham et al. 2010. Aldose reductase and AGE-RAGE pathways: central roles in the pathogenesis of vascular dysfunction in aging rats. Aging Cell 9(5): 776–84. doi: 10.1111/j.1474-9726.2010.00606.x.

Hamada, Y., J. Nakamura, K. Naruse, T. Komori, K. Kato, Y. Kasuya et al. 2000. Epalrestat, an aldose reductase inhibitor, reduces the levels of Nepsilon-(carboxymethyl)lysine protein adducts and their precursors in erythrocytes from diabetic patients. Diabetes Care 23(10): 1539–44.

Harja, E., D. X. Bu, B. I. Hudson, J. S. Chang, X. Shen, K. Hallam et al. 2008. Vascular and inflammatory stresses mediate atherosclerosis via RAGE and its ligands in apoE-/-mice. J Clin Invest 118(1): 183–94. doi: 10.1172/JCI32703.

Hartog, J. W., A. A. Voors, C. G. Schalkwijk, J. Scheijen, T. D. Smilde, K. Damman et al. 2007. Clinical and prognostic value of advanced glycation end-products in chronic heart failure. Eur Heart J 28(23): 2879–85. doi: 10.1093/eurheartj/ehm486.

Hasuike, Y., T. Nakanishi, Y. Otaki, M. Nanami, T. Tanimoto, N. Taniguchi et al. 2002. Plasma 3-deoxyglucosone elevation in chronic renal failure is associated with increased aldose reductase in erythrocytes. Am J Kidney Dis 40(3): 464–71. doi: 10.1053/ajkd.2002.34884.

Hausenloy, D. J. and D. M. Yellon. 2013. Myocardial ischemia-reperfusion injury: a neglected therapeutic target. J Clin Invest 123(1): 92–100. doi: 10.1172/JCI62874.

Hofmann, M. A., S. Drury, C. Fu, W. Qu, A. Taguchi, Y. Lu et al. 1999. RAGE mediates a novel proinflammatory axis: a central cell surface receptor for S100/calgranulin polypeptides. Cell 97(7): 889–901.

Horakova, L., M. K. Strosova, C. M. Spickett and D. Blaskovic. 2013. Impairment of calcium ATPases by high glucose and potential pharmacological protection. Free Radic Res 47Suppl 1: 81–92. doi: 10.3109/10715762.2013.807923.

Hovnanian, A. 2007. SERCA pumps and human diseases. Subcell Biochem 45: 337–63.

Huang, J. S., J. Y. Guh, H. C. Chen, W. C. Hung, Y. H. Lai and L. Y. Chuang. 2001. Role of receptor for advanced glycation end-product (RAGE) and the JAK/STAT-signaling pathway in AGE-induced collagen production in NRK-49F cells. J Cell Biochem 81(1): 102–13.

Hudson, B. I., A. Z. Kalea, M. Del Mar Arriero, E. Harja, E. Boulanger, V. D'Agati et al. 2008. Interaction of the RAGE cytoplasmic domain with diaphanous-1 is required for ligand-stimulated cellular migration through activation of Rac1 and Cdc42. J Biol Chem 283(49): 34457–68. doi: 10.1074/jbc.M801465200.

Huttunen, H. J., J. Kuja-Panula and H. Rauvala. 2002. Receptor for advanced glycation end products (RAGE) signaling induces CREB-dependent chromogranin expression during neuronal differentiation. J Biol Chem 277(41): 38635–46. doi: 10.1074/jbc.M202515200.

Hwang, Y. C., S. Sato, J. Y. Tsai, S. Yan, S. Bakr, H. Zhang et al. 2002. Aldose reductase activation is a key component of myocardial response to ischemia. FASEB J 16(2): 243–5. doi: 10.1096/fj.01-0368fje.

Hwang, Y. C., M. Kaneko, S. Bakr, H. Liao, Y. Lu, E. R. Lewis et al. 2004. Central role for aldose reductase pathway in myocardial ischemic injury. FASEB J 18(11): 1192–9. doi: 10.1096/fj.03-1400com.

Hwang, Y. C., S. Shaw, M. Kaneko, H. Redd, M. B. Marrero and R. Ramasamy. 2005. Aldose reductase pathway mediates JAK-STAT signaling: a novel axis in myocardial ischemic injury. FASEB J 19(7): 795–7. doi: 10.1096/fj.04-2780fje.

Jensen, L. J., K. Munk, A. Flyvbjerg, H. E. Botker and M. Bjerre. 2015. Soluble receptor of advanced glycation end-products in patients with acute myocardial infarction treated with remote ischaemic conditioning. Clin Lab 61(3-4): 323–8.

John, W. G. and E. J. Lamb. 1993. The Maillard or browning reaction in diabetes. Eye (Lond) 7(Pt 2): 230–7. doi: 10.1038/eye.1993.55.

Johnson, B. F., R. W. Nesto, M. A. Pfeifer, W. R. Slater, A. I. Vinik, D. A. Chyun et al. 2004. Cardiac abnormalities in diabetic patients with neuropathy: effects of aldose reductase inhibitor administration. Diabetes Care 27(2): 448–54.

Kass, D. A., E. P. Shapiro, M. Kawaguchi, A. R. Capriotti, A. Scuteri, R. C. deGroof et al. 2001. Improved arterial compliance by a novel advanced glycation end-product crosslink breaker. Circulation 104(13): 1464–70.

Kislinger, T., C. Fu, B. Huber, W. Qu, A. Taguchi, S. Du Yan et al. 1999. N(epsilon)-(carboxymethyl)lysine adducts of proteins are ligands for receptor for advanced glycation end products that activate cell signaling pathways and modulate gene expression. J Biol Chem 274(44): 31740–9.

Koschinsky, T., C. J. He, T. Mitsuhashi, R. Bucala, C. Liu, C. Buenting et al. 1997. Orally absorbed reactive glycation products (glycotoxins): an environmental risk factor in diabetic nephropathy. Proc Natl Acad Sci U S A 94(12): 6474–9.

Koyama, Y., Y. Takeishi, T. Arimoto, T. Niizeki, T. Shishido, H. Takahashi et al. 2007. High serum level of pentosidine, an advanced glycation end product (AGE), is a risk factor of patients with heart failure. J Card Fail 13(3): 199–206. doi: 10.1016/j.cardfail.2006.11.009.

Lakatta, E. G. 2003. Arterial and cardiac aging: major shareholders in cardiovascular disease enterprises: Part III: cellular and molecular clues to heart and arterial aging. Circulation 107(3): 490–7.

Lakatta, E. G. and D. Levy. 2003a. Arterial and cardiac aging: major shareholders in cardiovascular disease enterprises: Part I: aging arteries: a "set up" for vascular disease. Circulation 107(1): 139–46.

Lakatta, E. G. and D. Levy. 2003b. Arterial and cardiac aging: major shareholders in cardiovascular disease enterprises: Part II: the aging heart in health: links to heart disease. Circulation 107(2): 346–54.

Libby, P. 2002. Inflammation in atherosclerosis. Nature 420(6917): 868–74. doi: 10.1038/nature01323.

Liu, J., M. R. Masurekar, D. E. Vatner, G. N. Jyothirmayi, T. J. Regan, S. F. Vatner et al. 2003. Glycation end-product cross-link breaker reduces collagen and improves cardiac function in aging diabetic heart. Am J Physiol Heart Circ Physiol 285(6): H2587–91. doi: 10.1152/ajpheart.00516.2003.

Lo, T. W., M. E. Westwood, A. C. McLellan, T. Selwood and P. J. Thornalley. 1994. Binding and modification of proteins by methylglyoxal under physiological conditions. A kinetic and mechanistic study with N alpha-acetylarginine, N alpha-acetylcysteine, and N alpha-acetyllysine, and bovine serum albumin. J Biol Chem 269(51): 32299–305.

Loske, C., A. Neumann, A. M. Cunningham, K. Nichol, R. Schinzel, P. Riederer et al. 1998. Cytotoxicity of advanced glycation endproducts is mediated by oxidative stress. J Neural Transm 105(8-9): 1005–15.

McLellan, A. C., S. A. Phillips and P. J. Thornalley. 1993. The assay of S-D-lactoylglutathione in biological systems. Anal Biochem 211(1): 37–43. doi: 10.1006/abio.1993.1229.

McLellan, A. C., P. J. Thornalley, J. Benn and P. H. Sonksen. 1994. Glyoxalase system in clinical diabetes mellitus and correlation with diabetic complications. Clin Sci (Lond) 87(1): 21–9.

Miyata, T., Y. Ueda, T. Shinzato, Y. Iida, S. Tanaka, K. Kurokawa et al. 1996. Accumulation of albumin-linked and free-form pentosidine in the circulation of uremic patients with end-stage renal failure: renal implications in the pathophysiology of pentosidine. J Am Soc Nephrol 7(8): 1198–206.

Miyata, T., Y. Wada, Z. Cai, Y. Iida, K. Horie, Y. Yasuda et al. 1997. Implication of an increased oxidative stress in the formation of advanced glycation end products in patients with end-stage renal failure. Kidney Int 51(4): 1170–81.

Mozaffarian, D., E. J. Benjamin, A. S. Go, D. K. Arnett, M. J. Blaha, M. Cushman et al. 2015. Heart disease and stroke statistics—2015 update: a report from the American Heart Association. Circulation 131(4): e29–322. doi: 10.1161/CIR.0000000000000152.

Munch, G., B. Westcott, T. Menini and A. Gugliucci. 2012. Advanced glycation endproducts and their pathogenic roles in neurological disorders. Amino Acids 42(4): 1221–36. doi: 10.1007/s00726-010-0777-y.

Nelson, M. B., A. C. Swensen, D. R. Winden, J. S. Bodine, B. T. Bikman and P. R. Reynolds. 2015. Cardiomyocyte mitochondrial respiration is reduced by receptor for advanced glycation end-product signaling in a ceramide-dependent manner. Am J Physiol Heart Circ Physiol 309(1): H63–9. doi: 10.1152/ajpheart.00043.2015.

Nicholl, I. D. and R. Bucala. 1998. Advanced glycation endproducts and cigarette smoking. Cell Mol Biol (Noisy-le-grand) 44(7): 1025–33.

Nielsen, J. M., S. B. Kristiansen, R. Norregaard, C. L. Andersen, L. Denner, T. T. Nielsen et al. 2009. Blockage of receptor for advanced glycation end products prevents development of cardiac dysfunction in db/db type 2 diabetic mice. Eur J Heart Fail 11(7): 638–47. doi: 10.1093/eurjhf/hfp070.

Nishikawa, T., D. Edelstein, X. L. Du, S. Yamagishi, T. Matsumura, Y. Kaneda et al. 2000. Normalizing mitochondrial superoxide production blocks three pathways of hyperglycaemic damage. Nature 404(6779): 787–90. doi: 10.1038/35008121.

Niwa, H., A. Takeda, M. Wakai, T. Miyata, Y. Yasuda, T. Mitsuma et al. 1998. Accelerated formation of N epsilon-(carboxymethyl) lysine, an advanced glycation end product, by glyoxal and 3-deoxyglucosone in cultured rat sensory neurons. Biochem Biophys Res Commun 248(1): 93–7.

Niwa, T. 1999. 3-Deoxyglucosone: metabolism, analysis, biological activity, and clinical implication. J Chromatogr B Biomed Sci Appl 731(1): 23–36.

O'Brien, J. and P. A. Morrissey. 1989. Nutritional and toxicological aspects of the Maillard browning reaction in foods. Crit Rev Food Sci Nutr 28(3): 211–48. doi: 10.1080/10408398909527499.

Pamplona, R., M. Portero-Otin, M. J. Bellmun, R. Gredilla and G. Barja. 2002. Aging increases Nepsilon-(carboxymethyl)lysine and caloric restriction decreases Nepsilon-(carboxyethyl)lysine and Nepsilon-(malondialdehyde)lysine in rat heart mitochondrial proteins. Free Radic Res 36(1): 47–54.

Petrova, R., Y. Yamamoto, K. Muraki, H. Yonekura, K. Sakurai, T. Watanabe et al. 2002. Advanced glycation endproduct-induced calcium handling impairment in mouse cardiac myocytes. J Mol Cell Cardiol 34(10): 1425–31.

Rabbani, N. and P. J. Thornalley. 2015. Dicarbonyl stress in cell and tissue dysfunction contributing to ageing and disease. Biochem Biophys Res Commun 458(2): 221–6. doi: 10.1016/j.bbrc.2015.01.140.

Rai, V., F. Toure, S. Chitayat, R. Pei, F. Song, Q. Li et al. 2012. Lysophosphatidic acid targets vascular and oncogenic pathways via RAGE signaling. J Exp Med 209(13): 2339–50. doi: 10.1084/jem.20120873.

Ramasamy, R., P. J. Oates and S. Schaefer. 1997. Aldose reductase inhibition protects diabetic and nondiabetic rat hearts from ischemic injury. Diabetes 46(2): 292–300.

Ramasamy, R., S. F. Yan and A. M. Schmidt. 2010. Advanced glycation endproducts: from precursors to RAGE: round and round we go. Amino Acids. doi: 10.1007/s00726-010-0773-2.

Ramasamy, R., S. F. Yan and A. M. Schmidt. 2011. Receptor for AGE (RAGE): signaling mechanisms in the pathogenesis of diabetes and its complications. Ann N Y Acad Sci 1243: 88–102. doi: 10.1111/j.1749-6632.2011.06320.x.

Raposeiras-Roubin, S., B. K. Rodino-Janeiro, L. Grigorian-Shamagian, M. Moure-Gonzalez, A. Seoane-Blanco, A. Varela-Roman et al. 2010. Soluble receptor of advanced glycation end products levels are related to ischaemic aetiology and extent of coronary disease in chronic heart failure patients, independent of advanced glycation end products levels: New Roles for Soluble RAGE. Eur J Heart Fail 12(10): 1092–100. doi: 10.1093/eurjhf/hfq117.

Raposeiras-Roubin, S., B. K. Rodino-Janeiro, B. Paradela-Dobarro, L. Grigorian-Shamagian, J. M. Garcia-Acuna, P. Aguiar-Souto et al. 2013. Fluorescent advanced glycation end products and their soluble receptor: the birth of new plasmatic biomarkers for risk stratification of acute coronary syndrome. PLoS One 8(9): e74302. doi: 10.1371/journal.pone.0074302.

Roger, V. L., A. S. Go, D. M. Lloyd-Jones, E. J. Benjamin, J. D. Berry, W. B. Borden et al. 2012. Heart disease and stroke statistics—2012 update: a report from the American Heart Association. Circulation 125(1): e2–e220. doi: 10.1161/CIR.0b013e31823ac046.

Sakaguchi, T., S. F. Yan, S. D. Yan, D. Belov, L. L. Rong, M. Sousa et al. 2003. Central role of RAGE-dependent neointimal expansion in arterial restenosis. J Clin Invest 111(7): 959–72. doi: 10.1172/JCI17115.

Saunderson, C. E., R. A. Brogan, A. D. Simms, G. Sutton, P. D. Batin and C. P. Gale. 2014. Acute coronary syndrome management in older adults: guidelines, temporal changes and challenges. Age Ageing 43(4): 450–5. doi: 10.1093/ageing/afu034.

Savu, O., V. G. Sunkari, I. R. Botusan, J. Grunler, A. Nikoshkov and S. B. Catrina. 2011. Stability of mitochondrial DNA against reactive oxygen species (ROS) generated in diabetes. Diabetes Metab Res Rev 27(5): 470–9. doi: 10.1002/dmrr.1203.

Schleicher, E. D., E. Wagner and A. G. Nerlich. 1997. Increased accumulation of the glycoxidation product N(epsilon)-(carboxymethyl)lysine in human tissues in diabetes and aging. J Clin Invest 99(3): 457–68. doi: 10.1172/JCI119180.

Schmidt, A. M., M. Vianna, M. Gerlach, J. Brett, J. Ryan, J. Kao et al. 1992. Isolation and characterization of two binding proteins for advanced glycosylation end products from bovine lung which are present on the endothelial cell surface. J Biol Chem 267(21): 14987–97.

Schmidt, A. M., O. Hori, J. Brett, S. D. Yan, J. L. Wautier and D. Stern. 1994. Cellular receptors for advanced glycation end products. Implications for induction of oxidant stress and cellular dysfunction in the pathogenesis of vascular lesions. Arterioscler Thromb 14(10): 1521–8.

Schmidt, A. M., S. D. Yan, J. L. Wautier and D. Stern. 1999. Activation of receptor for advanced glycation end products: a mechanism for chronic vascular dysfunction in diabetic vasculopathy and atherosclerosis. Circ Res 84(5): 489–97.

Schmidt, A. M., S. D. Yan, S. F. Yan and D. M. Stern. 2001. The multiligand receptor RAGE as a progression factor amplifying immune and inflammatory responses. J Clin Invest 108(7): 949–55. doi: 10.1172/JCI14002.

Schmidt, A. M. 2015. Soluble RAGEs—prospects for treating & tracking metabolic and inflammatory disease. Vascul Pharmacol. doi: 10.1016/j.vph.2015.06.011.

Shamsi, F. A., A. Partal, C. Sady, M. A. Glomb and R. H. Nagaraj. 1998. Immunological evidence for methylglyoxal-derived modifications *in vivo*. Determination of antigenic epitopes. J Biol Chem 273(12): 6928–36.

Shang, L., R. Ananthakrishnan, Q. Li, N. Quadri, M. Abdillahi, Z. Zhu et al. 2010. RAGE modulates hypoxia/reoxygenation injury in adult murine cardiomyocytes via JNK and GSK-3beta signaling pathways. PLoS One 5(4): e10092. doi: 10.1371/journal.pone.0010092.

Shinohara, M., P. J. Thornalley, I. Giardino, P. Beisswenger, S. R. Thorpe, J. Onorato et al. 1998. Overexpression of glyoxalase-I in bovine endothelial cells inhibits intracellular advanced glycation endproduct formation and prevents hyperglycemia-induced increases in macromolecular endocytosis. J Clin Invest 101(5): 1142–7. doi: 10.1172/JCI119885.

Simm, A., C. Casselmann, A. Schubert, S. Hofmann, A. Reimann and R. E. Silber. 2004. Age associated changes of AGE-receptor expression: RAGE upregulation is associated with human heart dysfunction. Exp Gerontol 39(3): 407–13. doi: 10.1016/j.exger.2003.12.006.

Singh, R., A. Barden, T. Mori and L. Beilin. 2001. Advanced glycation end-products: a review. Diabetologia 44(2): 129–46. doi: 10.1007/s001250051591.

Suzuki, D., T. Miyata, N. Saotome, K. Horie, R. Inagi, Y. Yasuda et al. 1999. Immunohistochemical evidence for an increased oxidative stress and carbonyl modification of proteins in diabetic glomerular lesions. J Am Soc Nephrol 10(4): 822–32.

Taguchi, A., D. C. Blood, G. del Toro, A. Canet, D. C. Lee, W. Qu et al. 2000. Blockade of RAGE-amphoterin signalling suppresses tumour growth and metastases. Nature 405(6784): 354–60. doi: 10.1038/35012626.

Tanaka, S., G. Avigad, B. Brodsky and E. F. Eikenberry. 1988. Glycation induces expansion of the molecular packing of collagen. J Mol Biol 203(2): 495–505.

Tang, W. H., G. M. Kravtsov, M. Sauert, X. Y. Tong, X. Y. Hou, T. M. Wong et al. 2010. Polyol pathway impairs the function of SERCA and RyR in ischemic-reperfused rat hearts by increasing oxidative modifications of these proteins. J Mol Cell Cardiol 49(1): 58–69. doi: 10.1016/j.yjmcc.2009.12.003.

Tao, J., Y. F. Jin, Z. Yang, L. C. Wang, X. R. Gao, L. Lui et al. 2004. Reduced arterial elasticity is associated with endothelial dysfunction in persons of advancing age: comparative study of noninvasive pulse wave analysis and laser Doppler blood flow measurement. Am J Hypertens 17(8): 654–9. doi: 10.1016/j.amjhyper.2004.03.678.

Thornalley, P. J. 1993. The glyoxalase system in health and disease. Mol Aspects Med 14(4): 287–371.

Thornalley, P. J. 1998. Cell activation by glycated proteins. AGE receptors, receptor recognition factors and functional classification of AGEs. Cell Mol Biol (Noisy-le-grand) 44(7): 1013–23.

Thornalley, P. J. 2003. Glyoxalase I—structure, function and a critical role in the enzymatic defence against glycation. Biochem Soc Trans 31(Pt 6): 1343–8. doi: 10.1042/.

Tracey, W. R., W. P. Magee, C. A. Ellery, J. T. MacAndrew, A. H. Smith, D. R. Knight et al. 2000. Aldose reductase inhibition alone or combined with an adenosine A(3) agonist reduces ischemic myocardial injury. Am J Physiol Heart Circ Physiol 279(4): H1447–52.

Turk, Z. 2010. Glycotoxines, carbonyl stress and relevance to diabetes and its complications. Physiol Res 59(2): 147–56.

Uribarri, J., M. Peppa, W. Cai, T. Goldberg, M. Lu, C. He et al. 2003. Restriction of dietary glycotoxins reduces excessive advanced glycation end products in renal failure patients. J Am Soc Nephrol 14(3): 728–31.

Uribarri, J., W. Cai, O. Sandu, M. Peppa, T. Goldberg and H. Vlassara. 2005. Diet-derived advanced glycation end products are major contributors to the body's AGE pool and induce inflammation in healthy subjects. Ann N Y Acad Sci 1043: 461–6. doi: 10.1196/annals.1333.052.

Vaitkevicius, P. V., M. Lane, H. Spurgeon, D. K. Ingram, G. S. Roth, J. J. Egan et al. 2001. A cross-link breaker has sustained effects on arterial and ventricular properties in older rhesus monkeys. Proc Natl Acad Sci U S A 98(3): 1171–5. doi: 10.1073/pnas.98.3.1171.

Vlassara, H., Y. M. Li, F. Imani, D. Wojciechowicz, Z. Yang, F. T. Liu et al. 1995. Identification of galectin-3 as a high-affinity binding protein for advanced glycation end products (AGE): a new member of the AGE-receptor complex. Mol Med 1(6): 634–46.

Vlassara, H., W. Cai, J. Crandall, T. Goldberg, R. Oberstein, V. Dardaine et al. 2002. Inflammatory mediators are induced by dietary glycotoxins, a major risk factor for diabetic angiopathy. Proc Natl Acad Sci U S A 99(24): 15596–601. doi: 10.1073/pnas.242407999.

Vulesevic, B., R. W. Milne and E. J. Suuronen. 2014. Reducing methylglyoxal as a therapeutic target for diabetic heart disease. Biochem Soc Trans 42(2): 523–7. doi: 10.1042/BST20130254.

Wang, J., Y. Song, Q. Wang, P. M. Kralik and P. N. Epstein. 2006. Causes and characteristics of diabetic cardiomyopathy. Rev Diabet Stud 3(3): 108–17. doi: 10.1900/RDS.2006.3.108.

Wang, X. L., W. B. Lau, Y. X. Yuan, Y. J. Wang, W. Yi, T. A. Christopher et al. 2010. Methylglyoxal increases cardiomyocyte ischemia-reperfusion injury via glycative inhibition of thioredoxin activity. Am J Physiol Endocrinol Metab 299(2): E207–14. doi: 10.1152/ajpendo.00215.2010.

Wang, L. J., L. Lu, F. R. Zhang, Q. J. Chen, R. De Caterina and W. F. Shen. 2011. Increased serum high-mobility group box-1 and cleaved receptor for advanced glycation endproducts levels and decreased endogenous secretory receptor for advanced glycation endproducts levels in diabetic and non-diabetic patients with heart failure. Eur J Heart Fail 13(4): 440–9. doi: 10.1093/eurjhf/hfq231.

Wang, C. H., E. T. Wu, M. S. Wu, M. S. Tsai, Y. H. Ko, R. W. Chang et al. 2014. Pyridoxamine protects against mechanical defects in cardiac ageing in rats: studies on load dependence of myocardial relaxation. Exp Physiol 99(11): 1488–98. doi: 10.1113/expphysiol.2014.082008.

Wautier, M. P., O. Chappey, S. Corda, D. M. Stern, A. M. Schmidt and J. L. Wautier. 2001. Activation of NADPH oxidase by AGE links oxidant stress to altered gene expression via RAGE. Am J Physiol Endocrinol Metab 280(5): E685–94.

Wendt, T., L. Bucciarelli, W. Qu, Y. Lu, S. F. Yan, D. M. Stern et al. 2002. Receptor for advanced glycation endproducts (RAGE) and vascular inflammation: insights into the pathogenesis of macrovascular complications in diabetes. Curr Atheroscler Rep 4(3): 228–37.

Wu, E. T., J. T. Liang, M. S. Wu and K. C. Chang. 2011. Pyridoxamine prevents age-related aortic stiffening and vascular resistance in association with reduced collagen glycation. Exp Gerontol 46(6): 482–8. doi: 10.1016/j.exger.2011.02.001.

Yan, S. D., X. Chen, J. Fu, M. Chen, H. Zhu, A. Roher et al. 1996. RAGE and amyloid-beta peptide neurotoxicity in Alzheimer's disease. Nature 382(6593): 685–91. doi: 10.1038/382685a0.

Yan, S. F., R. Ramasamy, Y. Naka and A. M. Schmidt. 2003. Glycation, inflammation, and RAGE: a scaffold for the macrovascular complications of diabetes and beyond. Circ Res 93(12): 1159–69. doi: 10.1161/01.RES.0000103862.26506.3D.

Yan, S. F., R. Ramasamy and A. M. Schmidt. 2010a. The RAGE axis: a fundamental mechanism signaling danger to the vulnerable vasculature. Circ Res 106(5): 842–53. doi: 10.1161/CIRCRESAHA.109.212217.

Yan, S. F., R. Ramasamy and A. M. Schmidt. 2010b. Soluble RAGE: therapy and biomarker in unraveling the RAGE axis in chronic disease and aging. Biochem Pharmacol 79(10): 1379–86. doi: 10.1016/j.bcp.2010.01.013.

Yan, D., X. Luo, Y. Li, W. Liu, J. Deng, N. Zheng et al. 2014. Effects of advanced glycation end products on calcium handling in cardiomyocytes. Cardiology 129(2): 75–83. doi: 10.1159/000364779.

Yeh, C. H., L. Sturgis, J. Haidacher, X. N. Zhang, S. J. Sherwood, R. J. Bjercke et al. 2001. Requirement for p38 and p44/p42 mitogen-activated protein kinases in RAGE-mediated nuclear factor-kappaB transcriptional activation and cytokine secretion. Diabetes 50(6): 1495–504.

Zhang, L., D. Huang, D. Shen, C. Zhang, Y. Ma, S. A. Babcock et al. 2014. Inhibition of protein kinase C betaII isoform ameliorates methylglyoxal advanced glycation endproduct-induced cardiomyocyte contractile dysfunction. Life Sci 94(1): 83–91. doi: 10.1016/j.lfs.2013.11.011.

Zieman, S. J., V. Melenovsky and D. A. Kass. 2005. Mechanisms, pathophysiology, and therapy of arterial stiffness. Arterioscler Thromb Vasc Biol 25(5): 932–43. doi: 10.1161/01.ATV.0000160548.78317.29.

9

Advanced Glycation Endproducts in Aging Skin

Paraskevi Gkogkolou[a] and *Markus Böhm**

INTRODUCTION

Aging is the progressive accumulation damage over time, leading to disturbed cellular function, tissue homeostasis and ultimately to disease and death (Viña et al. 2007).

Skin, as the boundary between an organism and the environment, is not only exposed to the internal aging process but is also influenced by many environmental and external stressors. Thus, skin aging is traditionally divided into intrinsic and extrinsic aging. Both forms share some common characteristics but are distinguished by different pathogenetic mechanisms. Intrinsic aging refers to changes accumulating over time as part of the inevitable internal aging process. It is typically observed in skin areas protected from external stressors, such as the inner side of the arms or the gluteal regions. In extrinsic skin aging, on the other hand, environmental stressors like UV irradiation (Fisher et al. 2002), tobacco, chemicals and pollution (Makrantonaki and Zouboulis 2007, Zouboulis and Makrantonaki 2011) add up to the internal aging processes.

Intrinsic aged skin shows macroscopically fine wrinkles, dryness with scaling, loss of elasticity, decreased rigor and thickness as well as increased fragility and vulnerability to external traumata. Decreased proliferative capacity and cellular senescence are cardinal features. In the basal layer of the epidermis, irregular in size and volume keratinocytes with decreased mitotic activity and increased migration time to the stratum corneum lead to epidermal atrophy (thinning of the epidermis)

Department of Dermatology, Laboratory of Neuroendocrinology of the Skin and Interdisciplinary Endocrinology University of Münster, Münster, 48149, Germany.

[a] Email: paraskevi.gkogkolou@ukmuenster.de

* Corresponding author: bohmm@uni-muenster.de

(Makrantonaki and Zouboulis 2007). In the dermis, atrophy of the dermal extracellular matrix with decreased collagen and elastin fibers as well as of its cellular components, mainly fibroblasts is observed resulting in change of the biomechanical properties of the connective tissue. Sparse vessels with thin walls lead to decreased nutrient supply. Moreover, decreased number and impaired functionality of Langerhans cells leads to disturbed immune response and susceptibility to infections (Makrantonaki and Zouboulis 2007).

Extrinsic aged skin, on the contrary, is mainly characterized by a thickened epidermis with disturbed proliferation and desquamation of keratinocytes. Increased and irregular deposition of elastic fibers (solar elastosis) and reduced collagen content are prominent characteristics (Zouboulis and Makrantonaki 2011). Disorganized vasculature with thickened vessel walls as well as infiltration of inflammatory cells which produce and secrete proteolytic enzymes and reactive oxygen species are other characteristics.

The above mentioned changes due to intrinsic and extrinsic aging processes alter the structural and functional characteristics of skin, leading to an aged appearance as well as disturbed homeostasis. Aged skin shows increased fragility to external traumata, disturbed skin permeability, angiogenesis, decreased lipid and sweat production, decreased expression of antimicrobial peptides and immune function, decreased vitamin D synthesis, manifesting among others as impaired wound healing, atrophy, vulnerability to external stimuli and development of several benign and malignant diseases (reviewed in Zouboulis and Makrantonaki 2011).

Key mechanisms implicated in the pathogenesis of skin aging include alterations in DNA repair, cellular senescence, mitochondrial dysfunction and oxidative stress. In the last few years, Advanced Glycation Endproducts (AGEs) have emerged as important players in aging and age-related diseases. AGEs are products of non-enzymatic reactions between reducing sugars like glucose and proteins, lipids or nucleic acids. Details on formation and biochemistry of AGEs will not be provided in this chapter as this aspect is already addressed in a previous chapter.

Are AGEs Biomarkers for Skin Aging?

Skin, due to its immediate accessibility, offers an excellent opportunity for non-invasive and minimal invasive studies to assess AGE accumulation. To date, non-invasive "AGE readers", which measure *in vivo* the skin content of AGEs, based on their characteristic autofluorescence (Mulder et al. 2006), but also more accurate methods based on invasive skin sampling (Kawabata et al. 2011, Fan et al. 2010) exist. Within the wide spectrum of different AGEs many of them have been detected in the skin (Table 1).

Glycation-associated skin autofluorescence is increased in diabetic subjects in comparison to non-diabetic controls. However, AGEs also increase and correlate with chronological aging in healthy subjects (Corstjens et al. 2008). Moreover, skin autofluorescence positively correlates with various diabetes- and age-related complications such as renal disease, age-related macular degeneration and overall mortality (Mulder et al. 2006). Skin glycation can even serve as a predictor for the development of diabetic complications (Genuth et al. 2005).

Table 1. Detected AGEs in skin.

Type of AGE	Skin compartment	Targets of glycation	Methods of detection
CML	Epidermis (Kawabata et al. 2011) Dermis (Dyer et al. 1993, Fan et al. 2010, Jeanmaire et al. 2001) Photoaging - actinic elastosis (Jeanmaire et al. 2001, Mizutari et al. 1997)	Epidermis (SC-CK10, SS, SG) (Kawabata et al. 2011) Collagen (Dyer et al. 1993, Fan et al. 2010, Jeanmaire et al. 2001) Vimentin (Kueper et al. 2007) Elastin (Jeanmaire et al. 2001, Mizutari et al. 1997)	LC–ESI-TOF-MS, IF, IB (Kawabata et al. 2011) SIM/GC-MS (Dyer et al. 1993) IHC (Fan et al. 2010, Jeanmaire et al. 2001) IHC, ELISA, confocal microscopy (Jeanmaire et al. 2001, Mizutari et al. 1997)
Pentosidin	Dermis (Dyer et al. 1993, Yu et al. 2006)	Collagen (Dyer et al. 1993, Yu et al. 2006)	Reversed-phase HPLC (Dyer et al. 1993, Yu et al. 2006)
GO	Aged dermis (Fan et al. 2010)	Collagen (Fan et al. 2010)	LC/MS (Fan et al. 2010)
MGO	Aged dermis (Fan et al. 2010)	Collagen (Fan et al. 2010)	LC/MS (Fan et al. 2010)
Glucosepane	Aged dermis (Fan et al. 2010)	Collagen (Fan et al. 2010)	LC/MS (Fan et al. 2010)
Fructoselysine	Aged dermis (Fan et al. 2010)	Collagen (Fan et al. 2010)	LC/MS (Fan et al. 2010)
CEL	Aged dermis (Fan et al. 2010)	Collagen (Fan et al. 2010)	SIM/GC-MS (Fan et al. 2010)
GOLD	Aged dermis (Frye et al. 1998)	Collagen (Frye et al. 1998)	LC/MS (Frye et al. 1998)
MOLD	Aged dermis (Frye et al. 1998)	Collagen (Frye et al. 1998)	LC/MS (Frye et al. 1998)

As glycation is a slow procedure, accumulation of AGEs is mainly dependent on protein turnover rate. Therefore, dermal long-lived proteins like collagens type I and IV and fibronectin represent the major targets of AGE-modification (Verzijl et al. 2000, Dyer et al. 1993). Specifically, the appearance of glycated collagen is being first observed in the age of 20. With an accumulation rate of about 3.7% per year, it reaches 30–50% at 80 years of age (Dunn et al. 1991). Recently, N-ε-CML was detected in keratin 10 in the epidermis of healthy donors, pointing to a potential role of more short-lived proteins in glycation processes (Kawabata et al. 2011). Accordingly, in an *in vitro* reconstructed organ skin model, both epidermis and dermis were modified by glycation (Pageon et al. 2008).

AGE accumulation is also increased in photoaged skin, with 29.7% in sun-exposed versus 1.34% in sun-protected skin of young individuals (Jeanmaire et al. 2001). Areas with solar elastosis showed particular high amounts of AGEs, indicating that UV irradiation may promote formation of AGEs *in vivo* and that AGEs may play a crucial role in the various structural and functional modifications during photoaging (Mizutari et al. 1997). Smoking appears to be another aggravating factor of external skin aging by precipitating formation of AGEs. Smokers exhibit increased serum levels of AGEs compared to non-smokers (Cerami et al. 1997). The role of diet in skin aging and aging in general has been controversially discussed for years. Interestingly, about 10–30% of ingested AGEs are being absorbed in the circulation. Moreover, dietary AGEs directly correlate with serum levels of AGEs and inflammatory mediators (Uribarri et al. 2007, Vlassara et al. 2002).

It was initially believed that AGEs, once formed, can be only removed when the modified proteins degrade, with cathepsins D and L representing major enzymes for the intracellular degradation of endocytosed AGE-modified proteins (Grimm et al. 2012). However, it is now known, that many cells have intrinsic AGE-detoxifying pathways. The glutathione-dependent glyoxalase system consists of Glo I and II (Glyoxalase I and II), and uses reduced GSH to catalyze the conversion of glyoxal, methylglyoxal and other α-oxoaldehydes to the less toxic D-lactate (Xue et al. 2011). Fructosamine kinases are intracellular enzymes which phosphorylate and destabilize Amadori products leading to their spontaneous breakdown (Van Schaftingen et al. 2012). FN3K (Fructosamine-3-kinase), one of the most studied enzymes in this system, is almost ubiquitously expressed in human tissues, including the skin (Conner et al. 2005). Interestingly, decreased activity of such defense systems against AGEs has been reported during aging (Ramasamy et al. 2005). These age-related changes may increase the extent of deposited AGEs over time.

"RAGE" – Receptors for AGEs

AGEs exert their deleterious actions in part by interaction with a specific cell surface receptor, RAGE. RAGE is a multiligand member of the immunoglobulin superfamily. It is a promiscuous receptor that binds various molecules including AGEs, S-100/calgranulins, amphoterine and amyloid β-peptides. The binding of these ligands to RAGE stimulates various signaling pathways including the MAPKs and ERK 1 and 2. Stimulation of RAGE results in activation of the transcription factor NF-κB and subsequently of many proinflammatory genes. Notably, RAGE activation can directly induce oxidative stress by activating NADPH oxidase, decreasing activity of SOD and catalase (Ramasamy et al. 2005), and indirectly by reducing cellular antioxidant defenses, like GSH and ascorbic acid (Bierhaus et al. 2005). The reduction of GSH futher leads to decreased activity of Glo I, the major cellular defense system against α-oxoaldehydes, therefore, supporting further production of AGEs.

In the skin, RAGE is being expressed in both epidermis and dermis. RAGE expression is higher in sun-exposed than sun-protected areas (Lohwasser et al. 2006). Keratinocytes, fibroblasts, endothelial cells as well as immune cells like dendritic cells and monocytes express RAGE *in vitro* and *in vivo*.

AGE-R1 (oligosaccharyl transferase-48), AGE-R2 (80K-H phosphoprotein) and AGE-R3 (galectin-3) are additional receptors for AGEs. They can counteract RAGE signaling via regulation of endocytosis and degradation of AGEs (Lu et al. 2004). Such "intrinsic" RAGE-antagonism is also mediated by soluble RAGE (sRAGE) and endogenous secretory RAGE (esRAGE). Both are products of alternative gene splicing and post-translational proteolysis of the "full-length" RAGE and contain the ligand-binding domain but not the transmembrane domain. They do not have signaling properties and are considered as decoy receptors of RAGE (Vazzana et al. 2009).

Impact of AGEs in skin cells in vitro and in vivo

Advanced glycation targets amino-acids, lipids and nucleic acids in the intracellular as well as in the extracellular compartments. Therefore, AGEs can be formed almost ubiquitously in the organism and affect numerous physiological processes. Modifications of biological molecules by AGEs alter their biochemical and functional properties while interaction with RAGE leads to proinflammatory intracellular downstream signaling with pleotropic effects.

Extracellular matrix proteins

Extracellular Matrix (ECM) proteins serve as a supportive framework for cells and tissues but also directly interact with cell regulating various cellular functions like migration, differentiation and proliferation. Due to their long turnover rate they represent major targets for glycation.

Glycation creates intermolecular crosslinks of adjacent collagen fibers, thus increasing its stiffness, decreasing its flexibility and rendering it more prone to mechanical stimuli (Avery and Bailey 2006). AGE-mediated modifications on side chains of collagen cover contact-sites and impair interactions with cells and other components of the ECM (Haitoglou et al. 1992). Moreover, glycated collagen impairs actin polymerization and migration of immune cells (Haucke et al. 2014). The assembly of monomers into the triple helix and subsequently of collagen macromolecules can be affected, leading to an abnormal collagen network and decreased tissue permeability. Finally, AGE-modified collagen fibers become resistant against degradation by MMPs resulting in altered turnover and replacement with *de novo* synthesized collagen (DeGroot et al. 2001).

Elastin is another important target of glycation. Large and irregular aggregates of CML-modified elastin are found almost exclusively in sites of actinic elastosis in sun-exposed skin (Jeanmaire et al. 2001, Mizutari et al. 1997). These have decreased elastic properties and are resistant to proteolytic degradation (Yoshinaga et al. 2012). Mechanistically, UV-irradiation stimulates glycation of elastin in the presence of sugars *in vitro*. In addition, glycation of another ECM protein, fibronectin, leads to impaired interaction with integrin $\alpha v\beta 1$ of epidermal cells and delayed wound healing (Jacobsen et al. 2010).

As a consequence of these AGE-mediated changes on ECM proteins *in vitro*-glycated skin samples display impaired biomechanical properties. In accordance with this *in vivo* elasticity of skin from patients with diabetes is impaired in comparison to that of healthy controls (Yoon et al. 2002).

Intracellular proteins

Various intracellular proteins can be targets of glycation. Cytoskeletal proteins like vimentin in fibroblasts (Kueper et al. 2007) and keratin 10 in keratinocytes (Kawabata et al. 2011) can be modified by AGEs, thus compromising the stability of cytoskeleton and numerous cellular functions such as migration and cellular division. Basic Fibroblas Growth Factor (bFGF) displays impaired mitogenic activity in endothelial cells (Giardino et al. 1994), while glycation and crosslinking of the EGFR blocks its phosphorylation and subsequent activation of phospholipase C and ERKs

(Portero-Otín et al. 2002). Glycation of enzymes of the ubiquitin proteasome system and of the lysosomal proteolytic system inhibits their function (Uchiki et al. 2012), while antioxidant enzymes such as Cu-Zn-superoxide dismutase (Cu-Zn-SOD) can be inactivated (Ukeda et al. 1997). Glycation of DNA can lead to faulty DNA regulation and epigenetic modulation with detrimental effects (Baynes 2002).

RAGE-mediated signaling and effects

As mentioned above, AGE/RAGE-interaction initiates a cascade of signals influencing cell cycle and proliferation, gene expression, inflammation and extracellular matrix synthesis (reviewed in Bierhaus et al. 2005). In the skin, a wide variety of cell types express RAGE. *In vivo*, it is highly abundant in sites of solar elastosis, and its expression is induced by AGEs and proinflammatory cytokines like TNF-α (Lohwasser et al. 2006). In skin cells, RAGE decreases cell proliferation, induces apoptosis and increases MMP production, at least in part via NF-κB activation (Zhu et al. 2012).

Effects of AGEs on resident skin cells

AGEs affect various functions of skin cells *in vitro* (Table 2). In human dermal fibroblasts, they decrease proliferation, inhibit collagen- and ECM protein synthesis, induce premature senescence and even apoptosis, the latter via activation of RAGE, NF-κB and the caspases-3, -8 and -9 (Alikhani et al. 2005, Ravelojaona et al. 2009). In epidermal keratinocytes, AGEs decrease cell viability and migration, lead to senescence and induce the expression of proinflammatory mediators and MMPs (Zhu et al. 2011, Berge et al. 2007). Glycation of growth factors and their receptors like bFGF and EGFR may partly mediate these effects. Moreover, AGEs seem to potentiate the cytotoxic properties of UV-light, as UVA irradiated fibroblasts (Masaki et al. 1997) and keratinocytes (Wondrak et al. 2002) exhibit decreased viability after exposure to AGEs. Finally, AGEs increase the expression of adhesion molecules like ICAM-1 in vascular endothelial cells, resulting in adhesion and invasion of inflammatory cell in the skin (Schmidt et al. 1995).

The role of oxidative stress

Reactive Oxygen Species (ROS) can be generated from AGEs via various pathways. Firstly, during their crosslinking reactions AGEs act as electron donors, leading to formation of superoxide anion (Yim et al. 2001). Secondly, AGEs stimulate ROS-generating enzymes such as NADPH oxidase while thirdly intrinsic antioxidant enzymes such as Cu-Zn-SOD are functionally impaired (Ukeda et al. 1997). Thus, AGEs-mediated ROS generation seems to play a crucial role in photoaging. *In vitro* exposure of AGEs to UVA irradiation leads to formation of ROS, such as superoxide anion, hydrogen peroxide and hydroxyl radicals (Masaki et al. 1997). UV irradiation of human keratinocytes and fibroblasts in the presence of AGEs led to increased ROS formation and decreased proliferation *in vitro* (Wondrak et al. 2002). Finally, AGEs together with chromophores such as pentosidine act as endogenous photosensitizers leading to increased ROS formation after UVA irradiation of human skin (Fisher et al. 2002).

Table 2. Biological effects of AGEs and RAGE activator on skin.

Keratinocytes	Proliferation ↓ (Zhu et al. 2011)	Cell renewal↓
	Apoptosis ↑ (Zhu et al. 2012)	Epidermal homeostasis↓
	Oxidative stress↑ (Wondrak et al. 2002)	
	MMP-9↑, TIPM ↓ (Zhu et al. 2011)	
	Senescence ↑ (Berge et al. 2007)	
	NF-κB, proinflammatory mediators ↑ (Uchiki et al. 2012)	
	α2β1-integrin ↓ (Zhu et al. 2011)	
Fibroblasts	Proliferation ↓ (Alikhani et al. 2005)	Cell renewal↓
	Apoptosis↑ (Alikhani et al. 2005)	Dermal homeostasis↓
	ECM synthesis ↓ (Molinari et al. 2008)	Skin contractile function↓
	MMP↑ (Molinari et al. 2008)	
	Senescence ↑ (Ravelojaona et al. 2009)	
	NF-κB ↑ (Alikhani et al. 2005)	
	ROS ↑ (Wondrak et al. 2002)	
	Contractile properties ↓ (Kueper et al. 2007)	
Immune cells	Activation, differentiation ↑(Miyata et al. 1994, Chen et al. 2008)	Induction and propagation of inflammation
	Proinflammatory mediators ↑ (Miyata et al. 1994)	
ECM proteins (collagen, elastin, fibronectin)	Crosslinking (Dyer et al. 1993, Jeanmaire et al. 2001, Mizutari et al. 1997)	Elasticity ↓
	Resistance to MMP degradation (DeGroot et al. 2001)	Increased stiffness↑
	Impaired assembly of macromolecules to normal 3D structures (Yoshinaga et al. 2012, Reihsner et al. 2000)	Resistance to repair mechanisms
	Defect cross-talking to cells (Haitoglou et al. 1992)	Tissue permeability↓
Vascular endothelial cells	VCAM, ICAM, E-selectin ↑ (Schmidt et al. 1995)	Induction of proinflammatory mediators and recruitment of immune cells
	Permeability↑(Schmidt et al. 1995)	
	TNF-α, IL-6↑(Schmidt et al. 1995)	
	MCP-1↑(Schmidt et al. 1995)	

Strategies against the Impact of AGEs on Skin Aging

Due to the unraveled role of AGEs in diabetes- and age-related diseases, special efforts have been put on the development of anti-AGE-strategies. Various agents capable of inhibiting the formation of AGEs, of disrupting already formed AGEs and of antagonizing RAGE signaling have been identified. Some of them have already been tested in clinical trials (Elosta et al. 2012).

Aminoguanidine was one of the first substances that could reduce the formation of AGEs (Edelstein and Brownlee 1992). Aminoguanidine is a nucleophilic hydrazine which traps early glycation products such as carbonyl intermediate compounds and inhibits their further reaction with biological molecules. This molecule does not affect more advanced stages of glycation. Although it attenuated collagen glycation in an *in vitro* skin aging model (Pageon et al. 2008), its effects against AGE-mediated collagen modification *in vivo* still remain to be investigated.

Pyridoxamine, a naturally occurring vitamin B_6 isoform, is another potential agent against AGEs. Pyridoxamine traps reactive carbonyl intermediates, scavenges ROS and inhibits post-Amadori stages of AGEs formation. Oral intake of pyridoxamine resulted in potent inhibition of skin collagen CML formation in diabetic rats (Degenhardt et al. 2002).

Chemical substances and enzymes able to recognize and break the Maillard reaction crosslinks have also been identified. Such chemical "AGE-breakers" are dimethyl-3-phenayl-thiazolium chloride (ALT-711), N-phenacylthiazolium (PTB) and N-phenacyl-4,5-dimethylthiazolium (PMT), which can break the prototypical Maillard reaction crosslink via a thiazolium structure (Vasan et al. 2003). In a rat model, ALT-711 showed some promising results on skin hydration (Vasan et al. 2003).

Interference with intrinsic AGE-detoxification enzymes like FAOXs (Fructosyl-amine oxidases), FN3K and the enzymatic system of Glo is another interesting strategy to remove AGEs, as enzymes recognize specific substrates and may be associated with fewer side-effects (Wu and Monnier 2003). As noted above, decreased Glo I activity and increased accumulation of AGEs with age has been shown in many tissues and animals (Xue et al. 2011). Interestingly, it has been recently shown that Glo I is transcriptionally controlled by Nrf2, and that pharmacological Nrf2 activators increase Glo I mRNA and protein levels as well as its activity (Xue et al. 2012). Pharmacological induction of such enzymes may become a further future anti-AGE strategy.

Since oxidation steps are crucially involved in formation of many AGEs substances with antioxidative or metal chelating properties may have antiglycating effects as well. A lot of interest has therefore been directed towards nutrients and vitamins, so called "nutraceuticals", as natural weapons against AGEs (Elosta et al. 2012). The continuously growing list of these natural and often phytochemical agents includes ascorbic acid, α-tocopherol, niacinamide, riboflavin, zinc, α-lipoic acid, spices and herbs like ginger, cinnamon, marjoram and rosemary (Dearlove et al. 2008). Other promising compounds include naturally occurring flavonoids, such as luteolin, quercetin and rutin (Wu and Yen 2005), which can inhibit various stages of AGE formation. Interestingly, in a recent study, blueberry extract, an AGE-inhibitor and C-xyloside, (GAG) synthesis stimulator, were tested for 12 weeks in female diabetic subjects and showed significant improvement on firmness, wrinkles and hydration, although they failed to show a significant decrease in the cutaneous content of AGEs (Draelos et al. 2009).

Reduction of exogenously consumed AGEs may be perhaps the easiest way to reduce the impact of AGEs on the organism. AGE-rich diet correlates with serum levels of several proinflammatory mediators like TNF-α and C-reactive protein (Vlassara et al.

2002). AGE-poor diet, on the other hand, results in reduced levels of circulating AGEs and inflammatory biomarkers in patients with diabetes and renal failure (Yamagishi et al. 2007). It may even have beneficial effect on wound healing at least in a mouse model (Peppa et al. 2003). There are no studies addressing the impact of AGE-poor diet on skin aging in man. However, skin collagen glycation positively correlates with blood glucose levels in diabetes and decreases after treatment of hyperglycemia (Monnier et al. 1999), suggesting that a diet low in AGEs may have a beneficial effect on skin glycation and skin aging.

Inhibition of RAGE signaling via specific anti-RAGE antibodies and small molecular inhibitors is another potential strategy against excessive accumulation of AGEs. Promising effects in various systems have been shown *in vitro* and *in vivo* (Hudson et al. 2003). Potential protective effects of sRAGE have been shown in several diabetes and inflammatory models (Ramasamy et al. 2005, Yan et al. 2010). sRAGE could attenuate impaired wound healing in diabetic mice (Goova et al. 2001), although a potential effect on skin aging has not been investigated yet.

Finally, exploitation of molecular chaperones like carnosine and carninine may be a novel anti-AGE strategy with already promising results based on some skin aging parameters (Babizhayev et al. 2012).

Conclusions

Increasing evidence supports an important role of AGEs not only in diabetes but also in skin aging. Non-invasive methods can be employed to detect the amount of AGEs in skin and to monitor to extent of skin aging. Several strategies exist against the deleterious effect of AGEs. It is now time to start clinical trials to assess the effectiveness and feasibility of such strategies in the context of skin aging.

Keywords: Advanced glycation end products, skin, aging, photoaging, RAGE, AGEs

References

Alikhani, Z., M. Alikhani, C. M. Boyd, K. Naga, P. C. Trackman and D. T. Graves. 2005. Advanced glycation end products enhance expression of pro-apoptotic genes and stimulate fibroblast apoptosis through cytoplasmic and mitochondrial pathways. J Biol Chem 280: 12087–12095.

Avery, N. C. and A. J. Bailey. 2006. The effects of the Maillard reaction on the physical properties and cell interactions of collagen. Pathol Biol (Paris) 54: 387–395.

Babizhayev, M. A., A. I. Deyev, E. L. Savel'yeva, V. Z. Lankin and Y. E. Yegorov. 2012. Skin beautification with oral non-hydrolized versions of carnosine and carcinine: Effective therapeutic management and cosmetic skincare solutions against oxidative glycation and free-radical production as a causal mechanism of diabetic complications and skin aging. J Dermatolog Treat 23: 345–384.

Baynes, J. W. 2002. The Maillard hypothesis on aging: time to focus on DNA. Ann NY Acad Sci 959: 360–367.

Berge, U., J. Behrens and S. I. Rattan. 2007. Sugar-induced premature aging and altered differentiation in human epidermal keratinocytes. Ann N Y Acad Sci 1100: 524–529.

Bierhaus, A., P. M. Humpert, M. Morcos, T. Wendt, T. Chavakis, B. Arnold et al. 2005. Understanding RAGE, the receptor for advanced glycation end products. J Mol Med (Berl) 83: 876–886.

Cerami, C., H. Founds, I. Nicholl, T. Mitsuhashi, D. Giordano, S. Vanpatten et al. 1997. Tobacco smoke is a source of toxic reactive glycation products. Proc Natl Acad Sci USA 94: 13915–13920.

Chen, Y., E. M. Akirav, W. Chen, O. Henegariu, B. Moser, D. Desai et al. 2008. RAGE ligation affects T cell activation and controls T cell differentiation. J Immunol 181: 4272–4278.

Conner, J. R., P. J. Beisswenger and B. S. Szwergold. 2005. Some clues as to the regulation, expression, function, and distribution of fructosamine-3-kinase and fructosamine-3-kinase-related protein. Ann NY Acad Sci 1043: 824–836.

Corstjens, H., D. Dicanio, N. Muizzuddin, A. Neven, R. Sparacio, L. Declercq et al. 2008. Glycation associated skin autofluorescence and skin elasticity are related to chronological age and body mass index of healthy subjects. Exp Gerontol 43: 663–667.

Degenhardt, T. P., N. L. Alderson, D. D. Arrington, R. J. Beattie, J. M. Basgen, M. W. Steffes et al. 2002. Pyridoxamine inhibits early renal disease and dyslipidemia in the streptozotocin-diabetic rat. Kidney Int 61: 939–950.

DeGroot, J., N. Verzijl, M. J. Wenting-Van Wijk, R. A. Bank, F. P. Lafeber, J. W. Bijlsma et al. 2001. Age-related decrease in susceptibility of human articular cartilage to matrix metalloproteinase-mediated degradation: the role of advanced glycation end products. Arthritis Rheum 44: 2562–2571.

Dearlove, R. P., P. Greenspan, D. K. Hartle, R. B. Swanson and J. L. Hargrove. 2008. Inhibition of protein glycation by extracts of culinary herbs and spices. J Med Food 11: 275–281.

Draelos, Z. D., M. Yatskayer, S. Raab and C. Oresajo. 2009. An evaluation of the effect of a topical product containing C-xyloside and blueberry extract on the appearance of type II diabetic skin. J Cosmet Dermatol 8: 147–151.

Dunn, J. A., D. R. McCance, S. R. Thorpe, T. J. Lyons and J. W. Baynes. 1991. Age-dependent accumulation of N epsilon-(carboxymethyl)lysine and N epsilon-(carboxymethyl)hydroxylysine in human skin collagen. Biochemistry 30: 1205–1210.

Dyer, D. G., J. A. Dunn, S. R. Thorpe, K. E. Bailie, T. J. Lyons, D. R. McCance et al. 1993. Accumulation of Maillard reaction products in skin collagen in diabetes and aging. J Clin Invest 91: 2463–2469.

Edelstein, D. and M. Brownlee. 1992. Mechanistic studies of advanced glycosylation end product inhibition by aminoguanidine. Diabetes 41: 26–29.

Elosta, A., T. Ghous and N. Ahmed. 2012. Natural products as anti-glycation agents: possible therapeutic potential for diabetic complications. Curr Diabetes Rev 8: 92–108.

Fan, X., D. R. Sell, J. Zhang, I. Nemet, M. Theves, J. Lu et al. 2010. Anaerobic vs. aerobic pathways of carbonyl and oxidant stress in human lens and skin during aging and in diabetes: A comparative analysis. Free Radic Biol Med 49: 847–856.

Fisher, G. J., S. Kang, J. Varani, Z. Bata-Csorgo, Y. Wan, S. Datta et al. 2002. Mechanisms of photoaging and chronological skin aging. Arch Dermatol 138: 1462–1470.

Frye, E. B., T. B. Degenhardt, S. R. Thorpe and J. W. Baynes. 1998. Role of the Maillard reaction in aging of tissue proteins. Advanced glycation end product-dependent increase in imidazolium cross-links in human lens proteins. J Biol Chem 273: 18714–18719.

Genuth, S., W. Sun, P. Cleary, D. R. Sell, W. Dahms, J. Malone et al. 2005. Glycation and carboxymethyllysine levels in skin collagen predict the risk of future 10-year progression of diabetic retinopathy and nephropathy in the diabetes control and complications trial and epidemiology of diabetes interventions and complications participants with type 1 diabetes. Diabetes 54: 3103–3111.

Giardino, I., D. Edelstein and M. Brownlee. 1994. Nonenzymatic glycosylation *in vitro* and in bovine endothelial cells alters basic fibroblast growth factor activity. A model for intracellular glycosylation in diabetes. J Clin Invest 94: 110–117.

Goova, M. T., J. Li, T. Kislinger, W. Qu, Y. Lu, L. G. Bucciarelli et al. 2001. Blockade of receptor for advanced glycationend-products restores effective wound healing in diabetic mice. Am J Pathol 159: 513–525.

Grimm, S., M. Horlacher, B. Catalgol, A. Hoehn, T. Reinheckel and T. Grune. 2012. Cathepsins D and L reduce the toxicity of advanced glycation end products. Free Radic Biol Med 52: 1011–1023.

Haitoglou, C. S., E. C. Tsilibary, M. Brownlee and A. S. Charonis. 1992. Altered cellular interactions between endothelial cells and nonenzymatically glucosylated laminin/type IV collagen. J Biol Chem 267: 12404–12407.

Haucke, E., A. Navarrete-Santos, A. Simm, R. E. Silber and B. Hofmann. 2014. Glycation of extracellular matrix proteins impairs migration of immune cells. Wound Repair Regen 22: 239–245.

Hudson, B. I., L. G. Bucciarelli, T. Wendt, T. Sakaguchi, E. Lalla, W. Qu et al. 2003. Blockade of receptor for advanced glycation end products: a new target for therapeutic intervention in diabetic complications and inflammatory disorders. Arch Biochem Biophys 419: 80–88.

Jacobsen, J. N., B. Steffensen, L. Häkkinen, K. A. Krogfelt and H. S. Larjava. 2010. Skin wound healing in diabetic β6 integrin-deficient mice. APMIS 118: 753–764.

Jeanmaire, C., L. Danoux and G. Pauly. 2001. Glycation during human dermal intrinsic and actinic ageing: an *in vivo* and *in vitro* model study. Br J Dermatol 145: 10–18.

Kawabata, K., H. Yoshikawa, K. Saruwatari, Y. Akazawa, T. Inoue, T. Kuze et al. 2011. The presence of N(ε)-(Carboxymethyl) lysine in the human epidermis. Biochim Biophys Acta 1814: 1246–1252.

Kueper, T., T. Grune, S. Prahl, H. Lenz, V. Welge, T. Biernoth et al. 2007. Vimentin is the specific target in skin glycation. Structural prerequisites, functional consequences, and role in skin aging. J Biol Chem 282: 23427–23433.

Lohwasser, C., D. Neureiter, B. Weigle, T. Kirchner and D. Schuppan. 2006. The receptor for advanced glycation end products is highly expressed in the skin and upregulated by advanced glycation end products and tumor necrosis factor-alpha. J Invest Dermatol 126: 291–299.

Lu, C., J. C. He, W. Cai, H. Liu, L. Zhu and H. Vlassara. 2004. Advanced glycation end product (AGE) receptor 1 is a negative regulator of the inflammatory response to AGE in mesangial cells. Proc Natl Acad Sci USA 101: 11767–11772.

Makrantonaki, E. and C. C. Zouboulis. 2007. Molecular mechanisms of skin aging: state of the art. Ann NY Acad Sci 1119: 40–50.

Masaki, H., Y. Okano and H. Sakurai. 1997. Generation of active oxygen species from advanced glycation end products (AGE) under ultraviolet light A (UVA) irradiation. Biochem Biophys Res Commun 235: 306–310.

Miyata, T., R. Inagi, Y. Iida, M. Sato, N. Yamada, O. Oda, K. et al. 1994. Involvement of beta 2-microglobulin modified with advanced glycation end products in the pathogenesis of hemodialysis-associated amyloidosis. Induction of human monocyte chemotaxis and macrophage secretion of tumor necrosis factor-alpha and interleukin-1. J Clin Invest 93: 521–528.

Mizutari, K., T. Ono, K. Ikeda, K. Kayashima and S. Horiuchi. 1997. Photo-enhanced modification of human skin elastin in actinic elastosis by N(epsilon)-(carboxymethyl)lysine, one of the glycoxidation products of the Maillard reaction. J Invest. Dermatol 108: 797–802.

Molinari, J., E. Ruszova, V. Velebny, L. Robert. 2008. Effect of advanced glycation endproducts on gene expression profiles of human dermal fibroblasts. Biogerontology 9: 177–182.

Monnier, V. M., O. Bautista, D. Kenny, D. R. Sell, J. Fogarty, W. Dahms et al. 1999. Skin collagen glycation, glycoxidation, and crosslinking are lower in subjects with long-term intensive versus conventional therapy of type 1 diabetes: relevance of glycated collagen products versus HbA1c as markers of diabetic complications. DCCT Skin Collagen Ancillary Study Group. Diabetes Control and Complications Trial Diabetes 48: 870–880.

Mulder, D. J., T. V. Water, H. L. Lutgers, R. Graaff, R. O. Gans, F. Zijlstra et al. 2006. Skin autofluorescence, a novel marker for glycemic and oxidative stress-derived advanced glycation end products: an overview of current clinical studies, evidence, and limitations. Diabetes Technol Ther 8: 523–535.

Pageon, H., M. P. Técher and D. Asselineau. 2008. Reconstructed skin modified by glycation of the dermal equivalent as a model for skin aging and its potential use to evaluate anti-glycation molecules. Exp Gerontol 43: 584–588.

Peppa, M., H. Brem, P. Ehrlich, J. G. Zhang, W. Cai, Z. Li et al. 2003. Adverse effects of glycotoxins on wound healing in genetically diabetic mice. Diabetes 52: 2805–2813.

Portero-Otín, M., R. Pamplona, M. J. Bellmunt, M. C. Ruiz, J. Prat, R. Salvayre et al. 2002. Advanced glycation end product precursors impair epidermal growth factor receptor signaling. Diabetes 51: 1535–1542.

Ramasamy, R., S. J. Vannucci, S. S. Yan, K. Herold, S. F. Yan and A. M. Schmidt. 2005. Advanced glycation end products and RAGE: a common thread in aging, diabetes, neurodegeneration, and inflammation. Glycobiology 15: 16R–28R.

Ravelojaona, V., A. M. Robert and L. Robert. 2009. Expression of senescence-associated beta-galactosidase (SA-beta-Gal) by human skin fibroblasts, effect of advanced glycation end-products and fucose or rhamnose-rich polysaccharides. Arch Gerontol Geriatr 48: 151–154.

Reihsner, R., M. Melling, W. Pfeiler and E. J. Menzel. 2000. Alterations of biochemical and two-dimensional biomechanical properties of human skin in diabetes mellitus as compared to effects of *in vitro* non-enzymatic glycation. Clin Biomech 15: 379–386.

Schmidt, A. M., O. Hori, J. Chen, J. F. Li, J. Crandall, J. Zhang et al. 1995. Advanced glycation end products interacting with their endothelial receptor induce expression of vascular cell adhesion molecule-1 (VCAM-1): a potential mechanism for the accelerated vasculopathy of diabetes. J Clin Invest 96: 1395–1403.

Uchiki, T., K. A. Weikel, W. Jiao, F. Shang, A. Caceres, D. Pawlak et al. 2012. Glycation-altered proteolysis as a pathobiologic mechanism that links dietary glycemic index, aging, and age-related disease (in non diabetics). Aging Cell 11: 1–13.

Ukeda, H., Y. Hasegawa, T. Ishi and M. Sawamiura. 1997. Inactivation of Cu, Zn-superoxide dismutase by intermediates of Maillard reaction and glycolytic pathway and some sugars. Biosci Biotechnol Biochem 61: 2039–2042.

Uribarri, J., W. Cai, M. Peppa, S. Goodman, L. Ferrucci, G. Striker et al. 2007. Circulating glycotoxins and dietary advanced glycation end products: two links to inflammatory response, oxidative stress, and aging. J Gerontol A Biol Sci Med Sci 62: 427–433.

Van Schaftingen, E., F. Collard, E. Wiame and M. Veiga-da-Cunha. 2012. Enzymatic repair of amadori products. Amino Acids 42: 1143–1150.

Vasan, S., P. Foiles and H. Founds. 2003. Therapeutic potential of breakers of advanced glycation end product-protein crosslinks. Arch Biochem Biophys 419: 89–96.

Vazzana, N., F. Santilli, C. Cuccurullo and G. Davi. 2009. Soluble forms of RAGE in internal medicine. Intern Emerg Med 4: 389–401.

Verzijl, N., J. DeGroot, S. R. Thorpe, R. A. Bank, J. N. Shaw, T. J. Lyons et al. 2000. Effect of collagen turnover on the accumulation of advanced glycation end products. J Biol Chem 275: 39027–39031.

Viña, J., C. Borrás and J. Miquel. 2007. Theories of ageing. IUBMB Life 59: 249–254.

Vlassara, H., W. Cai, J. Crandall, T. Goldberg, R. Oberstein, V. Dardaine et al. 2002. Inflammatory mediators are induced by dietary glycotoxins, a major risk factor for diabetic angiopathy. Proc Natl Acad Sci USA 99: 15596–15601.

Wondrak, G. T., M. J. Roberts, M. K. Jacobson and E. L. Jacobson. 2002. Photosensitized growth inhibition of cultured human skin cells: Mechanism and suppression of oxidative stress from solar irradiation of glycated proteins. J Invest Dermatol 119: 489–498.

Wu, C. H. and G. C. Yen. 2005. Inhibitory effect of naturally occurring flavonoids on the formation of advanced glycation end products. J Agric Food Chem 53: 3167–3173.

Wu, X. and V. M. Monnier. 2003. Enzymatic deglycation of proteins. Arch Biochem Biophys 419: 16–24.

Xue, M., N. Rabbani and P. J. Thornalley. 2011. Glyoxalase in ageing. Semin. Cell Dev Biol 22: 293–301.

Xue, M., N. Rabbani, H. Momiji, P. Imbasi, M. M. Anwar, N. Kitteringham et al. 2012. Transcriptional control of glyoxalase 1 by Nrf2 provides a stress-responsive defence against dicarbonylglycation. Biochem J 443: 213–222.

Yamagishi, S., S. Ueda and S. Okuda. 2007. Food-derived advanced glycation end products (AGEs): a novel therapeutic target for various disorders. Curr Pharm Des 13: 2832–2836.

Yan, S. F., R. Ramasamy and A. M. Schmidt. 2010. Soluble RAGE: therapy and biomarker in unraveling the RAGE axis in chronic disease and aging. Biochem Pharmacol 79: 1379–1386.

Yim, M. B., H. S. Yim, C. Lee, S. O. Kang and P. B. Chock. 2001. Protein glycation: creation of catalytic sites for free radical generation. Ann NY Acad Sci 928: 48–53.

Yoon, H. S., S. H. Baik and C. H. Oh. 2002. Quantitative measurement of desquamation and skin elasticity in diabetic patients. Skin Res Technol 8: 250–254.

Yoshinaga, E., A. Kawada, K. Ono, E. Fujimoto, H. Wachi, S. Harumiya et al. 2012. N(ε)-(carboxymethyl) lysine modification of elastin alters its biological properties: implications for the accumulation of abnormal elastic fibers in actinic elastosis. J Invest Dermatol 132: 315–323.

Yu, Y., S. R. Thorpe, A. J. Jenkins, J. N. Shaw, M. A. Sochaski, D. McGee et al. 2006. Advanced glycation end-products and methionine sulphoxide in skin collagen of patients with type 1 diabetes. Diabetologia 49: 2488–2498.

Zhu, P., C. Yang, L. H. Chen, M. Ren, G. J. Lao and L. Yan. 2011. Impairment of human keratinocyte mobility and proliferation by advanced glycation end products-modified BSA. Arch Dermatol Res 303: 339–350.

Zhu, P., M. Ren, C. Yang, Y. X. Hu, J. M. Ran and L. Yan. 2012. Involvement of RAGE, MAPK and NF-κB pathways in AGEs-induced MMP-9 activation in HaCaT keratinocytes. Exp Dermatol 21: 123–129.

Zouboulis, C. C. and E. Makrantonaki. 2011. Clinical aspects and molecular diagnostics of skin aging. Clin Dermatol 29: 3–14.

10

Advanced Glycation Endproducts and Neurological Diseases

Teresita Menini

INTRODUCTION

AGEs are associated with long-lived proteins in all tissues and contribute to the aging process by interfering with the biological function of the affected proteins (Monnier and Cerami 1981). They also greatly contribute to the development of diseases such as atherosclerosis, neuropathy, nephropathy and neurodegeneration (Gasser and Forbes 2008).

It has been known for some time that in the aging brain and in the brains of patients suffering from neurodegenerative processes, AGEs accumulate intracellularly and extracellularly (Smith et al. 1995, Yan et al. 1994, Vitek et al. 1994). Using SDS-PAGE and immunoblotting with anti-AGE antibodies in tissues from AD patients, these products were identified in hippocampal pyramidal neurons, in extracellular neuropil and, in particular, pentosidine was found in senile plaques (Horie et al. 1997). AGEs deposits were less prominent in the neuropil of aged brains than in AD brains. Only faint deposits were found in brains of younger subjects, but no deposits were found in patients younger than 17 years of age (Takedo et al. 1996). Taking into consideration all their data, these authors hypothesized a direct relationship between the increased presence of AGEs and the process of neuronal aging and degeneration.

Accumulation of AGEs in the brain would be responsible for: (a) formation of insoluble protein deposits; (b) increased oxidative stress; (c) induction of inflammatory responses by stimulation of microglia and astrocytes (Kuhla et al. 2015, Jo et al. 2014, Schmidt et al. 2007, Munch et al. 2003).

Professor and Assistant Dean for Clinical Faculty Development Touro University – California 1310 Club Drive, Vallejo, 94592 CA.
Email: teresita.menini@tu.edu

In this chapter we will delineate the latest advances, in our knowledge, of the involvement of AGEs in the pathological process underlying the more frequent neurological diseases.

AGEs and Alzheimer's Disease (AD)

AD is the most common age-related degenerative disease of the nervous system characterized by severe diffuse cerebral atrophy evolving over a few years, which leads to progressive dementia due to neuronal death. Neuronal degeneration is mostly localized in the entorhinal cortex, hippocampus, and medial temporal lobe areas. Histologically, AD is characterized by three microscopic changes: (1) Intracellular deposits of thick fibers in the form of coils, loops or tangles of insoluble phosphorylated tau protein called neurofibrillary tangles (Braak and Braak 1988); (2) Deposits of amorphous material dispersed throughout the cerebral cortex composed of the protein, amyloid, which is surrounded by degenerating nerve terminals: amyloid plaques. The β-amyloid, a 39–43 amino acid peptide, the principal component in the amyloid plaques, is formed by intracellular proteolytic degradation of the transmembrane β-amyloid precursor protein (Tienari et al. 1997); (3) Neuronal degeneration, which is more evident in the hippocampus and is thought to reflect, in part, a defect in the phagocytosis of degraded proteins.

Most patients affected with AD have the sporadic, non-familial form of the disease with an incidence that increases after age 60. However, about 1% of cases are familial occurrences with a dominant inheritance and onset at a younger age. Mutations in three genes have been implicated in these familial cases: the gene encoding for the amyloid precursor protein, the presinilin 1 and the presinilin 2 genes. These mutations are believed to influence the rate of amyloid plaque formation, and this rate would be one of the parameters that determines the time of onset and progression of the disease (Holcomb et al. 1998).

The connection between AD and AGEs is provided by the fact that intracellular fibrillary tangles and extracellular amyloid plaques are mostly insoluble and long-lived, which makes them excellent targets for glycation and ultimately AGEs formation. The observations that metal ions such as zinc, copper and iron are elevated near the plaques environment prompted the study of these metals' participation in the AGEs formation process. It was found that exposure to trace amounts of transition metal ions accelerates the cross-linking of the Aβ peptide induced by AGEs (Loske et al. 2000).

AGEs attach to Aβ plaques, neurofibrillary tangles and senile plaques and have been implicated in the progression of Alzheimer's disease (Ahmed et al. 2005, Munch et al. 2003, Smith et al. 1996). Numerous studies have suggested that the formation of Aβ-AGE is more neurotoxic than the presence of Aβ alone, perhaps because of the increased cross linking and binding to the RAGE receptor. Aβ is one of the RAGE ligand and its affinity for the receptor is enhanced when Aβ contains AGEs (Li et al. 2013, Kuhla et al. 2004, Munch et al. 2003). The RAGE receptor is a well-characterized transmembrane multiligand protein of the immunoglobulin super family expressed in brain cells including neurons, microglia, astrocytes and endothelial vascular cells. The receptor extracellular domain binds different ligands: AGE, Aβ, S100/calgranulins and HMGB1, and two C-type immunoglobulin domains. Intracellularly, a short cytoplasmic domain is required for the mediating intracellular signaling pathways (Lv et al. 2015, Park et al. 2010). The evidence of RAGE participation in the process has been demonstrated. Accumulation of Aβ in the brains of AD patients produces increased RAGE expression in neurons, glial, and endothelial vascular cells and this over expression becomes greater as the disease progresses (Choi et al. 2014, Miller et al. 2008).

AGEs participate in tau hyperphosphorylation (Li et al. 2012b, Baki et al. 2004). Tau phosphorylation is increased by activation of protein kinases or inhibition of protein phosphatases. The most relevant of protein kinases promoting tau phosphorylation is GSK-3. The activity of GSK-3 is decreased by the PI3K-Akt intracellular signaling pathway, which promotes phosphorylation/inactivation of GSK-3 (Baki et al. 2004). Accumulation of extracellular AGEs upregulates RAGE expression, activates GSK-3, and inactivates Akt, all events leading lead to hyperphosphorylation of tau. When RAGE were blocked with antibodies, tau hyperphosphorylation decreased, suggesting that the AGE-mediated hyperphosphorylation of tau could be triggered by AGE-RAGE interactions, which are known to activate intracellular transduction pathways (Li et al. 2012a). In this way, both the accumulation of extracellular Aβ-AGE and the intracellular hyperphosphorylated tau protein, through AGE-RAGE mediated pathways, could inhibit synaptic LTP, producing the memory impairment demonstrated *in vivo* and *in vitro* studies (Chen et al. 2014, Li et al. 2013, Origlia et al. 2009, Chen et al. 2002). Other forms of synaptic plasticity have also been found to be altered. Accumulation of Aβ is able to disrupt LTD, which is dependent on the NMDA receptor activity on neurons and microglia (Origlia et al. 2010, Li et al. 2009, Snyder et al. 2005, Kim et al. 2001). However, LTP disruption starts before deficits in basal synaptic transmission and LTD, with increasing accumulation of Aβ. The disruption of synaptic transmission appears to be an earlier event than the histological changes described in AD (Oddo et al. 2003).

Glycation of proteins can also increase the production of free radicals, which in turn augments membrane peroxidation (Mullarkey et al. 1990). The aggregation of hyperphosphorylated tau protein to form neurofibrillary tangles is associated with high reactive products of lipid peroxidation (Perez et al. 2000). Oxidative stress is particularly relevant in the brain because of the increased energy needs and oxygen requirements in an environment with low antioxidant defense systems, compared to other organs. It has been demonstrated that lipid peroxidation precedes the formation of the amyloid plaques, making oxidative stress a strong factor in the pathogenesis of AD (Munch et al. 2003, Pratico et al. 2001). Methylglyoxal, a reactive dicarbonyl compound, which is an intermediate product of cell metabolism, is believed to be one of the main sources of AGEs in tissues. Methylglyoxal may contribute to neurodegeneration by triggering oxidative stress, which reflects an imbalance between reactive oxygen species generation and their detoxification by endogenous systems, in particular, glutathione and glutathione peroxidase. Methylglyoxal would directly inactivate glutathione peroxidase, resulting in an increased level of intracellular peroxides and subsequent cellular damage (Amicarelli et al. 2003). Using a neuroblastoma cell model, Deuther-Conrad et al. demonstrated that the presence of Aβ-AGE significantly increased the proportion of oxidized to reduced glutathione and that the generation of superoxide and hydrogen peroxide preceded these changes. These AGE-induced changes were attenuated when membrane permeable antioxidants were used, suggesting that they could be effective as therapeutic agents to normalized glutathione status (Deuther-Conrad et al. 2001).

The Aβ peptide has also been found circulating in the blood of AD patients (Shibata et al. 2000). Deane et al. have proposed a model for the transport of Aβ through the BBB. LRP–1 would mediate the transport of Aβ from the brain to the blood across the BBB while RAGEs would mediate the transport of circulating Aβ into the brain (Deane et al. 2003). RAGE would also mediate the transport of Aβ intracellularly contributing to Aβ-mediated neurotoxicity. The transport of Aβ by RAGE and the binding of AGE to the RAGE receptor activates transcription factors such as NF-kappa-B and the expression of pro-inflammatory cytokines such as IL-1, IL-6 and TNF-α (Berbaum et al. 2008, Deane et al. 2003).

Neuronal damage in AD has been postulated to occur also as a consequence of neurotoxicity related to the activation of microglia (Casal et al. 2002). Microglia would secrete pro-inflammatory molecules such as interleukin–1β, interleukin 6, TNF-α, NO and others, when activated by amyloid plaque Aβ-AGE accumulation and their binding to the RAGEs. These pro-inflammatory molecules may also induce oxidative stress and accelerate the neuronal demise (Munch et al. 1997). Origlia et al. in their 2010 paper confirm previous findings pointing, in particular, to interleukin-1b as one of the most potent factors responsible for inducing synaptic depression and disturbances in LTD. The intracellular pathways activated by the Aβ-RAGE interaction are protein kinase cascades, in particular JNK and p38MAPK that would be involved in the synaptic dysfunction (Origlia et al. 2010).

Recent studies have focused on the neuroprotective effects of geniposide, the pharmacological active component of the gardenia fruit. Lv et al. found that geniposide could decrease the inflammatory response triggered by Aβ-RAGE interaction by blocking the interaction, suppressing RAGE expression and decreasing the Aβ deposits by improving the expression of the enzymes that participate in Aβ degradation. Geniposide may, in this way, improve synaptic functioning and thus learning and memory in AD patients (Lv et al. 2015). It has also been demonstrated that geniposide reduces tau phosphorylation by decreasing GSK3β hyperactivity and lowers neuronal apoptosis in the strptozotocin-induced Alzheimer rat model (Gao et al. 2014). These studies point to a very promising agent with potential beneficial effects on AD which needs to be investigated in human studies.

AGEs and Amyotrophic Lateral Sclerosis (ALS)

ALS is a progressive neurodegenerative disease affecting primarily the upper and lower motor neurons in the brain and spinal cord. It is one of the most devastating neurological diseases, producing paralysis of voluntary body muscles and leading to death in 3 to 6 years. There is no primary therapy for ALS and riluzole, the only approved drug for use in these patients, only prolongs life for a few months. Symptomatic and psychological support are the main pillars of management in ALS. Ninety percent of the patients present with the sporadic form of the disease (SALS) and only 10% of cases are hereditary (FALS). The ultimate mechanisms responsible for the neuronal degeneration in ALS is the subject of intense investigation, however, it is known that the FALS patients show mutations in the SOD–1. The fact that SOD –1 is one of the enzymes involved in the quenching of superoxide radicals makes the hypothesis of oxidative stress involvement in the initiation and progression of ALS very plausible (Kikuchi et al. 2003, Kikuchi et al. 2002). Because most patients present have the sporadic form of the disease, the etiology of ALS is considered multifactorial. Some of the factors implicated in the dysfunction of the motor neurons are: excessive excitatory stimulation, protein misfolding, altered axonal transport and oxidative stress (Pasinelli and Brown 2006). Pathologically, both sporadic and hereditary cases present accumulations in the neuronal cytoplasm of neurofibrillary-derived Lewy-body-like hyaline inclusions (Gros-Louis et al. 2006, Kato et al. 2000).The presence of CML and non-CML AGEs was reported in the spinal cord anterior horn cell bodies of atrophic motor neurons and microglia of patients with ALS (Kikuchi et al. 2003, Kikuchi et al. 2002). Similar results were found by studying the presence of immunoreactivity for CML, pyrraline, carboxyethyl lysine and argpyrimidine in mutant SOD–1 transgenic mice compared with controls, and also in patients with an SOD–1 mutation (Shibata et al. 2002a), as well as patients affected by the sporadic form of ALS (Shibata et al. 2002b). The findings from their studies suggest that in patients with FALS, protein glycation and not lipid peroxidation was enhanced, while in SALS both protein glycoxidation and lipid peroxidation were enhanced. CML has also been found in the cerebrospinal fluid of patients affected with ALS (Kaufmann et al. 2004). Because of the known participation of AGE-RAGE interactions in the generation of reactive oxygen species through upregulation of pro-inflammatory pathways, several authors have investigated the presence of RAGEs, particularly the levels of the sRAGE in the serum of ALS patients, compared with a control group. A significant decrease in the serum levels of sRAGEs were found in ALS patients (Ilzecka 2009). The significance of these findings is unclear. Some authors postulate a protective effect of sRAGEs against oxidative stress and inflammation by consuming RAGE ligands (Geroldi et al. 2006, Falcone et al. 2005). In this way, lower levels of circulating sRAGE found in ALS and other neurodegenerative diseases could be associated with enhanced inflammatory responses and neurodegeneration (Sternberg et al. 2008, Emanuele et al. 2005).

All these findings clearly point towards a link between glycation and oxidative stress as responsible in the misfolding of proteins and accumulation of AGE in both sporadic and familial forms of ALS.

AGEs and Multiple Sclerosis (MS)

MS is a neurodegenerative autoimmune disease characterized by the presence of an inflammatory process with activation of microglial, recruitment of immunocompetent cells, and production of cytokines leading to demyelination of axons in the central nervous system, neuronal dysfunction and degeneration. Even if the ultimate molecular mechanisms responsible for disease progression are unclear, there have been important advances in the understanding of the disease. Here again, protein glycation with AGEs formation and enhanced oxidative stress leading to lipid and protein damage have been identified as possible mechanisms involved in the disease perpetuation (Sternberg et al. 2010, Gonsette 2008). As we discussed earlier in the AD section, AGEs could enhance the inflammatory process through upregulation of and interaction with the RAGE receptor, thus inducing microglial activation and oxidative stress (Schmidt et al. 2001). The RAGE receptor has been found upregulated in the brains of MS patients compared to controls, and the same results were found in animal models of MS (Andersson et al. 2008). One of the RAGE receptor ligands is HMGB1. This is a non-histone nuclear protein that may act as a cytokine when released from stimulated monocytes and macrophages, participating in the pathogenesis of inflammatory diseases (Gardella et al. 2002). HMGB1 has been shown to have an increased expression in active MS lesions and in animal model of MS, but it was expressed at normal levels in inactive lesions. This increased expression of HMGB1 was correlated with an increased expression of RAGE and other HMGB1 receptors in lesions and in the CSF from MS patients suggesting that HMGB1 and its receptors could also participate in the initiation and perpetuation of the inflammatory process (Andersson et al. 2008). These authors also demonstrated that macrophages and microglia were the major sources of HMGB1 in MS lesions from autopsy brains and in an immune-mediated animal model of MS. MS is accompanied by excessive production of reactive oxygen species and reactive nitrogen species, leading to oxidative and nitrosative stress. It is not clear yet if these processes could be initiators of the disease or if they are a consequence of the chronic inflammation in MS (Sadowska-Bartosz et al. 2013, Ljubisavljevic et al. 2013). Sternberg et al. measured the plasma levels of CML and CEL in MS patients and a healthy control group and found that CEL levels were higher in MS patients than in controls, while CML had similar levels in patients and controls. However, they also found that both CEL and CML levels increased with clinical relapse, even if the correlation was not statistically significant, probably due to the small number of relapsing patients. They also found that the plasma levels of CEL were decreased in patients administered with DMT such as interferon beta, glatiramer acetate and natalizumab, compared to patients not treated with DMT. Plasma CEL levels in these patients were more affected than CML levels (Sternberg et al. 2010). These authors also suggest that AGEs could be useful serum biomarkers in MS and that the use of AGE inhibitors or AGE-breaking agents could be part of new therapeutic modalities. In their latest paper, they studied the serum levels of esRAGE in MS patients and healthy controls to determine if those levels were modified by the chronic use of DMTs. They found that esRAGE levels are significantly higher in patients treated with DMTs and they hypothesized that this may reflect a reduction of the amounts of RAGE ligands (Sternberg et al. 2014). Because these studies have been conducted using small numbers of patients, a larger study is warranted to determine the clinical usefulness of these results.

Conclusion

Although there is no direct evidence that glycation and AGEs formation are causative per se, they play a significant role in the initiation and evolution of neurological diseases. The mechanisms by which AGEs produce their effects are multiple. AGEs would produce neuronal toxicity by: triggering aggregation of cross-linked, insoluble proteins and through the generation of oxidative stress and inflammatory processes by activation of the RAGE receptor. Multiple therapeutic opportunities are now open and should be explored in clinical trials. AGE-inhibitors, downregulation of RAGE expression, inactivation of RAGE ligands, anti-oxidation and inhibition of AGE-RAGE interaction are some of the targets that deserve further studies.

Keywords: AGE, RAGE, Alzheimer's, ALS, MS, neurodegeneration

References

Ahmed, N., U. Ahmed, P. J. Thornalley, K. Hager, G. Fleischer and G. Munch. 2005. Protein glycation, oxidation and nitration adduct residues and free adducts of cerebrospinal fluid in Alzheimer's disease and link to cognitive impairment. J Neurochem 92(2): 255–63.

Amicarelli, F., S. Colafarina, F. Cattani, A. Cimini, C. Di Ilio, M. P. Ceru et al. 2003. Scavenging system efficiency is crucial for cell resistance to ROS-mediated methylglyoxal injury. Free Radic Biol Med 35(8): 856–71.

Andersson, A., R. Covacu, D. Sunnemark, A. I. Danilov, A. Dal Bianco, M. Khademi et al. 2008. Pivotal advance: HMGB1 expression in active lesions of human and experimental multiple sclerosis. J Leukoc Biol 84(5): 1248–55.

Baki, L., J. Shioi, P. Wen, Z. Shao, A. Schwarzman, M. Gama-Sosa et al. 2004. PS1 activates PI3K thus inhibiting GSK-3 activity and tau overphosphorylation: effects of FAD mutations. EMBO J 23(13): 2586–96.

Berbaum, K., K. Shanmugam, G. Stuchbury, F. Wiede, H. Korner and G. Munch. 2008. Induction of novel cytokines and chemokines by advanced glycation endproducts determined with a cytometric bead array. Cytokine 41(3): 198–203.

Braak, H. and E. Braak. 1988. Neuropil threads occur in dendrites of tangle-bearing nerve cells. Neuropathol Appl Neurobiol 14(1): 39–44.

Casal, C., J. Serratosa and J. M. Tusell. 2002. Relationship between beta-AP peptide aggregation and microglial activation. Brain Res 928(1-2): 76–84.

Chen, C., X. H. Li, Y. Tu, H. T. Sun, H. Q. Liang, S. X. Cheng et al. 2014. Abeta-AGE aggravates cognitive deficit in rats via RAGE pathway. Neuroscience 257: 1–10.

Chen, Q. S., W. Z. Wei, T. Shimahara and C. W. Xie. 2002. Alzheimer amyloid beta-peptide inhibits the late phase of long-term potentiation through calcineurin-dependent mechanisms in the hippocampal dentate gyrus. Neurobiol Learn Mem 77(3): 354–71.

Choi, B. R., W. H. Cho, J. Kim, H. J. Lee, C. Chung, W. K. Jeon et al. 2014. Increased expression of the receptor for advanced glycation end products in neurons and astrocytes in a triple transgenic mouse model of Alzheimer's disease. Exp Mol Med 46: e75.

Deane, R., S. Du Yan, R. K. Submamaryan, B. LaRue, S. Jovanovic, E. Hogg et al. 2003. RAGE mediates amyloid-beta peptide transport across the blood-brain barrier and accumulation in brain. Nat Med 9(7): 907–13.

Deuther-Conrad, W., C. Loske, R. Schinzel, R. Dringen, P. Riederer and G. Munch. 2001. Advanced glycation end products change glutathione redox status in SH-SY5Y human neuroblastoma cells by a hydrogen peroxide dependent mechanism. Neurosci Lett 312(1): 29–32.

Emanuele, E., A. D'Angelo, C. Tomaino, G. Binetti, R. Ghidoni, P. Politi et al. 2005. Circulating levels of soluble receptor for advanced glycation end products in Alzheimer disease and vascular dementia. Arch Neurol 62(11): 1734–6.

Falcone, C., E. Emanuele, A. D'Angelo, M. P. Buzzi, C. Belvito, M. Cuccia et al. 2005. Plasma levels of soluble receptor for advanced glycation end products and coronary artery disease in nondiabetic men. Arterioscler Thromb Vasc Biol 25(5): 1032–7.

Gao, J., J. Teng, H. Liu, X. Han, B. Chen and A. Xie. 2014. Association of RAGE gene polymorphisms with sporadic Parkinson's disease in Chinese Han population. Neurosci Lett 559: 158–62.

Gardella, S., C. Andrei, D. Ferrera, L. V. Lotti, M. R. Torrisi, M. E. Bianchi et al. 2002. The nuclear protein HMGB1 is secreted by monocytes via a non-classical, vesicle-mediated secretory pathway. EMBO Rep 3(10): 995–1001.

Gasser, A. and J. M. Forbes. 2008. Advanced glycation: implications in tissue damage and disease. Protein Pept Lett 15(4): 385–91.

Geroldi, D., C. Falcone and E. Emanuele. 2006. Soluble receptor for advanced glycation end products: from disease marker to potential therapeutic target. Curr Med Chem 13(17): 1971–8.

Gonsette, R. E. 2008. Neurodegeneration in multiple sclerosis: the role of oxidative stress and excitotoxicity. J Neurol Sci 274(1-2): 48–53.

Gros-Louis, F., C. Gaspar and G. A. Rouleau. 2006. Genetics of familial and sporadic amyotrophic lateral sclerosis. Biochim Biophys Acta 1762(11-12): 956–72.

Holcomb, L., M. N. Gordon, E. McGowan, X. Yu, S. Benkovic, P. Jantzen et al. 1998. Accelerated Alzheimer-type phenotype in transgenic mice carrying both mutant amyloid precursor protein and presenilin 1 transgenes. Nat Med 4(1): 97–100.

Horie, K., T. Miyata, T. Yasuda, A. Takeda, Y. Yasuda, K. Maeda et al. 1997. Immunohistochemical localization of advanced glycation end products, pentosidine, and carboxymethyllysine in lipofuscin pigments of Alzheimer's disease and aged neurons. Biochem Biophys Res Commun 236(2): 327–32.

Ilzecka, J. 2009. Serum-soluble receptor for advanced glycation end product levels in patients with amyotrophic lateral sclerosis. Acta Neurol Scand 120(2): 119–22.

Jo, W. K., A. C. Law and S. K. Chung. 2014. The neglected co-star in the dementia drama: the putative roles of astrocytes in the pathogeneses of major neurocognitive disorders. Mol Psychiatry 19(2): 159–67.

Kato, S., S. Horiuchi, J. Liu, D. W. Cleveland, N. Shibata, K. Nakashima et al. 2000. Advanced glycation endproduct-modified superoxide dismutase-1 (SOD1)-positive inclusions are common to familial amyotrophic lateral sclerosis patients with SOD1 gene mutations and transgenic mice expressing human SOD1 with a G85R mutation. Acta Neuropathol 100(5): 490–505.

Kaufmann, E., B. O. Boehm, S. D. Sussmuth, R. Kientsch-Engel, A. Sperfeld, A. C. Ludolph et al. 2004. The advanced glycation end-product N epsilon-(carboxymethyl)lysine level is elevated in cerebrospinal fluid of patients with amyotrophic lateral sclerosis. Neurosci Lett 371(2-3): 226–9.

Kikuchi, S., K. Shinpo, A. Ogata, S. Tsuji, M. Takeuchi, Z. Makita et al. 2002. Detection of N epsilon-(carboxymethyl)lysine (CML) and non-CML advanced glycation end-products in the anterior horn of amyotrophic lateral sclerosis spinal cord. Amyotroph Lateral Scler Other Motor Neuron Disord 3(2): 63–8.

Kikuchi, S., K. Shinpo, M. Takeuchi, S. Yamagishi, Z. Makita, N. Sasaki et al. 2003. Glycation—a sweet tempter for neuronal death. Brain Res Brain Res Rev 41(2-3): 306–23.

Kim, J. H., R. Anwyl, Y. H. Suh, M. B. Djamgoz and M. J. Rowan. 2001. Use-dependent effects of amyloidogenic fragments of (beta)-amyloid precursor protein on synaptic plasticity in rat hippocampus *in vivo*. J Neurosci 21(4): 1327–33.

Kuhla, A., S. C. Ludwig, B. Kuhla, G. Munch and B. Vollmar. 2015. Advanced glycation end products are mitogenic signals and trigger cell cycle reentry of neurons in Alzheimer's disease brain. Neurobiol Aging 36(2): 753–61.

Kuhla, B., C. Loske, S. Garcia De Arriba, R. Schinzel, J. Huber and G. Munch. 2004. Differential effects of "Advanced glycation endproducts" and beta-amyloid peptide on glucose utilization and ATP levels in the neuronal cell line SH-SY5Y. J Neural Transm 111(3): 427–39.

Li, S., S. Hong, N. E. Shepardson, D. M. Walsh, G. M. Shankar and D. Selkoe. 2009. Soluble oligomers of amyloid Beta protein facilitate hippocampal long-term depression by disrupting neuronal glutamate uptake. Neuron 62(6): 788–801.

Li, X. H., B. L. Lv, J. Z. Xie, J. Liu, X. W. Zhou and J. Z. Wang. 2012a. AGEs induce Alzheimer-like tau pathology and memory deficit via RAGE-mediated GSK-3 activation. Neurobiol Aging 33(7): 1400–10.

Li, X. H., J. Z. Xie, X. Jiang, B. L. Lv, X. S. Cheng, L. L. Du et al. 2012b. Methylglyoxal induces tau hyperphosphorylation via promoting AGEs formation. Neuromolecular Med 14(4): 338–48.

Li, X. H., L. L. Du, X. S. Cheng, X. Jiang, Y. Zhang, B. L. Lv et al. 2013. Glycation exacerbates the neuronal toxicity of beta-amyloid. Cell Death Dis 4: e673. doi: 10.1038/cddis.2013.180.

Ljubisavljevic, S., I. Stojanovic, S. Vojinovic, D. Stojanov, S. Stojanovic, T. Cvetkovic et al. 2013. The patients with clinically isolated syndrome and relapsing remitting multiple sclerosis show different levels of advanced protein oxidation products and total thiol content in plasma and CSF. Neurochem Int 62(7): 988–97.

Loske, C., A. Gerdemann, W. Schepl, M. Wycislo, R. Schinzel, D. Palm et al. 2000. Transition metal-mediated glycoxidation accelerates cross-linking of beta-amyloid peptide. Eur J Biochem 267(13): 4171–8.

Lv, C., L. Wang, X. Liu, S. Yan, S. S. Yan, Y. Wang et al. 2015. Multi-faced neuroprotective effects of geniposide depending on the RAGE-mediated signaling in an Alzheimer mouse model. Neuropharmacology 89: 175–84.

Miller, M. C., R. Tavares, C. E. Johanson, V. Hovanesian, J. E. Donahue, L. Gonzalez et al. 2008. Hippocampal RAGE immunoreactivity in early and advanced Alzheimer's disease. Brain Res 1230: 273–80.

Monnier, V. M. and A. Cerami. 1981. Nonenzymatic browning in vivo: possible process for aging of long-lived proteins. Science 211(4481): 491–3.

Mullarkey, C. J., D. Edelstein and M. Brownlee. 1990. Free radical generation by early glycation products: a mechanism for accelerated atherogenesis in diabetes. Biochem Biophys Res Commun 173(3): 932–9.

Munch, G., J. Thome, P. Foley, R. Schinzel and P. Riederer. 1997. Advanced glycation endproducts in ageing and Alzheimer's disease. Brain Res Brain Res Rev 23(1-2): 134–43.

Munch, G., B. Kuhla, H. J. Luth, T. Arendt and S. R. Robinson. 2003. Anti-AGEing defences against Alzheimer's disease. Biochem Soc Trans 31(Pt 6): 1397–9.

Oddo, S., A. Caccamo, J. D. Shepherd, M. P. Murphy, T. E. Golde, R. Kayed et al. 2003. Triple-transgenic model of Alzheimer's disease with plaques and tangles: intracellular Abeta and synaptic dysfunction. Neuron 39(3): 409–21.

Origlia, N., S. Capsoni, A. Cattaneo, F. Fang, O. Arancio, S. D. Yan et al. 2009. Abeta-dependent Inhibition of LTP in different intracortical circuits of the visual cortex: the role of RAGE. J Alzheimers Dis 17(1): 59–68.

Origlia, N., C. Bonadonna, A. Rosellini, E. Leznik, O. Arancio, S. S. Yan et al. 2010. Microglial receptor for advanced glycation end product-dependent signal pathway drives beta-amyloid-induced synaptic depression and long-term depression impairment in entorhinal cortex. J Neurosci 30(34): 11414–25.

Park, H., F. G. Adsit and J. C. Boyington. 2010. The 1.5 Å crystal structure of human receptor for advanced glycation endproducts (RAGE) ectodomains reveals unique features determining ligand binding. J Biol Chem 285(52): 40762–70.

Pasinelli, P. and R. H. Brown. 2006. Molecular biology of amyotrophic lateral sclerosis: insights from genetics. Nat Rev Neurosci 7(9): 710–23.

Perez, M., R. Cuadros, M. A. Smith, G. Perry and J. Avila. 2000. Phosphorylated, but not native, tau protein assembles following reaction with the lipid peroxidation product, 4-hydroxy-2-nonenal. FEBS Lett 486(3): 270–4.

Pratico, D., K. Uryu, S. Leight, J. Q. Trojanoswki and V. M. Lee. 2001. Increased lipid peroxidation precedes amyloid plaque formation in an animal model of Alzheimer amyloidosis. J Neurosci 21(12): 4183–7.

Sadowska-Bartosz, I., M. Adamczyk-Sowa, S. Galiniak, S. Mucha, K. Pierzchala and G. Bartosz. 2013. Oxidative modification of serum proteins in multiple sclerosis. Neurochem Int 63(5): 507–16.

Schmidt, A. M., S. D. Yan, S. F. Yan and D. M. Stern. 2001. The multiligand receptor RAGE as a progression factor amplifying immune and inflammatory responses. J Clin Invest 108(7): 949–55.

Schmidt, A., B. Kuhla, K. Bigl, G. Munch and T. Arendt. 2007. Cell cycle related signaling in Neuro2a cells proceeds via the receptor for advanced glycation end products. J Neural Transm 114(11): 1413–24.

Shibata, M., S. Yamada, S. R. Kumar, M. Calero, J. Bading, B. Frangione et al. 2000. Clearance of Alzheimer's amyloid-ss(1-40) peptide from brain by LDL receptor-related protein-1 at the blood-brain barrier. J Clin Invest 106(12): 1489–99.

Shibata, N., A. Hirano, E. T. Hedley-Whyte, M. C. Dal Canto, R. Nagai, K. Uchida et al. 2002a. Selective formation of certain advanced glycation end products in spinal cord astrocytes of humans and mice with superoxide dismutase-1 mutation. Acta Neuropathol 104(2): 171–8.

Shibata, N., H. Oda, A. Hirano, Y. Kato, M. Kawaguchi, M. C. Dal Canto et al. 2002b. Molecular biological approaches to neurological disorders including knockout and transgenic mouse models. Neuropathology 22(4): 337–49.

Smith, M. A., V. M. Monnier, L. M. Sayre and G. Perry. 1995. Amyloidosis, advanced glycation end products and Alzheimer disease. Neuroreport 6(12): 1595–6.

Smith, M. A., L. M. Sayre and G. Perry. 1996. Diabetes mellitus and Alzheimer's disease: glycation as a biochemical link. Diabetologia 39(2): 247.

Snyder, E. M., Y. Nong, C. G. Almeida, S. Paul, T. Moran, E. Y. Choi et al. 2005. Regulation of NMDA receptor trafficking by amyloid-beta. Nat Neurosci 8(8): 1051–8.

Sternberg, Z., B. Weinstock-Guttman, D. Hojnacki, P. Zamboni, R. Zivadinov, K. Chadha et al. 2008. Soluble receptor for advanced glycation end products in multiple sclerosis: a potential marker of disease severity. Mult Scler 14(6): 759–63.

Sternberg, Z., C. Hennies, D. Sternberg, P. Wang, P. Kinkel, D. Hojnacki et al. 2010. Diagnostic potential of plasma carboxymethyllysine and carboxyethyllysine in multiple sclerosis. J Neuroinflammation 7: 72.

Sternberg, Z., D. Sternberg, A. Drake, T. Chichelli, J. Yu and D. Hojnacki. 2014. Disease modifying drugs modulate endogenous secretory receptor for advanced glycation end-products, a new biomarker of clinical relapse in multiple sclerosis. J Neuroimmunol 274(1-2): 197–201.

Takedo, A., T. Yasuda, T. Miyata, K. Mizuno, M. Li, S. Yoneyama et al. 1996. Immunohistochemical study of advanced glycation end products in aging and Alzheimer's disease brain. Neurosci Lett 221(1): 17–20.

Tienari, P. J., N. Ida, E. Ikonen, M. Simons, A. Weidemann, G. Multhaup et al. 1997. Intracellular and secreted Alzheimer beta-amyloid species are generated by distinct mechanisms in cultured hippocampal neurons. Proc Natl Acad Sci U S A 94(8): 4125–30.

Vitek, M. P., K. Bhattacharya, J. M. Glendening, E. Stopa, H. Vlassara, R. Bucala et al. 1994. Advanced glycation end products contribute to amyloidosis in Alzheimer disease. Proc Natl Acad Sci U S A 91(11): 4766–70.

Yan, S. D., X. Chen, A. M. Schmidt, J. Brett, G. Godman, Y. S. Zou et al. 1994. Glycated tau protein in Alzheimer disease: a mechanism for induction of oxidant stress. Proc Natl Acad Sci U S A 91(16): 7787–91.

11

Advanced Glycation Endproducts for Age-at-Death Estimation

Luis L. Cabo,[1] Christian Thomas[2] and Sara C. Zapico[2],*

INTRODUCTION

Age-at-death estimation is one of the fundamental parameters in the forensic sciences to create a biological profile leading to the identification of deceased individuals (Zapico and Ubelaker 2013). Age can be estimated very accurately in childhood using anthropological methods. In contrast, in adulthood, estimation of age-at-death is influenced by endogenous and exogenous factors, which make an accurate determination difficult (Zapico and Ubelaker 2013). For that reason, new methodologies for age-at-death estimation are being developed based on the natural process of aging, which leads to alterations of tissues and organs at different biochemical levels. One of these alterations can be seen in the study of Advanced Glycation endproducts. Although few studies were conducted with forensic purposes in mind, this chapter will give an overview of different works that correlate AGEs with aging, pointing out their application to forensic science research.

AGEs and Aging

The Maillard reaction is a complex series of reactions between reducing sugars and amino groups of proteins, which lead to browning, fluorescence and cross-linking

[1] Mercyhurst University Department of Applied Forensic Sciences Zurn 119C, 501 E.38th St. Erie, PA 16546, USA.
  Email: lcabo@mercyhurst.edu
[2] Smithsonian Institution, National Museum of Natural History Anthropology Department 10th and Constitution Ave, NW, PO Box 37012 Washington, DC 20560, USA.
  Email: crf.thomas@gmail.com
* Corresponding author: saiczapico@gmail.com

of protein (Baynes 2001). AGEs products, formed during the later stages of the Maillard reaction, accumulate in long-lived tissue proteins and may contribute to the development of complications in aging (Thorpe and Baynes 1996). Human aging is associated with a stiffening of tissues that are rich in extracellular matrix and long-lived proteins, such as skeletal muscle, tendons, joints, bone, heart, arteries, lung, skin and lens (Monnier et al. 2005). GO, MGO, and deoxyglucosones belong to a series of dicarbonyl compounds, identified as intermediates in the Maillard reaction. GO and MGO react with lysine and arginine residues in protein yielding compounds such as the N-(carboxy-alkyl)lysines, CML and CEL, and imidazolones and dehydroimidazolones (Ahmed et al. 1997, Ahmed et al. 1986, Henle et al. 1994, Lo et al. 1994). Glucosepane appears to be the most important cross-linking AGE in human tissues, as well as MGO and pentoside. Cross-linking of collagen and other proteins by AGEs affects the mechanical properties of tissues, especially of the vasculature (Monnier et al. 2005).

Different studies were developed to determine the implication of AGEs on aging. Semba et al. studied population-based cohorts of aging in the United States and Europe. These studies suggested that older adults with elevated circulating CML are at greater risk of arterial stiffness, chronic kidney disease, anemia, poor skeletal muscle strength and physical and cardiovascular performance; all of which are causes of mortality (Semba et al. 2009c, Dalal et al. 2009, Semba et al. 2010, Semba et al. 2008a, Semba et al. 2008b, Semba et al. 2009b, Semba et al. 2009a). CML and CEL were found to increase in skin collagen and lens proteins with age (Ahmed et al. 1997, Dunn et al. 1991, Dunn et al. 1989, Dyer et al. 1993, Kumar et al. 2007). Besides, MOLD and GOLD correlated significantly with age in human lens proteins. Recent studies suggest that AGEs accumulate in erythrocytes and alter their deformability (Iwata et al. 2004).

In most mature tissues, collagen molecules have a long life-time (> 200 years), making them susceptible to NEG via the Maillard reaction. Evidence is emerging suggesting that the accumulation of AGEs in bone contributes to disturbed bone modeling and deterioration of bone tissue quality (Hein 2006). AGEs concentrations increase in cortical and trabecular bone with age and are negatively associated with bone density and mineralization (Odetti et al. 2005). Accumulation of AGEs in the collagen matrix of bone alters the mechanical properties of bone, increasing stiffness and fragility (Tang et al. 2007).

In skeletal muscle, there is an increase in cross-linking of collagen and deposition of AGEs in older adults (Haus et al. 2007). The accumulation of NEG products is supported by the color change of articular cartilage from bluish in young age to yellow/brownish in the elderly. There was an increase in the level of pentosidine in cartilage with advancing age (Eyre et al. 1984, Sell et al. 1991, Vlassara et al. 1994, Verzijl et al. 2002, Bank et al. 1998). This has been shown to increase the stiffness of human articular cartilage (Verzijl et al. 2002).

Apart from the tissues described above, there is strong evidence for the role of AGEs in atherosclerosis (Basta et al. 2004). With aging, AGEs are deposited in arterial walls, especially the elastic membrane and intimal extracellular matrix, and the process appears to be accelerated in diabetes (Nerlich and Schleicher 1999).

AGEs accumulate in the human brain as age increases. This accumulation was found in hippocampal pyramidal neurons (Kimura et al. 1996). With the forensic sciences in mind, Sato et al. in 2001 described the same increase of AGEs with age in hippocampal pyramidal neurons using an immunohistochemical approach, with an antibody against AGEs. They did not find an AGE signal in young individuals; however this signal increased with age, with higher amounts detected in older individuals. This correlation also was found in cases of death by fire, so the analysis of AGEs may be useful in such cases and likely correlates with the age at death.

Moving on from the studies above, RAGE, AGEs receptor, seems to play a key role in the aging process. RAGE accelerates and perpetuates the aging process and might shorten lifespan. The most direct mechanism for this process is through its interaction with AGEs, triggering the induction of increased reactive oxygen species, activating NADPH oxidase, increasing expression of adhesion molecules, and upregulating inflammation through NF-κB and other signaling pathways; leading to a perpetuated pro-inflammatory phenotype, which is associated with age-related neurodegenerative, musculoskeletal and vascular diseases (Bierhaus and Nawroth 2009, Bierhaus et al. 2001, Coughlan et al. 2009). Moreover, it has recently been observed that activation of RAGE can lead to epigenetic alterations. The remodeling of chromatin through posttranslational modifications of histone proteins is emerging as a viable determinant of the aging process. Further recent studies of the ligation of RAGE have shown its ability to directly induce modifications of chromatin through the thioredoxin-interacting protein, an endogenous inhibitor of antioxidant thioredoxin. RAGE mediates epigenetic remodeling of histone H_3 lysine K9 by decreasing its trimethylation and increasing its acetylation, which in turn amplifies and sustains pro-inflammatory gene activation (Perrone et al. 2009).

AGEs for Age-at-Death Estimation in Forensic Sciences

A study conducted by Sato et al. in 2001 for forensic purposes, pointed out the potential utility of AGEs for age-at-death estimation even in cases of death by fire. Other authors studied the relationship between AGEs and age based on the color changes causes by these compounds. In this line of research, Pilin et al. (Pilin et al. 2007a,b,c) found age-related color changes in intervertebral discs excisions, Achilles tendons and rib cartilage as a result of accumulation of AGEs. The color changes and correlation coefficients were different in different tissues. They were more conspicuous in the rib cartilage then in the intervertebral discs, while in the Achilles tendon nearly no relation with age was registered. These studies found reliable age estimation based on color changes up to the age of 45 years. After this age, the stretch of values increases, which makes the usability of this method low for this age range.

Likewise, the color changes caused by aging and the utilization of this effect for age estimation were studied on hard dental tissues. The change of dentin color and tooth enamel was used as one of the morphological parameters for age estimation (Solheim 1993, 1988). In fact, it was reported that teeth in higher age groups become more yellow with age (de Jonge 1950, Brudevold et al. 1957). Ten Cate et al. (Ten Cate et al. 1977), using a transmission densitometer, found a correlation between root color density and age, with root color becoming more yellow with increasing age in both sexes. Recently, Solheim confirmed these results, using reflective spectrophotometry, reporting that crown and root dentin color was related to age. Lackovic and Wood

(Lackovic and Wood 2000) found that mesial surfaces of the roots were less yellow than the other three surfaces. There was a significant difference in the percentage of yellowness between two of the four root surfaces. In spite of these differences, they found a positive increase in the percentage of the measured color with age, suggesting that this could be due to continued cementum deposition through life or due to some undefined intrinsic change in the dentin. Using the spectroradiometry technique to measure dental color changes, Martín-de las Heras et al. (Martín-de las Heras et al. 2003) analyzed different colorimetric variables, obtaining the highest correlation coefficient with whiteness of the tooth, with a calibration error averaging 13.7 years. However, they found differences in dental color depending on the postmortem interval, with this relation weaker in postmortem extracts than in fresh extracted teeth. So, this technique is not reliable in extended postmortem intervals.

Conclusions

There are several studies that point out the increase of AGEs with aging in different tissues. In spite of these works, only few were developed with forensic sciences purposes in mind, indicating that AGEs could be a new and interesting approach for age-at-death estimation. In fact, there is a possibility that use of AGEs in burnt bodies has an advantage with respect to other methodologies that are affected by high temperatures, such as aspartic acid racemization. Therefore, further research is needed to clarify this line of inquiry.

Keywords: Age-at-death estimation, anthropological methods, Advanced Glycation Endproducts (AGEs), aging, Maillard reaction, CML, pentoside, lens, bone, cartilage, RAGE, hippocampal pyramidal neurons, intervertebral discs, color changes, teeth

References

Ahmed, M. U., S. R. Thorpe and J. W. Baynes. 1986. Identification of N epsilon-carboxymethyllysine as a degradation product of fructoselysine in glycated protein. J Biol Chem 261(11): 4889–94.

Ahmed, M. U., E. Brinkmann Frye, T. P. Degenhardt, S. R. Thorpe and J. W. Baynes. 1997. N-epsilon-(carboxyethyl)lysine, a product of the chemical modification of proteins by methylglyoxal, increases with age in human lens proteins. Biochem J 324(Pt 2): 565–70.

Bank, R. A., M. T. Bayliss, F. P. Lafeber, A. Maroudas and J. M. Tekoppele. 1998. Ageing and zonal variation in post-translational modification of collagen in normal human articular cartilage. The age-related increase in non-enzymatic glycation affects biomechanical properties of cartilage. Biochem J 330(Pt 1): 345–51.

Basta, G., A. M. Schmidt and R. De Caterina. 2004. Advanced glycation end products and vascular inflammation: implications for accelerated atherosclerosis in diabetes. Cardiovasc Res 63(4): 582–92. doi: 10.1016/j.cardiores.2004.05.001.

Baynes, J. W. 2001. The role of AGEs in aging: causation or correlation. Exp Gerontol 36(9): 1527–37.

Bierhaus, A., S. Schiekofer, M. Schwaninger, M. Andrassy, P. M. Humpert, J. Chen et al. 2001. Diabetes-associated sustained activation of the transcription factor nuclear factor-kappaB. Diabetes 50(12): 2792–808.

Bierhaus, A. and P. P. Nawroth. 2009. Multiple levels of regulation determine the role of the receptor for AGE (RAGE) as common soil in inflammation, immune responses and diabetes mellitus and its complications. Diabetologia 52(11): 2251–63. doi: 10.1007/s00125-009-1458-9.

Brudevold, F., J. W. Hein, J. F. Bonner, R. B. Nevin, B. G. Bibry and H. C. Hodge. 1957. Reaction of tooth surfaces with one ppm of fluoride as sodium fluoride. J Dent Res 36(5): 771–9.

Coughlan, M. T., D. R. Thorburn, S. A. Penfold, A. Laskowski, B. E. Harcourt, K. C. Sourris et al. 2009. RAGE-induced cytosolic ROS promote mitochondrial superoxide generation in diabetes. J Am Soc Nephrol 20(4): 742–52. doi: 10.1681/ASN.2008050514.

Dalal, M., L. Ferrucci, K. Sun, J. Beck, L. P. Fried and R. D. Semba. 2009. Elevated serum advanced glycation end products and poor grip strength in older community-dwelling women. J Gerontol A Biol Sci Med Sci 64(1): 132–7. doi: 10.1093/gerona/gln018.

de, Jonge Te. 1950. [Senescence of the dentition]. Paradentologie 4(2): 83–98.

Dunn, J. A., D. R. McCance, S. R. Thorpe, T. J. Lyons and J. W. Baynes. 1991. Age-dependent accumulation of N epsilon-(carboxymethyl)lysine and N epsilon-(carboxymethyl)hydroxylysine in human skin collagen. Biochemistry 30(5): 1205–10.

Dunn, J. A., J. S. Patrick, S. R. Thorpe and J. W. Baynes. 1989. Oxidation of glycated proteins: age-dependent accumulation of N epsilon-(carboxymethyl)lysine in lens proteins. Biochemistry 28(24): 9464–8.

Dyer, D. G., J. A. Dunn, S. R. Thorpe, K. E. Bailie, T. J. Lyons, D. R. McCance et al. 1993. Accumulation of Maillard reaction products in skin collagen in diabetes and aging. J Clin Invest 91(6): 2463–9. doi: 10.1172/JCI116481.

Eyre, D. R., M. A. Paz and P. M. Gallop. 1984. Cross-linking in collagen and elastin. Annu Rev Biochem 53: 717–48. doi: 10.1146/annurev.bi.53.070184.003441.

Haus, J. M., J. A. Carrithers, S. W. Trappe and T. A. Trappe. 2007. Collagen, cross-linking, and advanced glycation end products in aging human skeletal muscle. J Appl Physiol (1985) 103(6): 2068–76. doi: 10.1152/japplphysiol.00670.2007.

Hein, G. E. 2006. Glycation endproducts in osteoporosis—is there a pathophysiologic importance? Clin Chim Acta 371(1-2): 32–6. doi: 10.1016/j.cca.2006.03.017.

Henle, T., A. W. Walter and H. Klostermeyer. 1994. A simple method for the preparation of furosine and pyridosine reference material. Z Lebensm Unters Forsch 198(1): 66–7.

Iwata, H., H. Ukeda, T. Maruyama, T. Fujino and M. Sawamura. 2004. Effect of carbonyl compounds on red blood cells deformability. Biochem Biophys Res Commun 321(3): 700–6. doi: 10.1016/j.bbrc.2004.07.026.

Kimura, T., J. Takamatsu, K. Ikeda, A. Kondo, T. Miyakawa and S. Horiuchi. 1996. Accumulation of advanced glycation end products of the Maillard reaction with age in human hippocampal neurons. Neurosci Lett 208(1): 53–6.

Kumar, P. A., M. S. Kumar and G. B. Reddy. 2007. Effect of glycation on alpha-crystallin structure and chaperone-like function. Biochem J 408(2): 251–8. doi: 10.1042/BJ20070989.

Lackovic, K. P. and R. E. Wood. 2000. Tooth root colour as a measure of chronological age. J Forensic Odontostomatol 18(2): 37–45.

Lo, T. W., M. E. Westwood, A. C. McLellan, T. Selwood and P. J. Thornalley. 1994. Binding and modification of proteins by methylglyoxal under physiological conditions. A kinetic and mechanistic study with N alpha-acetylarginine, N alpha-acetylcysteine, and N alpha-acetyllysine, and bovine serum albumin. J Biol Chem 269(51): 32299–305.

Martin-de las Heras, S., A. Valenzuela, R. Bellini, C. Salas, M. Rubino and J. A. Garcia. 2003. Objective measurement of dental color for age estimation by spectroradiometry. Forensic Sci Int 132(1): 57–62.

Monnier, V. M., G. T. Mustata, K. L. Biemel, O. Reihl, M. O. Lederer, D. Zhenyu et al. 2005. Cross-linking of the extracellular matrix by the maillard reaction in aging and diabetes: an update on "a puzzle nearing resolution". Ann N Y Acad Sci 1043: 533–44. doi: 10.1196/annals.1333.061.

Nerlich, A. G. and E. D. Schleicher. 1999. N(epsilon)-(carboxymethyl)lysine in atherosclerotic vascular lesions as a marker for local oxidative stress. Atherosclerosis 144(1): 41–7.

Odetti, P., S. Rossi, F. Monacelli, A. Poggi, M. Cirnigliaro, M. Federici et al. 2005. Advanced glycation end products and bone loss during aging. Ann N Y Acad Sci 1043: 710–7. doi: 10.1196/annals.1333.082.

Perrone, L., T. S. Devi, K. Hosoya, T. Terasaki and L. P. Singh. 2009. Thioredoxin interacting protein (TXNIP) induces inflammation through chromatin modification in retinal capillary endothelial cells under diabetic conditions. J Cell Physiol 221(1): 262–72. doi: 10.1002/jcp.21852.

Pilin, A., F. Pudil and V. Bencko. 2007a. Changes in colour of different human tissues as a marker of age. Int J Legal Med 121(2): 158–62. doi: 10.1007/s00414-006-0136-4.

Pilin, A., F. Pudil and V. Bencko. 2007b. The image analysis of colour changes of different human tissues in the relation to the age. Part 1. Methodological approach. Soud Lek 52(2): 26–30.

Pilin, A., F. Pudil and V. Bencko. 2007c. The image analysis of colour changes of different human tissues in the relation to the age. Part 2: practical applicability. Soud Lek 52(3): 36–42.

Sato, Y., T. Kondo and T. Ohshima. 2001. Estimation of age of human cadavers by immunohistochemical assessment of advanced glycation end products in the hippocampus. Histopathology 38(3): 217–20.

Sell, D. R., R. H. Nagaraj, S. K. Grandhee, P. Odetti, A. Lapolla, J. Fogarty et al. 1991. Pentosidine: a molecular marker for the cumulative damage to proteins in diabetes, aging, and uremia. Diabetes Metab Rev 7(4): 239–51.

Semba, R. D., S. de Pee, M. O. Ricks, M. Sari and M. W. Bloem. 2008a. Diarrhea and fever as risk factors for anemia among children under age five living in urban slum areas of Indonesia. Int J Infect Dis 12(1): 62–70. doi: 10.1016/j.ijid.2007.04.011.

Semba, R. D., K. V. Patel, K. Sun, J. M. Guralnik, W. B. Ershler, D. L. Longo et al. 2008b. Association between serum carboxymethyl-lysine, a dominant advanced glycation end product, and anemia in adults: the Baltimore longitudinal study of aging. J Am Geriatr Soc 56(11): 2145–7. doi: 10.1111/j.1532-5415.2008.01968.x.

Semba, R. D., L. Ferrucci, J. C. Fink, K. Sun, J. Beck, M. Dalal et al. 2009a. Advanced glycation end products and their circulating receptors and level of kidney function in older community-dwelling women. Am J Kidney Dis 53(1): 51–8. doi: 10.1053/j.ajkd.2008.06.018.

Semba, R. D., L. Ferrucci, K. Sun, J. Beck, M. Dalal, R. Varadhan et al. 2009b. Advanced glycation end products and their circulating receptors predict cardiovascular disease mortality in older community-dwelling women. Aging Clin Exp Res 21(2): 182–90.

Semba, R. D., L. Ferrucci, K. Sun, K. V. Patel, J. M. Guralnik and L. P. Fried. 2009c. Elevated serum advanced glycation end products and their circulating receptors are associated with anaemia in older community-dwelling women. Age Ageing 38(3): 283–9. doi: 10.1093/ageing/afp011.

Semba, R. D., E. J. Nicklett and L. Ferrucci. 2010. Does accumulation of advanced glycation end products contribute to the aging phenotype? J Gerontol A Biol Sci Med Sci 65(9): 963–75. doi: 10.1093/gerona/glq074.

Solheim, T. 1988. Dental color as an indicator of age. Gerodontics 4(3): 114–8.

Solheim, T. 1993. A new method for dental age estimation in adults. Forensic Sci Int 59(2): 137–47.

Tang, S. Y., U. Zeenath and D. Vashishth. 2007. Effects of non-enzymatic glycation on cancellous bone fragility. Bone 40(4): 1144–51. doi: 10.1016/j.bone.2006.12.056.

Ten Cate, A. R., G. W. Thompson, J. B. Dickinson and H. A. Hunter. 1977. The estimation of age of skeletal remains from the colour of roots of teeth. Dent J 43(2): 83–6.

Thorpe, S. R. and J. W. Baynes. 1996. Role of the Maillard reaction in diabetes mellitus and diseases of aging. Drugs Aging 9(2): 69–77.

Verzijl, N., J. DeGroot, Z. C. Ben, O. Brau-Benjamin, A. Maroudas, R. A. Bank et al. 2002. Crosslinking by advanced glycation end products increases the stiffness of the collagen network in human articular cartilage: a possible mechanism through which age is a risk factor for osteoarthritis. Arthritis Rheum 46(1): 114–23. doi: 10.1002/1529-0131(200201)46:1<114::AID-ART10025>3.0.CO;2-P.

Vlassara, H., R. Bucala and L. Striker. 1994. Pathogenic effects of advanced glycosylation: biochemical, biologic, and clinical implications for diabetes and aging. Lab Invest 70(2): 138–51.

Zapico, S. C. and D. H. Ubelaker. 2013. Applications of physiological bases of ageing to forensic sciences. Estimation of age-at-death. Ageing Res Rev 12(2): 605–17. doi: 10.1016/j.arr.2013.02.002.

Section III
Telomeres

12

Introduction to Telomere Biology

Celia de Frutos[a] and Pablo Bermejo-Álvarez*

INTRODUCTION

Telomeres constitute one of the most reliable markers of cellular aging and were the first molecular feature associated to cellular senescence. The starting point for the understanding of cellular aging was the observation that cells cultured *in vitro* age and eventually stop their division (Hayflick and Moorhead 1961). This pioneer study proved that the replication ability of human fibroblasts was finite and that aging occurred at the cellular level. At that time, the notion of cellular aging constituted a breakthrough that contradicted the general consensus that aging was "an attribute of the multicellular body as a whole", supported by the experiments conducted by two Nobel prize winners who reported that chicken hearth fibroblasts could be cultured *in vitro* for 34 years (Carrel and Ebeling 1921). The concept introduced by Hayflick not only led to the fall of the so called "Myth of Immortal Normal Chick Cell" (Witkowski 1979), but also introduced a novel concept in cell biology: the "replicometer".

The "replicometer" was a cellular mechanism for counting cellular division, a mechanism by which cells kept a memory of their replicative history (Hayflick 1965). Experiments conducted with enucleated cells observed that the "replicometer" resided in the nucleus (Muggleton-Harris and Hayflick 1976, Wright and Hayflick 1975), but the mechanism and molecular nature of the "replicometer" remained unclear. Telomeres were actually discovered two decades before the concepts of cellular aging and "replicometer" were coined. In 1938, two independent studies observed that the end of the chromosomes in maize (Zea mays) (McClintock 1938) and fruit fly (Drosophila melanogaster) (Muller 1938) was different from the classic chromosome breaks, harbouring special properties that protects them from terminal fusions.

INIA, Department of Animal Reproduction Avda Puerta de Hierro 12 Local 10 Madrid, Madrid Spain, 28040.
[a] Email: cfbenitez@gmail.com
* Corresponding author: borrillobermejo@hotmail.com

Muller named these structures at the chromosomes end "telomeres" from the greek words *telos* (end) and *meros* (part).

Despite the early discovery of telomeres, its composition and structure, and how they exerted a protective role in the chromosome ends remained unknown for decades. The double helix tri-dimensional structure of DNA was discovered almost two decades later (Watson and Crick 1953b) and it was not until 1971 that Alexander Olovnikov associated the end of chromosomes with the cellular arrest described by Hayflick 10 years earlier (Olovnikov 1973). Olovnikov's theory, called DNA Marginotomy, proposed that the limited potential for cell replication was due to the shortening of a DNA replicate with respect to its template during cell division. This phenomenon was not considered in the classic DNA replication model (Watson and Crick 1953a) and it was later named as "the end replication problem" by James Watson (Watson 1972), a concept that refers to the inability of lineal chromosomes to replicate its 3' end, which thereby shortens on each mitotic cycle.

Olovnikov reasoned that it was not logical to lose essential genetic information on every cell division and, therefore, telomeres could be constituted by repeated nucleotide sequences containing no information (Olovnikov 1973). These disposable sequences, referred by Olovnikov as "*telogenes*", would serve as a buffer to prevent the loss of genes containing relevant information. The length of the telomeres would determine the number of cell divisions a cell would be able to overcome without losing essential information, thereby establishing a solid link between cellular senescence and telomere.

The molecular composition of telomeres was uncovered later. First, telomeres were found to be composed by a repeated sequence (TTGGG) in the ciliated protozoon *Tetrahymena thermophila* (Blackburn and Gall 1978). Subsequent experiments assayed the protective role of telomeres by transfecting yeast with a chimeric lineal plasmid, formed by fusing the end fragments of *Tetrahymena* ribosomal DNA to a lineal yeast plasmid (Szostak and Blackburn 1982). These hybrid lineal plasmids, flanked by *Tetrahymena* telomeres remained stable for several generations, proving the protective role of telomeres in the maintenance of chromosomes. Furthermore, the fact that yeast were able to recognize the telomeres from the protozoon *Tetrahymena* suggested that telomere function was conserved in eukaryotes (Szostak and Blackburn 1982). Later, human telomeres were also found to be composed by a repeated sequence, albeit different from protozoa (TTAGGG) (Moyzis et al. 1988).

The mechanism by which telomeres were formed remained elusive, but researchers reasoned that given that telomere shortening limited the cellular proliferative potential, cells with unlimited potential to divide, such as tumoral cells, should possess a mechanism for telomere elongation. The first mechanism responsible for telomere elongation described consisted of an unknown reverse transcriptase, which was initially named "telomere terminal transferase" in Tetrahymena (Greider and Blackburn 1985, 1987) and subsequently known as telomerase in human immortal cells (Morin 1989). The enzyme activity was detected not only in immortal cell lines, but also in most tumors (Kim et al. 1994), showing the role of telomere elongation on inducing unlimited cell division.

The first telomerase component described was its RNA subunit (TERC) (Greider and Blackburn 1989) and it was discovered that the enzyme used a template for the synthesis of telomere repetitions, as mutations on its sequence CAACCCCAA resulted in altered telomeric repeats (Yu et al. 1990). The gene encoding the catalytic subunit (TERT), was discovered 6 years later (Lingner et al. 1997, Nakamura et al. 1997). These findings further proved the role of telomerase in cellular immortalization by transfecting telomerase-negative cells with vectors encoding TERT (Bodnar et al. 1998).

The End Replication Problem

During cellular replication, the lineal chromosome of eukaryotes suffers from a telomere shortening on each mitosis estimated at 100 bp in humans, due to the DNA replication process. DNA replication in eukaryotes is a semiconservative process, which means that it is carried out by copying the parental DNA strand, as originally suggested by Watson and Crick model (Watson and Crick 1953a). DNA replication starts by separating both strands of the double helix of DNA, which leaves one strand oriented in 5'-3' sense and the other in reverse sense. The enzyme responsible for the *de novo* synthesis of DNA, the DNA polymerase, can only synthesize in the 5'-3' sense, requiring a small oligonucleotide sequence (RNA primer) to start replication. The strand synthesized in that sense, called leading strand, is continuously synthesized to the end of the template strand. The end replication problem arises in the complementary strand, oriented in 3'-5' sense. On this strand, the nascent DNA strand, called lagging strand, requires to be synthesized by small and discontinuous fragments, named Okazaki fragments, which are later bound by a ligase, replacing the RNA primers by DNA. As ligase requires a DNA strand in front of the primer, the last RNA primer is removed and therefore, DNA polymerase is unable to replicate the 3' end from the template strand and the sequence is shortened in each mitotic cycle (Fig. 1).

Telomeres Roles in Maintenance of Genome Stability

Chromosome ends present two problems: the end replication problem and the end protection problem. To avoid these problems, which would ultimately lead to the loss of genetic information and the development of chromosome aberrations, chromosome ends have been adapted genetically and structurally. The length of telomeric repeats compensate for the end replication problem, while the secondary structure of telomeres provides protection.

Telomeres are the DNA sequences suffering from shortening during cellular replication, thereby avoiding the loss of gene sequences (Hackett et al. 2001). In this perspective, the initial telomere length and the telomere shortening rate during replication determines the number of divisions a somatic cell may overcome before telomeres become critically short and lose their protective ability (Allsopp et al. 1992). The loss of telomeres' protective ability is associated with the entry into mitotic arrest (known as replicative senescence) or cell death (apoptosis) (O'Sullivan and Karlseder 2010). A finite telomere length, thereby, limits the replication of individual cells (Allsopp et al. 1992) and prevents cellular immortalization and the proliferation of aged cells with pre-cancer mutations (Bechter et al. 2004). The additional cost of the limit for cell proliferation in somatic tissues is the loss of regenerative ability, contributing to cell and organic aging (Cesare and Karlseder 2012).

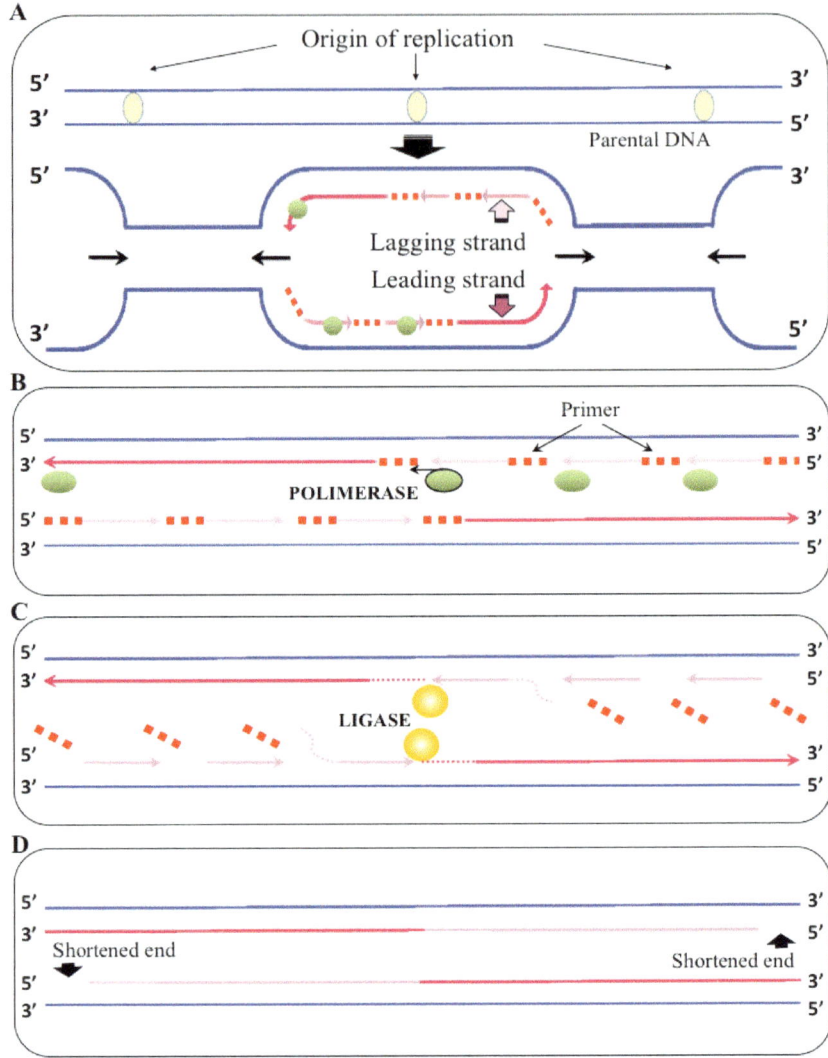

Figure 1. The end replication problem. During DNA replication lagging and leading, strands are synthesized (A) on 5'-3' sense using RNA primers (B). RNA primers are substituted by DNA (C), but RNA polymerase cannot replace the 5' RNA primer, thereby DNA polymerase is unable to replicate the 3' end from the template strand and the sequence is shortened in each mitotic cycle.

The role of telomeres in chromosome stability was initially proposed by Barbara McClintock, who realized that X-ray irradiated plant cells suffered from chromosomal breakage-fusion-bridge (McClintock 1938). This finding led to hypothesize that aberrant chromosome fusions will break on the next cellular divisions, resulting in an unequal and random distribution of the genetic material in daughter cells. Dysfunctional telomeres, that lose their protective ability, give way to the formation of the same chromosomal breakage-fusion-bridge described by McClintock, suggesting that the loss of chromosome end protection may be the base for genomic instability in mammals (O'Sullivan and Karlseder 2010).

Telomere protective role depends on its length and secondary structure. An excessive telomere shortening has been proved to result in chromosome instability. The fourth generation of telomerase null mice lacks detectable telomere repeats and presents aneuploidy and chromosomal abnormalities, including terminal fusions (Blasco et al. 1997). However, telomere length is not the only factor determining cellular proliferative potential (O'Sullivan and Karlseder 2010) and it is unclear the minimum telomere length required for telomere protection. Telomeres may lose the protective ability from cellular activities such as nucleases or DNA-break repairing proteins, even in telomeres of a normal length; functional alterations in shelterin complex cause telomere-associated senescence independently of telomere length (Karlseder et al. 2002, Martinez and Blasco 2010).

Apart from their role in the protection of chromosome ends, telomeres play other essential roles, such as chromosome positioning during mitosis (Perez-Morga et al. 2001) and meiosis (Zalenskaya and Zalensky 2002).

Molecular Base for the Protection provided by Telomeres: "Telomere Capping"

The chromosome ends display properties similar to DNA DSB and also telomeres contain a region of ssDNA (de Lange 2009). Both features make them prone to activate the DNA damage checkpoints that trigger the two signalling ways for DNA DSB: (1) ATM and 2) ATR. The activation of ATM or ATR results in cell cycle detention to avoid further DNA erosion (d'Adda di Fagagna et al. 2003). Apart from impeding the cell cycle blockage mediated by ATM or ATR, the chromosome ends must avoid the activation of the DSB repairing mechanisms: (1) HDR and (2) NHEJ. The activation of HDR would lead to a sequence exchange between two telomeres or one telomere and another part of the genome, resulting in chromosome rearrangements, terminal deletions or drastic changes in telomere length. Similarly, the activation of NHEJ would lead to terminal fusions resulting in dicentric chromosomes, which are unstable on mitosis. Telomeres must, therefore, avoid the activation of DSB signalling and DSB repair by avoiding DDR (d'Adda di Fagagna et al. 2004).

A crucial telomere mechanism to avoid DDR is the sequester of the 3' free end through the T-loop structure, that prevents it to be recognized as a DSB (Griffith et al. 1999). Telomere associated proteins also inhibit DSB signalling and repairing. Among the telomere-associated proteins, TRF2 prevents from NHEJ (van Steensel et al. 1998) and represses ATM and the overexpression of p53 by avoiding the phosphorylation of Chk2 kinase, an activator of ATM (Karlseder et al. 1999). Another telomere-associated protein, POT1, prevents the activation of ATR by impeding the phosphorylation of Chk1, a critical step in that pathway (Guo et al. 2007, Hockemeyer et al. 2006).

Telomere dysfunction, either caused by telomere shortening or the loss of structures protecting the chromosome end is associated to genomic instability, which triggers tumorogenesis and premature cell aging (Perera et al. 2008). To avoid telomere dysfunction by shortening, mammals use two mechanisms for telomere elongation:

Telomerase is the main factor responsible for telomere extension and maintenance, whereas some mammalian cells use a telomerase-independent mechanism named ALT.

Telomerase

Telomerase is one of the many mechanisms to lengthen telomeres that have appeared through eukaryotes evolution. It may have evolved from retrotransposition, as TERT holds great functional similarity with the transcriptase from non-LTR transposable elements (de Lange 2004). TR is a ribonucleoproteic holoenzyme formed by multiple subunits that includes the TERT, TERC, which provides the template for telomere DNA synthesis, and a set of species-specific proteins. TERT and TERC are sufficient for telomere synthesis *in vitro* (Bryan et al. 2003, Weinrich et al. 1997), whereas *in vivo* many other factors are required for enzyme assembly, activity and location (Smekalova et al. 2012). Indeed, revealing telomerase structure in vertebrates is particularly challenging due to the low expression levels and the cell cycle-dependent regulation of its subunits (Miracco et al. 2014). The most stable telomerase components in humans are TERT, TERC and dyskerin.

TERT is the catalytic subunit, more conserved across evolution than TERC, and it contains three domains: TRBD that interacts with TERC region CR4/CR5, RT and a CTE less conserved than the other two (Sandin and Rhodes 2014). In mammals, the central domain RT is flanked by the domain TEN, which displays a special affinity for ssDNA, is involved in the binding to the primer, and facilitates the addition of telomeric repeats.

TERC is the RNA component of telomerase and its function is not limited only to provide a template for the synthesis of telomere repeats; it also collaborates in the incorporation of nucleotides (Qiao and Cech 2008), in the translocation process (Qiao and Cech 2008, Cristofari et al. 2007), in the association and assembly of the telomerase complex (Theimer et al. 2007, Robart and Collins 2010), and regulates telomerase activity (Cairney and Keith 2008). All TERCs contain three essential elements: the template region for telomere repeat synthesis, the adjacent element pseudoknot—both essential for catalytic activity (Sandin and Rhodes 2014), and a distal loop (stem terminus element, STE) named CR4-CR5 (CR: conserved region) in vertebrates. These structural elements interact directly with TERT, whereas the 3' hairpin-hinge-hairpin-ACA (H/ACA) motif is essential for RNA folding and for holoenzyme assembly (Egan and Collins 2010).

DKC1, is a RNA binding protein that recognizes the H/ACA and is involved in the biogenesis of the ribonucleoprotein telomerase and in TERC stability (Mochizuki et al. 2004). Other potential components of the active telomerase include three dyskerin associated proteins evolutionary conserved: NOP10 ribonucleoprotein (NOP10), GAR1 ribonucleoprotein (GAR1) and NHP2 ribonucleoprotein (NHP2) (Dragon et al. 2000, Kiss et al. 2006, Fu and Collins 2007, Pogacic et al. 2000), SMG6 nonsense mediated mRNA decay factor (SMG6, also known as EST1A) (Snow et al. 2003) and the Cajal body localization protein WDR79/TCAB1 (Venteicher et al. 2009).

Telomerase activity is regulated during the cell cycle, being detected predominantly during the late S/G2 phase, coupled to telomere replication (Marcand et al. 2000). Being a reverse transcriptase, telomerase synthesize DNA using RNA as template, and it requires as primer the telomere 3' free end. In contrast with other RT, telomerase synthesize a small segment of its own RNA to dissociate from the newly synthesized DNA and relocate again close to its primer to start a new elongation cycle. The ability to use only a small region of its RNA (8–11 nt) for DNA synthesis is unusual, and requires a mechanism that facilitates enzyme translocation, which requires several steps and is yet mostly unknown (Smekalova et al. 2012).

Alternative Telomere Elongation (ALT)

ALT is a telomere lengthening mechanism that does not require telomerase and relies on genes involved in homologous recombination (RAD32 nuclease) (Lundblad and Blackburn 1993). ALT requires from DNA recombination, but the molecular mechanism remains unclear (Cesare and Reddel 2010) and two models have been proposed: (1) unequal T-SCE, resulting in lengthening of one telomere and shortening of the one in the sister chromatid, and (2) homologous recombination-dependent DNA replication, where another telomere is used as a template (Dunham et al. 2000). The template for telomere synthesis may be the telomere from a non-homologous chromosome

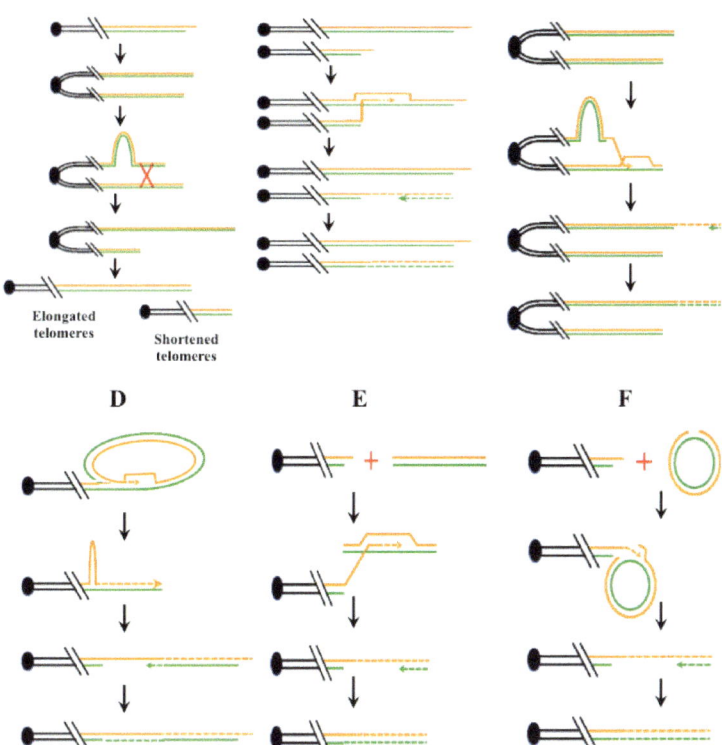

Figure 2. Two models have been proposed models for ALT: Unequal telomere sister chromatid exchange (A) or homologous recombination-dependent DNA replication (B). The recombination template may be a sister chromatid (C), a T-loop (D) or extrachromosomic telomeric lineal (E) or circular (F) fragments.

(intertelomeric recombination (Dunham et al. 2000)), the telomeric sequences from the same chromosome or a sister chromatid (intratelomeric recombination (Muntoni et al. 2009)) or extrachromosomic telomeric DNA (Henson et al. 2002) (Fig. 2).

Acknowledgements

CF was supported by a fellowship from INIA (Spain). PBA is supported by Grants RYC-2012-10193 and AGL2014-58739-R from the Spanish Ministry of Economy and Competitiveness.

Keywords: Replicometer, end replication problem, telomere stability, telomere capping, telomerase, TERT, TERC, alternative lenghtening of telomeres, intertelomeric recombination, intratelomeric recombination, T-loop

References

Allsopp, R. C., H. Vaziri, C. Patterson, S. Goldstein, E. V. Younglai, A. B. Futcher et al. 1992. Telomere length predicts replicative capacity of human fibroblasts. Proc Natl Acad Sci U S A 89(21): 10114–8.

Bechter, O. E., Y. Zou, W. Walker, W. E. Wright and J. W. Shay. 2004. Telomeric recombination in mismatch repair deficient human colon cancer cells after telomerase inhibition. Cancer Research 64(10): 3444–51.

Blackburn, E. H. and J. G. Gall. 1978. A tandemly repeated sequence at the termini of the extrachromosomal ribosomal RNA genes in Tetrahymena. Journal of Molecular Biology 120(1): 33–53.

Blasco, M. A., H. W. Lee, M. P. Hande, E. Samper, P. M. Lansdorp, R. A. DePinho et al. 1997. Telomere shortening and tumor formation by mouse cells lacking telomerase RNA. Cell 91(1): 25–34.

Bodnar, A. G., M. Ouellette, M. Frolkis, S. E. Holt, C. P. Chiu, G. B. Morin et al. 1998. Extension of life-span by introduction of telomerase into normal human cells. Science 279(5349): 349–52.

Bryan, T. M., K. J. Goodrich and T. R. Cech. 2003. Tetrahymena telomerase is active as a monomer. Molecular Biology of the Cell 14(12): 4794–804.

Cairney, C. J. and W. N. Keith. 2008. Telomerase redefined: integrated regulation of hTR and hTERT for telomere maintenance and telomerase activity. Biochimie 90(1): 13–23.

Carrel, A. and A. H. Ebeling. 1921. Age and multiplication of fibroblasts. Journal of Experimental Medicine 34(6): 599–623.

Cesare, A. J. and R. R. Reddel. 2010. Alternative lengthening of telomeres: models, mechanisms and implications. Nat Rev Genet 11(5): 319–30.

Cesare, A. J. and J. Karlseder. 2012. A three-state model of telomere control over human proliferative boundaries. Current Opinion in Cell Biology 24(6): 731–8.

Cristofari, G., E. Adolf, P. Reichenbach, K. Sikora, R. M. Terns, M. P. Terns et al. 2007. Human telomerase RNA accumulation in Cajal bodies facilitates telomerase recruitment to telomeres and telomere elongation. Mol Cell 27(6): 882–9.

d'Adda di Fagagna, F., P. M. Reaper, L. Clay-Farrace, H. Fiegler, P. Carr, T. Von Zglinicki et al. 2003. A DNA damage checkpoint response in telomere-initiated senescence. Nature 426(6963): 194–8.

d'Adda di Fagagna, F., S. H. Teo and S. P. Jackson. 2004. Functional links between telomeres and proteins of the DNA-damage response. Genes Dev 18(15): 1781–99.

de Lange, T. 2004. T-loops and the origin of telomeres. Nat Rev Mol Cell Biol 5(4): 323–9.

de Lange, T. 2009. How telomeres solve the end-protection problem. Science 326(5955): 948–52.

Dragon, F., V. Pogacic and W. Filipowicz. 2000. *In vitro* assembly of human H/ACA small nucleolar RNPs reveals unique features of U17 and telomerase RNAs. Mol Cell Biol 20(9): 3037–48.

Dunham, M. A., A. A. Neumann, C. L. Fasching and R. R. Reddel. 2000. Telomere maintenance by recombination in human cells. Nature Genetics 26(4): 447–50.

Egan, E. D. and K. Collins. 2010. Specificity and stoichiometry of subunit interactions in the human telomerase holoenzyme assembled *in vivo*. Mol Cell Biol 30(11): 2775–86.

Fu, D. and K. Collins. 2007. Purification of human telomerase complexes identifies factors involved in telomerase biogenesis and telomere length regulation. Mol Cell 28(5): 773–85.

Greider, C. W. and E. H. Blackburn. 1985. Identification of a specific telomere terminal transferase activity in Tetrahymena extracts. Cell 43(2 Pt 1): 405–13.

Greider, C. W. and E. H. Blackburn. 1987. The telomere terminal transferase of Tetrahymena is a ribonucleoprotein enzyme with two kinds of primer specificity. Cell 51(6): 887–98.

Greider, C. W. and E. H. Blackburn. 1989. A telomeric sequence in the RNA of Tetrahymena telomerase required for telomere repeat synthesis. Nature 337(6205): 331–7.

Griffith, J. D., L. Comeau, S. Rosenfield, R. M. Stansel, A. Bianchi, H. Moss et al. 1999. Mammalian telomeres end in a large duplex loop. Cell 97(4): 503–14.

Guo, X., Y. Deng, Y. Lin, W. Cosme-Blanco, S. Chan, H. He et al. 2007. Dysfunctional telomeres activate an ATM-ATR-dependent DNA damage response to suppress tumorigenesis. EMBO Journal 26(22): 4709–19.

Hackett, J. A., D. M. Feldser and C. W. Greider. 2001. Telomere dysfunction increases mutation rate and genomic instability. Cell 106(3): 275–86.

Hayflick, L. 1965. The limited *in vitro* lifetime of human diploid cell strains. Exp Cell Res 37: 614–36.

Hayflick, L. and P. S. Moorhead. 1961. The serial cultivation of human diploid cell strains. Exp Cell Res 25: 585–621.

Henson, J. D., A. A. Neumann, T. R. Yeager and R. R. Reddel. 2002. Alternative lengthening of telomeres in mammalian cells. Oncogene 21(4): 598–610.

Hockemeyer, D., J. P. Daniels, H. Takai and T. de Lange. 2006. Recent expansion of the telomeric complex in rodents: Two distinct POT1 proteins protect mouse telomeres. Cell 126(1): 63–77.

Karlseder, J., D. Broccoli, Y. Dai, S. Hardy and T. de Lange. 1999. p53- and ATM-dependent apoptosis induced by telomeres lacking TRF2. Science 283(5406): 1321–5.

Karlseder, J., A. Smogorzewska and T. de Lange. 2002. Senescence induced by altered telomere state, not telomere loss. Science 295(5564): 2446–9.

Kim, N. W., M. A. Piatyszek, K. R. Prowse, C. B. Harley, M. D. West, P. L. Ho et al. 1994. Specific association of human telomerase activity with immortal cells and cancer. Science 266(5193): 2011–5.

Kiss, T., E. Fayet, B. E. Jady, P. Richard and M. Weber. 2006. Biogenesis and intranuclear trafficking of human box C/D and H/ACA RNPs. Cold Spring Harbor Symposia on Quantitative Biology 71: 407–17.

Lingner, J., T. R. Hughes, A. Shevchenko, M. Mann, V. Lundblad and T. R. Cech. 1997. Reverse transcriptase motifs in the catalytic subunit of telomerase. Science 276(5312): 561–7.

Lundblad, V. and E. H. Blackburn. 1993. An alternative pathway for yeast telomere maintenance rescues est1—senescence. Cell 73(2): 347–60.

Marcand, S., V. Brevet, C. Mann and E. Gilson. 2000. Cell cycle restriction of telomere elongation. Curr Biol 10(8): 487–90.

Martinez, P. and M. A. Blasco. 2010. Role of shelterin in cancer and aging. Aging Cell 9(5): 653–66.

McClintock, B. 1938. The fusion of broken ends of sister half chromatids following chromatid breakage at meiotic anaphase. Mo Agric Exp Stn Res Bull 290: 1–48.

Miracco, E. J., J. Jiang, D. D. Cash and J. Feigon. 2014. Progress in structural studies of telomerase. Current Opinion in Structural Biology 24: 115–24.

Mochizuki, Y., J. He, S. Kulkarni, M. Bessler and P. J. Mason. 2004. Mouse dyskerin mutations affect accumulation of telomerase RNA and small nucleolar RNA, telomerase activity, and ribosomal RNA processing. Proc Natl Acad Sci U S A 101(29): 10756–61.

Morin, G. B. 1989. The human telomere terminal transferase enzyme is a ribonucleoprotein that synthesizes TTAGGG repeats. Cell 59(3): 521–9.

Moyzis, R. K., J. M. Buckingham, L. S. Cram, M. Dani, L. L. Deaven, M. D. Jones et al. 1988. A highly conserved repetitive DNA sequence, (TTAGGG)n, present at the telomeres of human chromosomes. Proc Natl Acad Sci U S A 85(18): 6622–6.

Muggleton-Harris, A. L. and L. Hayflick. 1976. Cellular aging studied by the reconstruction of replicating cells from nuclei and cytoplasms isolated from normal human diploid cells. Exp Cell Res 103(2): 321–30.

Muller, H. 1938. The re-making of chromosomes. Collecting Net, Woods Hole 13: 181–198.

Muntoni, A., A. A. Neumann, M. Hills and R. R. Reddel. 2009. Telomere elongation involves intra-molecular DNA replication in cells utilizing alternative lengthening of telomeres. Human Molecular Genetics 18(6): 1017–27.

Nakamura, T. M., G. B. Morin, K. B. Chapman, S. L. Weinrich, W. H. Andrews, J. Lingner et al. 1997. Telomerase catalytic subunit homologs from fission yeast and human. Science 277(5328): 955–9.

O'Sullivan, R. J. and J. Karlseder. 2010. Telomeres: protecting chromosomes against genome instability. Nat Rev Mol Cell Biol 11(3): 171–81.

Olovnikov, A. M. 1973. A theory of marginotomy. The incomplete copying of template margin in enzymic synthesis of polynucleotides and biological significance of the phenomenon. Journal of Theoretical Biology 41(1): 181–90.

Perera, S. A., R. S. Maser, H. Xia, K. McNamara, A. Protopopov, L. Chen et al. 2008. Telomere dysfunction promotes genome instability and metastatic potential in a K-ras p53 mouse model of lung cancer. Carcinogenesis 29(4): 747–53.

Perez-Morga, D., A. Amiguet-Vercher, D. Vermijlen and E. Pays. 2001. Organization of telomeres during the cell and life cycles of Trypanosoma brucei. Journal of Eukaryotic Microbiology 48(2): 221–6.

Pogacic, V., F. Dragon and W. Filipowicz. 2000. Human H/ACA small nucleolar RNPs and telomerase share evolutionarily conserved proteins NHP2 and NOP10. Mol Cell Biol 20(23): 9028–40.

Qiao, F. and T. R. Cech. 2008. Triple-helix structure in telomerase RNA contributes to catalysis. Nat Struct Mol Biol 15(6): 634–40.

Robart, A. R. and K. Collins. 2010. Investigation of human telomerase holoenzyme assembly, activity, and processivity using disease-linked subunit variants. Journal of Biological Chemistry 285(7): 4375–86.

Sandin, S. and D. Rhodes. 2014. Telomerase structure. Current Opinion in Structural Biology 25: 104–10.

Smekalova, E. M., O. S. Shubernetskaya, M. I. Zvereva, E. V. Gromenko, M. P. Rubtsova and O. A. Dontsova. 2012. Telomerase RNA biosynthesis and processing. Biochemistry 77(10): 1120–8.

Snow, B. E., N. Erdmann, J. Cruickshank, H. Goldman, R. M. Gill, M. O. Robinson et al. 2003. Functional conservation of the telomerase protein Est1p in humans. Curr Biol 13(8): 698–704.

Szostak, J. W. and E. H. Blackburn. 1982. Cloning yeast telomeres on linear plasmid vectors. Cell 29(1): 245–55.

Theimer, C. A., B. E. Jady, N. Chim, P. Richard, K. E. Breece, T. Kiss et al. 2007. Structural and functional characterization of human telomerase RNA processing and cajal body localization signals. Mol Cell 27(6): 869–81.

van Steensel, B., A. Smogorzewska and T. de Lange. 1998. TRF2 protects human telomeres from end-to-end fusions. Cell 92(3): 401–13.

Venteicher, A. S., E. B. Abreu, Z. Meng, K. E. McCann, R. M. Terns, T. D. Veenstra et al. 2009. A human telomerase holoenzyme protein required for Cajal body localization and telomere synthesis. Science 323(5914): 644–8.

Watson, J. D. 1972. Origin of concatemeric T7 DNA. Nat New Biol 239(94): 197–201.

Watson, J. D. and F. H. Crick. 1953a. Genetical implications of the structure of deoxyribonucleic acid. Nature 171(4361): 964–7.

Watson, J. D. and F. H. Crick. 1953b. Molecular structure of nucleic acids; a structure for deoxyribose nucleic acid. Nature 171(4356): 737–8.

Weinrich, S. L., R. Pruzan, L. Ma, M. Ouellette, V. M. Tesmer, S. E. Holt et al. 1997. Reconstitution of human telomerase with the template RNA component hTR and the catalytic protein subunit hTRT. Nature Genetics 17(4): 498–502.

Witkowski, J. A. 1979. Alexis Carrel and the mysticism of tissue culture. Medical History 23(3): 279–96.

Wright, W. E. and L. Hayflick. 1975. Nuclear control of cellular aging demonstrated by hybridization of anucleate and whole cultured normal human fibroblasts. Exp Cell Res 96(1): 113–21.

Yu, G. L., J. D. Bradley, L. D. Attardi and E. H. Blackburn. 1990. *In vivo* alteration of telomere sequences and senescence caused by mutated Tetrahymena telomerase RNAs. Nature 344(6262): 126–32.

Zalenskaya, I. A. and A. O. Zalensky. 2002. Telomeres in mammalian male germline cells. International Review of Cytology 218: 37–67.

13

Role of Telomeres in Aging

Christian Thomas[a] and *Sara C. Zapico**

INTRODUCTION

Telomeres, located at the end of chromosomes to maintain their integrity, are a major focus of aging research as (1) a small portion of telomeric DNA is lost with each cell division, (2) they are shortened by oxidative damage and (3) when telomere length reaches a critical limit, cells enter a senescence state and/or apoptosis (Simons 2015). Telomere length may serve as a biological clock to determine the lifespan of a cell and an organism (Shammas 2011), as pioneer experiments demonstrated that telomere length predicts the *in vitro* replicative capacity of human fibroblasts and that overexpressing telomerase immortalizes fibroblasts in cell cultures (Bodnar et al. 1998, Rudolph et al. 1999). It was later demonstrated that *in vivo* human telomeres shorten during aging (Muezzinler et al. 2013).

Telomere Length Regulation and Cellular Senescence

Telomere shortening occurs at each DNA replication and continuous shortening eventually leads to chromosomal degradation and cell death (Shin et al. 2006). Telomerase activity, which lengthens telomeres, is present in stem cells, gamete precursors and certain hematopoietic cells and is absent or partially repressed in somatic cells (Chiu et al. 1996, Counter et al. 1995, Hiyama et al. 1995, Frenck et al. 1998). The absence of telomerase determines the progressive shortening occurring during replication in somatic cells.

Smithsonian Institution, National Museum of Natural History, Anthropology Department 10th and Constitution Ave, NW, PO Box 37012 Washington, DC 20560 USA.
[a] Email: crf.thomas@gmail.com
* Corresponding author: saiczapico@gmail.com

Telomere loss is explained as a result of incomplete DNA replication, "the end replication problem", which accounts for a loss of approximately 100 bp telomere length at each population doubling (Oeseburg et al. 2010), but has been also explained by the processing of chromosome ends following replication (Watson 1972, Olovnikov 1973, Makarov et al. 1997, Lingner and Cech 1996). Oxidative damage of telomeric DNA has been proposed as the major cause of telomere shortening in human cells (von Zglinicki 2000). The guanine-rich nature of telomeric DNA makes it particularly vulnerable to oxidative damage (Henle et al. 1999, Oikawa et al. 2001). Telomeric proteins are known to actively suppress DNA damage responses in yeast (Michelson et al. 2005) and it is possible that certain lesions in telomeric DNA appear only during replication, resulting in variable losses of telomere repeats. Recently, two other mechanisms of telomere shortening have been proposed: the failure to correctly process higher order structures of G-rich telomeric DNA (Crabbe et al. 2004, Ding et al. 2004) and the deletion of T-loops by homologous recombination (Wang et al. 2004). The factors and pathways involved in the repair of replication forks that are stalled or have collapsed at telomeres are not well understood. Repair during S-phase could involve either telomerase or homologous recombination pathways, perhaps including proteins such as the Fanconi proteins, BRCA1 and BRCA2 (Aubert and Lansdorp 2008).

Regarding the role of telomere shortening in senescence, although telomere attrition might not be primarily involved in the induction of senescence (Chen et al. 2001), the cumulative burden of oxidative stress and cumulative telomere attrition might increase the likelihood of a cell to enter senescence (Kurz et al. 2004). Based on the "Hayflick limit", in mortality stage I, one or more telomeres become critically short, are recognized within the cell as chromosome breaks and the cell cycle is arrested irreversibly. The signal that triggers replicative senescence is not the telomere array length, but rather the inability to form a protective higher order complex in association with telomere-specific proteins, leading to dysfunctional telomeres (Blackburn 2000, Griffith et al. 1999, Chin et al. 1999). These activate cell cycle checkpoints and induce either replicative senescence or apoptosis. In the absence of such checkpoint systems, cells continue to proliferate and telomere erosion proceeds until nearly all telomeres reach a critical length at mortality stage II and enter crisis, resulting in chromosomal instability by erroneous DNA damage repair. At this point, the number of cell divisions is counterbalanced by an equal number of cell deaths and chromosomal end fusions and other cytological abnormalities accumulate. Rare cells can escape from this crisis by the activation of telomerase, ensuring propagation of this short telomere length and conferring immortality (Chan and Blackburn 2003). Apart from the direct role of telomere shortening in senescence, there is ample evidence that disruption of telomere binding proteins results in early senescence (Karlseder et al. 2002, van Steensel et al. 1998, Yang et al. 2005).

Determinants of telomere length

Before studying the role of telomeres in lifespan and aging disease, several determinants of telomere length should be taken into account. Average telomere length was shown to be a heritable trait in several studies (Rufer et al. 1999b, Jeanclos et al. 2000, Graakjaer et al. 2003, Slagboom et al. 1994). For instance, African American telomeres are longer in general than in White Americans (Hunt et al. 2008). Interestingly, several studies agree that paternal inheritance is stronger than maternal (Njajou et al. 2007), which may be the consequence of epigenetic inheritance mediated by spermatozoa telomeres. Sex is another telomere length determinant, as women exhibit longer telomeres than men (Hunt et al. 2008), a difference that may have its origin during preimplantation development (Bermejo-Alvarez et al. 2008).

Although telomere length is primarily a heritable trait, it is also affected by a combination of factors like genetic and epigenetic make-up and environment (Steinert et al. 2004, Munoz et al. 2005, Celli and de Lange 2005, Benetti et al. 2007), social and economic status (Cherkas et al. 2008, Adams et al. 2007), exercise (Cherkas et al. 2008), body weight (Nordfjall et al. 2008b, Nordfjall et al. 2008a) and smoking (Valdes et al. 2005). As described above, oxidative stress accelerates telomere erosion *in vitro* (Serra et al. 2000) and *in vivo* and antioxidant defenses can in turn slow down telomere attrition (Serra et al. 2003, von Zglinicki et al. 2000). In fact, sexual dimorphism in telomere length has been attributed to differences in hormonal status and lifestyle, linked to antioxidative defenses. It is also attributed to body size differences (Stindl 2004) and different telomere dynamics on the X and Y chromosome (Nawrot et al. 2004, Manestar-Blazic 2004).

Telomere Length during Aging

Telomere length varies between different species, tissues, cells and even between single chromosomes (Bekaert 2005). Further, it is highly variable between human individuals and highly variable between organs from one subject (Takubo et al. 2002). However, at any moment during extra uterine life, telomere length is a reflection of the mean telomere length set during intra uterine development and the telomere attrition rate, making it a valuable biomarker for human aging. Any chromosome end shows a large variation in the number of telomere repeats, although the average length was found to vary significantly between chromosome arms. Chromosome 17p, 13p and 19p showed relatively short telomeres, which suggests that these telomeres will be the first to become "uncapped" upon progressive telomere shortening with proliferation and age (Britt-Compton et al. 2006, Martens et al. 1998, Graakjaer et al. 2003). The rate of telomere attrition seems to vary between chromosome ends. For example, the inactivate X chromosomes in female cells accelerates telomere loss relative to autosomal and the active X chromosomes (Surralles et al. 1999). This suggests that differences in the average length of specific chromosomes are in part generated during proliferation and with age.

Telomere length per se is not able to predict lifespan, but the Telomere Rate of Change (TROC) in length does (Haussmann et al. 2003). There are reports related to telomere shortening indicating that this is gradual and constant, or increasing or decreasing with age (Rufer et al. 1999a, Frenck et al. 1998). In fact, a biphasic telomere attrition in leukocytes was described, with fast attrition from birth up to the age of 4–5, after which telomere length is relatively stable throughout childhood, preadolescent and adolescent years. Thereafter there is a slower and more gradual erosion, with increasing telomere attrition at very old age (Rufer et al. 1999b).

Telomere attrition in newborn and young individuals is shown to be significantly higher than in their older counterparts, up to four-fold faster during the first five years of life than in adults (Iwama et al. 1998, Manestar-Blazic 2004, Zeichner et al. 1999). Telomere attrition then levels off to a more gradual and constant telomere loss. As telomere length does not vary considerably between different cell populations in a given individual (Pommier et al. 1997) and as a large synchrony between tissues exists within fetuses and newborns (Okuda et al. 2002, Youngren et al. 1998), this difference in erosion rate can most probably be attributed to increased immune cell replication rates accompanying the characteristic changes of the immune system in newborns and infants.

In the elderly (older than 60 years), telomere attrition is significantly associated with higher mortality rates, both from infectious and cardiovascular diseases (Cawthon et al. 2003). In fact, an important aging hypothesis is that telomere shortening increases at the onset of disease.

Telomere Shortening and Aging-associated Diseases

Numerous links between aging-associated diseases and telomere length have been reported. Accelerated telomere attrition increases the risk of cancer (Artandi et al. 2000, Artandi and DePinho 2000, DePinho and Wong 2003), osteoarthritis (Martin and Buckwalter 2003), decreased immune function (Effros 2004), atherosclerosis (Samani et al. 2001), diabetes mellitus (Obana et al. 2003) and Alzheimer's disease (Panossian et al. 2003, Epel et al. 2004). Other correlations have been found in patients with chronic infections like chronic hepatitis and liver cirrhosis (Aikata et al. 2000, Kitada et al. 1995) and are ultimately linked to an 8-fold increased mortality rate in an elderly subpopulation (Cawthon et al. 2003), as described above.

Apart from the previously mentioned diseases, accelerated telomere attrition and telomere dysfunction has been implicated in the development of telomeric syndromes (Armanios and Blackburn 2012), as described below.

Mutations in telomerase, leading to telomere shortening, cause bone marrow failure syndromes such as constitutional AA and MDS. Subjects with aplastic anemia experience accelerated telomere shortening and die young (Marrone et al. 2004).

Dyskeratosis Congenital (DC) is an inherited genetic disorder produced by mutations in the Dyskerin (DKC1) gene, associated with the X-linked inheritance form of the disease (Heiss et al. 1998). Dyskerin is a nucleolar protein that has been involved in the modification of specific small RNA molecules, specifically ribosomal RNAs and the telomerase template RNA or hTERC. This disease can be caused by genetic factors other than DKC1, but the mutation always affects the telomerase complex, resulting in decreased telomerase stability and shorter telomeres (Mitchell et al. 1999). Patients with DC develop numerous different pathologies, including short stature, hypogonadism, infertility, bone marrow failure, abnormal skin pigmentation, hematopoietic defects and premature death.

Mutations in telomere related genes, distinct from telomerase components, are also responsible for human diseases. Fanconi Anemia (FA) and Ataxia Telangiectasia are autosomal recessive diseases caused by mutations in Fanconi genes (encoding any of 12 Fanconi anemia complementation group proteins) and the ataxia telangiectasia mutated genes (encoding the ATM protein). These proteins are implicated in DNA damage and repair pathways (Dokal 2006, Lavin and Kozlov 2007, Shiloh 1997). These diseases are associated with accelerated telomere shortening (Brummendorf et al. 2001, Leteurtre et al. 1999, Li et al. 2003, Metcalfe et al. 1996) and abnormalities in telomere replication or repair are thought to play a role in the pathogenesis, particularly in the progression of the disease to immunodeficiency and bone marrow failure, as well as in the increased susceptibility to malignancy in young adults.

Some diseases originating from mutations in genes of the DNA repair system also result in a phenotype characterized by accelerated telomere shortening and premature aging. HGPS is caused by point mutations in lamin A, a key component of nuclear scaffolding (Cao and Hegele 2003, Eriksson et al. 2003). Deficiency of this gene results in absence of hair, craniofacial deformities, emaciated and wrinkled appearance and cardiovascular defects that eventually lead to stroke or heart attack at a very young age. This disease is characterized by specific defects in nuclear shape (Scaffidi and Misteli 2006). However, although it is known that DNA damage responses in cells expressing mutant lamin A are abnormal, the role of telomeres in this disorder remains to be clarified (Manju et al. 2006).

Conclusions

Telomere rate of change, rather than telomere length per se, has been associated with lifespan. Telomere shortening increases with age and has been linked to several aging-associated diseases like cancer, diabetes, atherosclerosis and heart failure. However, the best examples of the role of telomeres in aging are telomeric syndromes and related diseases, caused by specific mutations in telomerase or telomere binding proteins, which prove the crucial role of telomere shortening in human aging and pathogenesis. Further research in this field, including the study of telomere length and telomerase activity at multiple time-points in large population based cohorts, may help in the development of novel epigenetic based therapies for aging and telomeric syndromes.

Keywords: Telomere length, telomerase, Hayflick limit, senescence, oxidative stress, telomere shortening (attrition), aging, heritable trait, telomere rate of change (TROC), lifespan, aging-associated diseases, progeroid syndromes

References

Adams, J., C. Martin-Ruiz, M. S. Pearce, M. White, L. Parker and T. von Zglinicki. 2007. No association between socio-economic status and white blood cell telomere length. Aging Cell 6(1): 125–8. doi: 10.1111/j.1474-9726.2006.00258.x.

Aikata, H., H. Takaishi, Y. Kawakami, S. Takahashi, M. Kitamoto, T. Nakanishi et al. 2000. Telomere reduction in human liver tissues with age and chronic inflammation. Exp Cell Res 256(2): 578–82. doi: 10.1006/excr.2000.4862.

Armanios, M. and E. H. Blackburn. 2012. The telomere syndromes. Nat Rev Genet 13(10): 693–704. doi: 10.1038/nrg3246.

Artandi, S. E., S. Chang, S. L. Lee, S. Alson, G. J. Gottlieb, L. Chin et al. 2000. Telomere dysfunction promotes non-reciprocal translocations and epithelial cancers in mice. Nature 406(6796): 641–5. doi: 10.1038/35020592.

Artandi, S. E. and R. A. DePinho. 2000. A critical role for telomeres in suppressing and facilitating carcinogenesis. Curr Opin Genet Dev 10(1): 39–46.

Aubert, G. and P. M. Lansdorp. 2008. Telomeres and aging. Physiol Rev 88(2): 557–79. doi: 10.1152/physrev.00026.2007.

Bekaert, S., T. De Meyer and P. Van Oostveldt. 2005. Telomere attrition as ageing biomarker. Anticancer Res 25(4): 3011–21.

Benetti, R., M. Garcia-Cao and M. A. Blasco. 2007. Telomere length regulates the epigenetic status of mammalian telomeres and subtelomeres. Nat Genet 39(2): 243–50. doi: 10.1038/ng1952.

Bermejo-Alvarez, P., D. Rizos, D. Rath, P. Lonergan and A. Gutierrez-Adan. 2008. Epigenetic differences between male and female bovine blastocysts produced *in vitro*. Physiol Genomics 32(2): 264–72. doi: 10.1152/physiolgenomics.00234.2007.

Blackburn, E. H. 2000. Telomere states and cell fates. Nature 408(6808): 53–6. doi: 10.1038/35040500.

Bodnar, A. G., M. Ouellette, M. Frolkis, S. E. Holt, C. P. Chiu, G. B. Morin et al. 1998. Extension of life-span by introduction of telomerase into normal human cells. Science 279(5349): 349–52.

Britt-Compton, B., J. Rowson, M. Locke, I. Mackenzie, D. Kipling and D. M. Baird. 2006. Structural stability and chromosome-specific telomere length is governed by cis-acting determinants in humans. Hum Mol Genet 15(5): 725–33. doi: 10.1093/hmg/ddi486.

Brummendorf, T. H., J. P. Maciejewski, J. Mak, N. S. Young and P. M. Lansdorp. 2001. Telomere length in leukocyte subpopulations of patients with aplastic anemia. Blood 97(4): 895–900.

Cao, H. and R. A. Hegele. 2003. LMNA is mutated in Hutchinson-Gilford progeria (MIM 176670) but not in Wiedemann-Rautenstrauch progeroid syndrome (MIM 264090). J Hum Genet 48(5): 271–4. doi: 10.1007/s10038-003-0025-3.

Cawthon, R. M., K. R. Smith, E. O'Brien, A. Sivatchenko and R. A. Kerber. 2003. Association between telomere length in blood and mortality in people aged 60 years or older. Lancet 361(9355): 393–5. doi: 10.1016/S0140-6736(03)12384-7.

Celli, G. B. and T. de Lange. 2005. DNA processing is not required for ATM-mediated telomere damage response after TRF2 deletion. Nat Cell Biol 7(7): 712–8. doi: 10.1038/ncb1275.

Chan, S. W. and E. H. Blackburn. 2003. Telomerase and ATM/Tel1p protect telomeres from nonhomologous end joining. Mol Cell 11(5): 1379–87.

Chen, Q. M., K. R. Prowse, V. C. Tu, S. Purdom and M. H. Linskens. 2001. Uncoupling the senescent phenotype from telomere shortening in hydrogen peroxide-treated fibroblasts. Exp Cell Res 265(2): 294–303. doi: 10.1006/excr.2001.5182.

Cherkas, L. F., J. L. Hunkin, B. S. Kato, J. B. Richards, J. P. Gardner, G. L. Surdulescu et al. 2008. The association between physical activity in leisure time and leukocyte telomere length. Arch Intern Med 168(2): 154–8. doi: 10.1001/archinternmed.2007.39.

Chin, L., S. E. Artandi, Q. Shen, A. Tam, S. L. Lee, G. J. Gottlieb et al. 1999. p53 deficiency rescues the adverse effects of telomere loss and cooperates with telomere dysfunction to accelerate carcinogenesis. Cell 97(4): 527–38.

Chiu, C. P., W. Dragowska, N. W. Kim, H. Vaziri, J. Yui, T. E. Thomas et al. 1996. Differential expression of telomerase activity in hematopoietic progenitors from adult human bone marrow. Stem Cells 14(2): 239–48. doi: 10.1002/stem.140239.

Counter, C. M., J. Gupta, C. B. Harley, B. Leber and S. Bacchetti. 1995. Telomerase activity in normal leukocytes and in hematologic malignancies. Blood 85(9): 2315–20.

Crabbe, L., R. E. Verdun, C. I. Haggblom and J. Karlseder. 2004. Defective telomere lagging strand synthesis in cells lacking WRN helicase activity. Science 306(5703): 1951–3. doi: 10.1126/science.1103619.

DePinho, R. A. and K. K. Wong. 2003. The age of cancer: telomeres, checkpoints, and longevity. J Clin Invest 111(7): S9–14.

Ding, H., M. Schertzer, X. Wu, M. Gertsenstein, S. Selig, M. Kammori et al. 2004. Regulation of murine telomere length by Rtel: an essential gene encoding a helicase-like protein. Cell 117(7): 873–86. doi: 10.1016/j.cell.2004.05.026.

Dokal, I. 2006. Fanconi's anaemia and related bone marrow failure syndromes. Br Med Bull 77-78: 37–53. doi: 10.1093/bmb/ldl007.

Effros, R. B. 2004. From Hayflick to Walford: the role of T cell replicative senescence in human aging. Exp Gerontol 39(6): 885–90. doi: 10.1016/j.exger.2004.03.004.

Epel, E. S., E. H. Blackburn, J. Lin, F. S. Dhabhar, N. E. Adler, J. D. Morrow et al. 2004. Accelerated telomere shortening in response to life stress. Proc Natl Acad Sci U S A 101(49): 17312–5. doi: 10.1073/pnas.0407162101.

Eriksson, M., W. T. Brown, L. B. Gordon, M. W. Glynn, J. Singer, L. Scott et al. 2003. Recurrent *de novo* point mutations in lamin A cause Hutchinson-Gilford progeria syndrome. Nature 423(6937): 293–8. doi: 10.1038/nature01629.

Frenck, R. W., Jr., E. H. Blackburn and K. M. Shannon. 1998. The rate of telomere sequence loss in human leukocytes varies with age. Proc Natl Acad Sci U S A 95(10): 5607–10.

Graakjaer, J., C. Bischoff, L. Korsholm, S. Holstebroe, W. Vach, V. A. Bohr et al. 2003. The pattern of chromosome-specific variations in telomere length in humans is determined by inherited, telomere-near factors and is maintained throughout life. Mech Ageing Dev 124(5): 629–40.

Griffith, J. D., L. Comeau, S. Rosenfield, R. M. Stansel, A. Bianchi, H. Moss et al. 1999. Mammalian telomeres end in a large duplex loop. Cell 97(4): 503–14.

Haussmann, M. F., D. W. Winkler, K. M. O'Reilly, C. E. Huntington, I. C. Nisbet and C. M. Vleck. 2003. Telomeres shorten more slowly in long-lived birds and mammals than in short-lived ones. Proc Biol Sci 270(1522): 1387–92. doi: 10.1098/rspb.2003.2385.

Heiss, N. S., S. W. Knight, T. J. Vulliamy, S. M. Klauck, S. Wiemann, P. J. Mason et al. 1998. X-linked dyskeratosis congenita is caused by mutations in a highly conserved gene with putative nucleolar functions. Nat Genet 19(1): 32–8. doi: 10.1038/ng0598-32.

Henle, E. S., Z. Han, N. Tang, P. Rai, Y. Luo and S. Linn. 1999. Sequence-specific DNA cleavage by $Fe2^+$-mediated fenton reactions has possible biological implications. J Biol Chem 274(2): 962–71.

Hiyama, K., Y. Hirai, S. Kyoizumi, M. Akiyama, E. Hiyama, M. A. Piatyszek et al. 1995. Activation of telomerase in human lymphocytes and hematopoietic progenitor cells. J Immunol 155(8): 3711–5.

Hunt, S. C., W. Chen, J. P. Gardner, M. Kimura, S. R. Srinivasan, J. H. Eckfeldt et al. 2008. Leukocyte telomeres are longer in African Americans than in whites: the National Heart, Lung, and Blood Institute Family Heart Study and the Bogalusa Heart Study. Aging Cell 7(4): 451–8. doi: 10.1111/j.1474-9726.2008.00397.x.

Iwama, H., K. Ohyashiki, J. H. Ohyashiki, S. Hayashi, N. Yahata, K. Ando et al. 1998. Telomeric length and telomerase activity vary with age in peripheral blood cells obtained from normal individuals. Hum Genet 102(4): 397–402.

Jeanclos, E., N. J. Schork, K. O. Kyvik, M. Kimura, J. H. Skurnick and A. Aviv. 2000. Telomere length inversely correlates with pulse pressure and is highly familial. Hypertension 36(2): 195–200.

Karlseder, J., A. Smogorzewska and T. de Lange. 2002. Senescence induced by altered telomere state, not telomere loss. Science 295(5564): 2446–9. doi: 10.1126/science.1069523.

Kitada, T., S. Seki, N. Kawakita, T. Kuroki and T. Monna. 1995. Telomere shortening in chronic liver diseases. Biochem Biophys Res Commun 211(1): 33–9. doi: 10.1006/bbrc.1995.1774.

Kurz, D. J., S. Decary, Y. Hong, E. Trivier, A. Akhmedov and J. D. Erusalimsky. 2004. Chronic oxidative stress compromises telomere integrity and accelerates the onset of senescence in human endothelial cells. J Cell Sci 117(Pt 11): 2417–26. doi: 10.1242/jcs.01097.

Lavin, M. F. and S. Kozlov. 2007. DNA damage-induced signalling in ataxia-telangiectasia and related syndromes. Radiother Oncol 83(3): 231–7. doi: 10.1016/j.radonc.2007.04.032.

Leteurtre, F., X. Li, P. Guardiola, G. Le Roux, J. C. Sergere, P. Richard et al. 1999. Accelerated telomere shortening and telomerase activation in Fanconi's anaemia. Br J Haematol 105(4): 883–93.

Li, X., F. Leteurtre, V. Rocha, P. Guardiola, R. Berger, M. T. Daniel et al. 2003. Abnormal telomere metabolism in Fanconi's anaemia correlates with genomic instability and the probability of developing severe aplastic anaemia. Br J Haematol 120(5): 836–45.

Lingner, J. and T. R. Cech. 1996. Purification of telomerase from Euplotes aediculatus: requirement of a primer 3' overhang. Proc Natl Acad Sci U S A 93(20): 10712–7.

Makarov, V. L., Y. Hirose and J. P. Langmore. 1997. Long G tails at both ends of human chromosomes suggest a C strand degradation mechanism for telomere shortening. Cell 88(5): 657–66.

Manestar-Blazic, T. 2004. Hypothesis on transmission of longevity based on telomere length and state of integrity. Med Hypotheses 62(5): 770–2. doi: 10.1016/j.mehy.2003.12.017.

Manju, K., B. Muralikrishna and V. K. Parnaik. 2006. Expression of disease-causing lamin A mutants impairs the formation of DNA repair foci. J Cell Sci 119(Pt 13): 2704–14. doi: 10.1242/jcs.03009.

Marrone, A., D. Stevens, T. Vulliamy, I. Dokal and P. J. Mason. 2004. Heterozygous telomerase RNA mutations found in dyskeratosis congenita and aplastic anemia reduce telomerase activity via haploinsufficiency. Blood 104(13): 3936–42. doi: 10.1182/blood-2004-05-1829.

Martens, U. M., J. M. Zijlmans, S. S. Poon, W. Dragowska, J. Yui, E. A. Chavez et al. 1998. Short telomeres on human chromosome 17p. Nat Genet 18(1): 76–80. doi: 10.1038/ng0198-018.

Martin, J. A. and J. A. Buckwalter. 2003. The role of chondrocyte senescence in the pathogenesis of osteoarthritis and in limiting cartilage repair. J Bone Joint Surg Am 85-A Suppl 2: 106–10.

Metcalfe, J. A., J. Parkhill, L. Campbell, M. Stacey, P. Biggs, P. J. Byrd et al. 1996. Accelerated telomere shortening in ataxia telangiectasia. Nat Genet 13(3): 350–3. doi: 10.1038/ng0796-350.

Michelson, R. J., S. Rosenstein and T. Weinert. 2005. A telomeric repeat sequence adjacent to a DNA double-stranded break produces an anticheckpoint. Genes Dev 19(21): 2546–59. doi: 10.1101/gad.1293805.

Mitchell, J. R., E. Wood and K. Collins. 1999. A telomerase component is defective in the human disease dyskeratosis congenita. Nature 402(6761): 551–5. doi: 10.1038/990141.

Muezzinler, A., A. K. Zaineddin and H. Brenner. 2013. A systematic review of leukocyte telomere length and age in adults. Ageing Res Rev 12(2): 509–19. doi: 10.1016/j.arr.2013.01.003.

Munoz, P., R. Blanco, J. M. Flores and M. A. Blasco. 2005. XPF nuclease-dependent telomere loss and increased DNA damage in mice overexpressing TRF2 result in premature aging and cancer. Nat Genet 37(10): 1063–71. doi: 10.1038/ng1633.

Nawrot, T. S., J. A. Staessen, J. P. Gardner and A. Aviv. 2004. Telomere length and possible link to X chromosome. Lancet 363(9408): 507–10. doi: 10.1016/S0140-6736(04)15535-9.

Njajou, O. T., R. M. Cawthon, C. M. Damcott, S. H. Wu, S. Ott, M. J. Garant et al. 2007. Telomere length is paternally inherited and is associated with parental lifespan. Proc Natl Acad Sci U S A 104(29): 12135–9. doi: 10.1073/pnas.0702703104.

Nordfjall, K., M. Eliasson, B. Stegmayr, S. Lundin, G. Roos and P. M. Nilsson. 2008a. Increased abdominal obesity, adverse psychosocial factors and shorter telomere length in subjects reporting early ageing; the MONICA Northern Sweden Study. Scand J Public Health 36(7): 744–52. doi: 10.1177/1403494808090634.

Nordfjall, K., M. Eliasson, B. Stegmayr, O. Melander, P. Nilsson and G. Roos. 2008b. Telomere length is associated with obesity parameters but with a gender difference. Obesity (Silver Spring) 16(12): 2682–9. doi: 10.1038/oby.2008.413.

Obana, N., S. Takagi, Y. Kinouchi, Y. Tokita, A. Sekikawa, S. Takahashi et al. 2003. Telomere shortening of peripheral blood mononuclear cells in coronary disease patients with metabolic disorders. Intern Med 42(2): 150–3.

Oeseburg, H., R. A. de Boer, W. H. van Gilst and P. van der Harst. 2010. Telomere biology in healthy aging and disease. Pflugers Arch 459(2): 259–68. doi: 10.1007/s00424-009-0728-1.

Oikawa, S., S. Tada-Oikawa and S. Kawanishi. 2001. Site-specific DNA damage at the GGG sequence by UVA involves acceleration of telomere shortening. Biochemistry 40(15): 4763–8.

Okuda, K., A. Bardeguez, J. P. Gardner, P. Rodriguez, V. Ganesh, M. Kimura et al. 2002. Telomere length in the newborn. Pediatr Res 52(3): 377–81. doi: 10.1203/00006450-200209000-00012.

Olovnikov, A. M. 1973. A theory of marginotomy. The incomplete copying of template margin in enzymic synthesis of polynucleotides and biological significance of the phenomenon. J Theor Biol 41(1): 181–90.

Panossian, L. A., V. R. Porter, H. F. Valenzuela, X. Zhu, E. Reback, D. Masterman et al. 2003. Telomere shortening in T cells correlates with Alzheimer's disease status. Neurobiol Aging 24(1): 77–84.

Pommier, J. P., L. Gauthier, J. Livartowski, P. Galanaud, F. Boue, A. Dulioust et al. 1997. Immunosenescence in HIV pathogenesis. Virology 231(1): 148–54. doi: 10.1006/viro.1997.8512.

Rudolph, K. L., S. Chang, H. W. Lee, M. Blasco, G. J. Gottlieb, C. Greider et al. 1999. Longevity, stress response, and cancer in aging telomerase-deficient mice. Cell 96(5): 701–12.

Rufer, N., T. H. Brummendorf, V. Dragowska, M. Shultzer, L. D. Wadsworth and P. M. Lansdorp. 1999a. Turnover of stem cells, naive and memory T lymphocytes, estimated from telomere fluorescence measurements. Cytotherapy 1(4): 342. doi: 10.1080/0032472031000141274.

Rufer, N., T. H. Brummendorf, S. Kolvraa, C. Bischoff, K. Christensen, L. Wadsworth et al. 1999b. Telomere fluorescence measurements in granulocytes and T lymphocyte subsets point to a high turnover of hematopoietic stem cells and memory T cells in early childhood. J Exp Med 190(2): 157–67.

Samani, N. J., R. Boultby, R. Butler, J. R. Thompson and A. H. Goodall. 2001. Telomere shortening in atherosclerosis. Lancet 358(9280): 472–3. doi: 10.1016/S0140-6736(01)05633-1.

Scaffidi, P. and T. Misteli. 2006. Lamin A-dependent nuclear defects in human aging. Science 312(5776): 1059–63. doi: 10.1126/science.1127168.

Serra, V., T. Grune, N. Sitte, G. Saretzki and T. von Zglinicki. 2000. Telomere length as a marker of oxidative stress in primary human fibroblast cultures. Ann N Y Acad Sci 908: 327–30.

Serra, V., T. von Zglinicki, M. Lorenz and G. Saretzki. 2003. Extracellular superoxide dismutase is a major antioxidant in human fibroblasts and slows telomere shortening. J Biol Chem 278(9): 6824–30. doi: 10.1074/jbc.M207939200.

Shammas, M. A. 2011. Telomeres, lifestyle, cancer, and aging. Curr Opin Clin Nutr Metab Care 14(1): 28–34. doi: 10.1097/MCO.0b013e32834121b1.

Shiloh, Y. 1997. Ataxia-telangiectasia and the Nijmegen breakage syndrome: related disorders but genes apart. Annu Rev Genet 31: 635–62. doi: 10.1146/annurev.genet.31.1.635.

Shin, J. S., A. Hong, M. J. Solomon and C. S. Lee. 2006. The role of telomeres and telomerase in the pathology of human cancer and aging. Pathology 38(2): 103–13. doi: 10.1080/00313020600580468.

Simons, M. J. 2015. Questioning causal involvement of telomeres in aging. Ageing Res Rev doi: 10.1016/j.arr.2015.08.002.

Slagboom, P. E., S. Droog and D. I. Boomsma. 1994. Genetic determination of telomere size in humans: a twin study of three age groups. Am J Hum Genet 55(5): 876–82.

Steinert, S., J. W. Shay and W. E. Wright. 2004. Modification of subtelomeric DNA. Mol Cell Biol 24(10): 4571–80.

Stindl, R. 2004. Tying it all together: telomeres, sexual size dimorphism and the gender gap in life expectancy. Med Hypotheses 62(1): 151–4.

Surralles, J., M. P. Hande, R. Marcos and P. M. Lansdorp. 1999. Accelerated telomere shortening in the human inactive X chromosome. Am J Hum Genet 65(6): 1617–22. doi: 10.1086/302665.

Takubo, K., N. Izumiyama-Shimomura, N. Honma, M. Sawabe, T. Arai, M. Kato et al. 2002. Telomere lengths are characteristic in each human individual. Exp Gerontol 37(4): 523–31.

Valdes, A. M., T. Andrew, J. P. Gardner, M. Kimura, E. Oelsner, L. F. Cherkas et al. 2005. Obesity, cigarette smoking, and telomere length in women. Lancet 366(9486): 662–4. doi: 10.1016/S0140-6736(05)66630-5.

van Steensel, B., A. Smogorzewska and T. de Lange. 1998. TRF2 protects human telomeres from end-to-end fusions. Cell 92(3): 401–13.

von Zglinicki, T. 2000. Role of oxidative stress in telomere length regulation and replicative senescence. Ann N Y Acad Sci 908: 99–110.

von Zglinicki, T., V. Serra, M. Lorenz, G. Saretzki, R. Lenzen-Grossimlighaus, R. Gessner et al. 2000. Short telomeres in patients with vascular dementia: an indicator of low antioxidative capacity and a possible risk factor? Lab Invest 80(11): 1739–47.

Wang, R. C., A. Smogorzewska and T. de Lange. 2004. Homologous recombination generates T-loop-sized deletions at human telomeres. Cell 119(3): 355–68. doi: 10.1016/j.cell.2004.10.011.

Watson, J. D. 1972. Origin of concatemeric T7 DNA. Nat New Biol 239(94): 197–201.

Yang, Q., Y. L. Zheng and C. C. Harris. 2005. POT1 and TRF2 cooperate to maintain telomeric integrity. Mol Cell Biol 25(3): 1070–80. doi: 10.1128/MCB.25.3.1070-1080.2005.

Youngren, K., E. Jeanclos, H. Aviv, M. Kimura, J. Stock, M. Hanna et al. 1998. Synchrony in telomere length of the human fetus. Hum Genet 102(6): 640–3.

Zeichner, S. L., P. Palumbo, Y. Feng, X. Xiao, D. Gee, J. Sleasman et al. 1999. Rapid telomere shortening in children. Blood 93(9): 2824–30.

14

Telomeres and Cardiovascular Diseases
Facts, Controversies and Limitations

Vicente Andrés,[1,*] *Beatriz Dorado*[1,a] and
Ioakim Spyridopoulos[2]

INTRODUCTION

Despite advances in diagnosis and treatment, CVD remains the major cause of morbidity and mortality in industrialized countries and is expected to become so worldwide by 2020. The economic, social and human costs associated with CVD treatment are immense. To improve diagnosis, prevention and treatment, it is thus urgent to refine classic CVD risk scores and identify new genetic, cellular and molecular mechanisms underlying disease initiation and early progression. This chapter examines advances in our understanding of the role of telomere shortening in the etiopathogenesis of CVD, and discusses limitations and perspectives for future research.

Telomere attrition, particularly the accumulation of critically short telomeres, is considered a marker of irreversible cell senescence and aging, the main CVD risk factor. Many *in vitro* and animal models and clinical studies have revealed associations between short telomeres and the appearance of disease symptoms. One of the most informative mouse models has been the late-generation Terc-KO mouse with critically short telomeres. These mice show symptoms of premature aging and associated anomalies including hair graying, alopecia, fertility loss, heart failure, and atrophy

[1] Centro Nacional de Investigaciones Cardiovasculares Carlos III (CNIC), c/Melchor Fernández Almagro, 3 28029 Madrid, Spain.
[a] Email: bjdorado@cnic.es
[2] Newcastle University, Institute of Genetic Medicine, Central Parkway Newcastle Upon Tyne, NE1 3BZ United Kingdom.
  Email: ioakim.spyridopoulos@ncl.ac.uk
* Corresponding author: vandres@cnic.es

of several tissues (Herrera et al. 1999, Leri et al. 2003, Wong et al. 2003). However, debate continues about whether telomere shortening is a cause or consequence of CVD in humans.

Telomere Dynamics in Different Hematopoietic Cell Populations

Most population-based studies of TL use peripheral blood leukocytes. The largest leukocyte fraction is composed of granulocytes, which reflect the TL of myeloid-derived stem and progenitor cells (Spyridopoulos et al. 2009). Isolated peripheral blood mononuclear cells are mostly lymphocytes, mainly T cells. Determination of TL separately for these populations requires the Flow-FISH method or prior cell sorting, and it can be assumed that most studies in unsorted peripheral blood leukocytes track myeloid TL. However, lymphocyte telomeres shorten faster during life, and a high content of short-telomeric T-cells can therefore decrease average LTL. Billions of new blood cells are produced every day from HSCs, but nonetheless, there is evidence that telomerase activity in HSCs cannot prevent age- and proliferation-associated telomere loss in this cell compartment (Aviv and Levy 2012). HSC telomere dynamics *in vivo* have been studied during the repopulation of the hematopoietic system in the course of bone marrow transplantation. TL in peripheral leukocytes of allogenic bone-marrow-transplant recipients is shorter than that of their donors, indicating that hematopoietic reconstitution places a greater proliferative demand on stem cells than they experience under normal hematopoiesis (Notaro et al. 1997). Granulocytes from young and old donors have similar rates of TL loss to myeloid bone marrow cells (Spyridopoulos et al. 2009). Moreover, telomere shortening in granulocytes and lymphocytes follows a cubic function during aging *in vivo*, marked by a significant drop in TL within the first year of life and a slower, steady decline thereafter (Rufer et al. 1999). The decline in TL is significantly faster in total lymphocytes than in granulocytes (53 bp/year vs. 39 bp/year) (Rufer et al. 1999, Spyridopoulos et al. 2008, Spyridopoulos et al. 2009). This difference might be attributable to the memory lymphocyte compartment. Memory T cells expressing CD4 and CD8 proteins both have much shorter telomeres than their naive counterparts (Rufer et al. 1999), and the increasing number of effector and memory T-cell subsets with age thus accelerates the decrease in lymphocyte TL (Koch et al. 2008). On the other hand, granulocyte telomeres shorten with age at the same rate as telomeres in naïve T cells (Rufer et al. 1999).

Telomere Length in Humans and Association with Cardiovascular Disease

Heritability and genetic determinants

TL is largely inherited and shows high inter-individual variability. Consistent with the observed age-dependent telomerase inactivation in somatic cells, telomeres in human leukocytes are longer at birth and undergo rapid shortening until adolescence, followed by slower shortening until old age (Aviv and Levy 2012). Telomere attrition can be strongly accelerated by genetic factors, including mutations in TERT, TERC, WRN, and SIRT1 (Armanios and Blackburn 2012, Aviv 2012).

A recent meta-analysis investigated the heritability of LTL, its mode of inheritance and the influence of parental age at birth (Broer et al. 2013). The study included six independent cohorts, totaling almost 20,000 subjects and including three cohorts with twins. The estimated LTL heritability was 0.70 (95% CI: 0.64–0.76). The authors found a strong correlation for monozygotic twins (r = 0.72), while much lower but similar correlations were found for dizygotic twins (r = 0.50) and other siblings (r = 0.49). The mother-offspring correlation (r = 0.42) was stronger than the father-offspring correlation (r = 0.33) and the authors also found a significant positive association with paternal age at offspring birth (β = 0.005). Interestingly, a significant and quite substantial correlation in LTL between spouses (r = 0.25) was seen, which seemed stronger in older spouse pairs (mean age $\geq$ 55 years; r = 0.31) than in younger couples (mean age < 55 years; r = 0.20), suggesting the influence of environmental factors. In summary, the results of this meta-analysis reveal a high and very consistent heritability estimate for mean LTL, evidence for a maternal inheritance component and a positive association with paternal age.

In another approach, GWAS were conducted on ≈40,000 individuals, with replication of selected variants in a further 11,000 individuals (Codd et al. 2013). This study identified SNPs in seven loci associated with mean LTL ($P < 5 \times 10^{-8}$). Five of the loci contained genes already implicated in telomere biology (TERC, TERT, NAF1, OBFC1, RTEL1). A genetic risk score analysis was conducted combining lead variants at all seven loci in over 22,233 cases of CAD and 64,762 controls. This analysis showed that alleles linked to shorter LTL were associated with a 21% higher risk of CAD (95% CI: 5–35% per standard deviation in LTL, p = 0.014). The authors suggested that the per allele effect of the different SNPs on LTL ranges from ~57 to 117 bp/year. Earlier Southern blotting studies showed a mean LTL attrition rate of about 30–50 bp/year (see previous section).

Telomeres and Cardiovascular Disease

Telomere attrition is associated with aging, the major risk factor for CVD. LTL shortening can be accelerated by other established cardiovascular risk factors, including endogenous factors (e.g., inflammation, oxidative stress, oxidized-LDL cholesterol, lack of estrogens, insulin-resistance) and environmental factors related to life-style (e.g., smoking, obesity, alcohol abuse, sedentary life-style or psychological stress) (Aviv 2012, Aviv and Levy 2012, Fuster and Andrés 2006). A recent meta-analysis of 24 studies, compiling data from over 40,000 participants and 8,000 CVD patients, found that the associations between short LTL and CVD was broadly consistent across all characteristics tested, including clinically relevant subgroups, such as different mean ages and sex distribution and across prospective and retrospective studies (Haycock et al. 2014). The pooled relative risk for CAD in all studies was 1.54 (95% CI: 1.30–1.83) and in prospective studies alone was 1.40 (95% CI: 1.15–1.70). The association between shorter telomeres and cerebrovascular disease was comparable to the association with CAD, but was not significant when restricting data to prospective studies. In spite of the observation that shorter LTL is associated with higher risk of CAD, independently of conventional cardiovascular risk factors, a causative link between telomere attrition and CAD in humans has not been established and conflicting results have been obtained in animal models (see below).

Telomere Length and Cardiovascular Disease in Premature Aging Syndromes

Premature aging disorders, such as HGPS or Werner syndrome, are characterized by precocious CVD and a severely reduced life span (Trigueros-Motos et al. 2011). Mechanisms thought to contribute to the etiopathogenesis of both diseases include transcriptional alterations, impaired DNA repair, accelerated telomere attrition and premature cellular senescence (Andrés and González 2009). In HGPS, mutations in the LMNA gene (encoding A-type lamins) lead to abnormal accumulation of progerin, which unlike mature wild-type lamin A remains permanently farnesylated. This accumulation disrupts the nuclear lamina and leads to a multitude of nuclear defects. Interestingly, in normal cells progressive telomere damage during cellular senescence activates progerin production, a mechanism through which low progerin expression might contribute to normal aging (Cao et al. 2011).

A recent study showed that association of TRF2 with interstitial telomeric sequences is stabilized by co-localization with A-type lamins (lamin A/C) and that HGPS-causing mutations or reduced lamin A/C levels lead to impaired interstitial t-loop formation and telomere loss (rather than telomere shortening) (Wood et al. 2014). Moreover, the interaction of LAP2α with telomeres is impaired in HGPS, and progerin-induced defects in cell proliferation and chromatin organization are rescued by increasing LAP2α levels, while lowering LAP2α levels has the opposite effect (Chojnowski et al. 2015). It is conceivable that telomere shortening in HGPS is the net result of normal length of some telomeres and the complete loss of others, yielding a 'shorter' mean telomere length per cell (metaphase cells contain 92 telomere foci). Atherosclerotic plaque generation in the setting of HGPS may be accelerated by progerin-dependent induction of VSMC senescence due to increased oxidative stress, DNA damage, inhibition of telomerase and abnormally high telomere shortening. However, heart failure in the absence of atherosclerosis is a common phenotype in other genetic disorders associated with telomere shortening, such as DKC. It would thus seem that short telomeres alone do not necessarily cause premature atherosclerosis, such as in DKC patients, but might lead to accelerated senescence and therefore accelerated atherosclerosis in the setting of HGPS and Werner Syndrome.

Telomerase and Telomeres in Cellular and Animal Models of Cardiovascular Disease

Studies in different patient populations have consistently revealed significant associations between short LTL and various forms of clinically-relevant CVD (e.g., myocardial infarction and heart failure) independently of traditional cardiovascular risk factors. However, cross-sectional studies of CVD-free middle-aged populations revealed that average LTL and short telomere load are not significant independent determinants of early subclinical atherosclerosis after adjusting for age or other well-known cardiovascular risk factors (De Meyer et al. 2009, Fernández-Alvira et al. 2016). Moreover, a causative link between telomerase inactivation/telomere attrition and CVD in humans has not been established, and a number of seemingly contradictory results remain unresolved. For example, at advanced stages, hypertension, heart failure and atherosclerosis share features of cellular senescence characterized by low or absent telomerase activity and short telomeres. Moreover, telomere attrition and the onset of senescence and/or apoptosis are induced by telomerase inhibition in cultured VSMCs, ECs and cardiomyocytes. Conversely, forced telomerase activation promotes the proliferation of cardiovascular cells and prolongs their lifespan (Fuster and Andrés 2006). Lack of telomerase activity and telomere attrition are therefore often assumed to be detrimental in the setting of CVD. However, some studies have revealed strong telomerase activity at early stages of hypertension and atherosclerosis, suggesting that telomerase activation may contribute to the onset of these disorders. Moreover, atherosclerosis is significantly attenuated in transgenic mice born with critically short telomeres (Poch et al. 2004). In this section we discuss studies supporting a biphasic model of telomerase activity, rather than a model of strictly progressive telomere attrition.

Heart disease

Strong evidence for the importance of telomere attrition in heart disease has come from studies in late-generation Terc-KO mice with critically short telomeres. These mice show weak proliferation and increased apoptosis of cardiomyocytes, ventricular dilation, myocardial thinning, cardiac dysfunction and sudden death (Leri et al. 2003). Recent studies in this mouse model have revealed that postnatal telomere dysfunction induces cardiomyocyte cell-cycle arrest through p21 activation (Aix et al. 2016). Conversely, transgenic mice with cardiac-specific over-expression of human TERT maintain telomerase activity and TL in adult heart, correlating with protection against heart damage after mechanical and ischemic injury (Oh et al. 2001, Oh et al. 2003). Likewise, cardiac-specific TERT expression mediated by adeno-associated viruses causes telomere elongation and a gene expression switch towards the regeneration signature of neonatal mice, and confers cardio-protection to the adult mouse heart after acute myocardial infarction (Bar et al. 2014).

The myocardium of end-stage heart failure patients shows TRF2 downregulation, activation of Chk2, and shortened telomeres and TRF2 inactivation in cultured cardiomyocytes causes rapid telomere shortening, Chk2 activation and apoptosis (Oh et al. 2003). Consistent with these findings, analysis of biopsies from heart-failure patients reveals shorter telomeres and increased cellular senescence and cell death (Chimenti et al. 2003). Telomere shortening may be a critical factor in the aging heart, where the number of senescent cardiomyocytes with short telomeres increases, possibly as a consequence of excessive oxidative stress generated by mitochondrial dysfunction (Moslehi et al. 2012). This loss of cardiomyocytes in the aging heart could lead to age-related heart diseases such as diastolic heart failure. In summary, reduced telomerase activity and short telomeres in cardiac tissue may trigger heart failure by promoting cardiomyocyte senescence and hampering tissue regeneration.

Hypertension

A study of SHR at ages before manifestion of hypertension revealed increased aortic TERT expression and telomerase activity (Cao et al. 2002). Moreover, compared with cells from control rats, primary cultures of medial VSMCs from the aorta of young SHR show increased telomerase activity and telomere length associated with augmented proliferation. However, in older SHRs, with established hypertension, EPCs show reduced telomerase and mitotic activity associated with increased cellular senescence. Similar findings were observed in EPCs from other hypertensive rat models (deoxycorticosterone acetate salt hypertensive rats and angiotensin II-infused hypertensive rats) and in the white blood cells and EPCs of hypertensive patients (Imanishi et al. 2005, Kobayashi et al. 2006). A causal relationship between oxidative stress, established hypertension and short telomeres is supported by the hypertension and high blood and urine levels of the endothelium-derived vasoconstrictor peptide endothelin-1 in late generation Terc-KO mice with critically short telomeres (Pérez-Rivero et al. 2006). In addition, embryonic fibroblasts from these mutant mice show enhanced production of reactive oxygen species and treatment with antioxidants reduces their expression of endothelin-converting enzyme. Similarly, inhibiting telomere function in ECs by over-expressing a TRF2 mutant diminishes endothelial nitric oxide synthase activity and induces a senescent phenotype (Minamino et al. 2002). VSMC proliferation supported by early telomerase activation may therefore contribute to the initial phases of vascular remodeling (e.g., medial hypertrophy) associated with the onset of hypertension, while insulin resistance or prolonged exposure to inflammation and oxidative stress may accelerate vascular cell telomere erosion and senescence (Fuster et al. 2007), characteristics of established hypertension.

Atherosclerosis

Studies in human and animal arteries have revealed excessive cellular proliferation at early stages of atherosclerosis and accumulation of senescent cells in advanced atherosclerotic lesions (Andrés and Castro 2003). Cellular senescence in atherosclerotic vessels can be caused by DNA damage, oxidative stress, epigenetic modifications and telomere attrition, a set of non-mutually exclusive molecular mechanisms able to feedback on one another. The possible causal relationship between telomere attrition or telomerase inactivation and atherosclerosis is debated, with some studies indicating telomere dysfunction as a cause and others as a consequence of the disease.

Senescent ECs have been demonstrated in aged human iliac, thoracic and coronary arteries, particularly in atherosclerotic lesions (Fuster and Andrés 2006, Aviv and Levy 2012). High hemodynamic stress in these vessels is thought to locally enhance EC turnover, with subsequent telomere shortening due to lack of telomerase activity (Chang and Harley 1995, Matthews et al. 2006). Telomerase inhibition in cultured ECs induces telomere attrition and the onset of senescence and/or apoptosis; conversely, telomerase or TRF2 over-expression in ECs extends their lifespan and inhibits the functional alterations related to senescence (Aviv and Levy 2012). Moreover, it has been suggested that shortened LTL predicts augmented atherosclerotic risk because the injurious component of atherosclerosis exceeds the repair capacity of HSC reserves, thus reducing the number of EPCs available to replace senescent ECs (Aviv and Levy 2012).

Excessive proliferation of neointimal VSMCs and macrophages contributes to the growth of atherosclerotic lesions (Andrés and Castro 2003, Robbins et al. 2013). TERT activation extends the lifespan of cultured VSMCs, whereas telomerase inhibition abrogates VSMC proliferation (Fuster and Andrés 2006). Liu et al. reported strong TERT expression and activity in atherosclerotic human epicardial coronary arteries (four-five fold increase vs. control) (Liu et al. 2005). They also found an association of TERT expression with actively proliferating cells in early lesions, but advanced lesions were characterized by a paucity of proliferating TERT-expressing cells. High levels of TERT expression have also been found in macrophages in human atherosclerotic lesions, and telomerase is activated during atherosclerosis development in low-density lipoprotein receptor-deficient mice (Ldlr-KO) (Gizard et al. 2011). Telomerase is likewise activated during the proliferative response in rat and mouse arteries subjected to mechanical injury (Ogawa et al. 2006, Bu et al. 2010).

The complexity of the role of telomerase/telomeres in atherosclerosis development is highlighted by late-generation mice doubly deficient for apolipoprotein E and Terc (apoE-Terc-DKO). These mice have significantly shorter telomeres in medial VSMCs than apoE-KO controls with intact TERC, and exhibit significantly below normal numbers of advanced atherosclerotic lesions (Poch et al. 2004). Primary lymphocytes and macrophages from apoE-Terc-DKO mice show significantly reduced mitogen-induced proliferation, suggesting that immunoreplicative senescence in this model may contribute to reduced atherosclerosis upon telomere shortening. In another study, plasma cholesterol level increased similarly in young (four-five month-old) and old (four-five year-old) rabbits fed an atherogenic diet for two months. However, aortic atherosclerotic lesions, composed mainly of neointimal macrophages, were more prominent in young animals, which had a higher DNA-binding activity of NF-kB in aortic tissue (Cortés et al. 2002). Remarkably, inflammatory stimuli cause telomerase activation in macrophages and VSMCs through the transcriptional activation of the TERT promoter by NF-kB and PPARγ (Gizard et al. 2011, Ogawa et al. 2006, Bu et al. 2010). NF-kB can also upregulate telomerase activity post-transcriptionally via TERT nuclear translocation through a physical association (Akiyama et al. 2003).

Final Remarks

TL in humans is largely inherited and declines with age, at least, in part due to rapid inactivation of telomerase after birth. The rate of telomere shortening appears to be influenced by several environmental factors, including classical cardiovascular risk factors. Indeed, multiple human population studies have consistently revealed that telomere ablation in circulating leukocytes is associated with clinically-relevant CVD and predicts mortality. However, in cross-sectional human studies with CVD-free middle-aged populations, LTL is not an independent determinant of early asymptomatic atherosclerosis after adjusting for age and other well-known cardiovascular risk factors (De Meyer et al. 2009, Fernandez-Alvira et al. 2016). Telomere dysfunction is associated with genomic instability, accumulation of DNA damage, mitochondrial dysfunction and stem-cell exhaustion, which have all been implicated in the aging process (Sahin and Depinho 2010). Nevertheless, it remains unresolved whether telomere attrition is a cause of aging and associated CVD or simply a consequence (e.g., readout of oxidative stress, DNA damage and cell senescence or increased cell turnover under systemic low grade inflammation).

The number of senescent cardiomyocytes with short telomeres increase with age and animal studies indicate that telomere ablation associated with defective telomerase activity hamper cardio-protection after acute myocardial infarction and increase the risk of heart failure. However, the situation is more complex for hypertension and atherosclerosis, which seem to exhibit a biphasic telomerase behavior. VSMC proliferation supported by early telomerase activation may contribute to the initial phases of vascular remodeling (e.g., medial hypertrophy) at pre-hypertension stages, whereas subsequent telomerase inactivation may contribute to the establishment of hypertension by accelerating telomere erosion and senescence of vascular cells. Likewise, prompt activation of telomerase in immune cells and VSMCs might contribute to early atherosclerosis development through increased cell proliferation. Moreover, subsequent telomere attrition resulting from telomerase inactivation may contribute to the accumulation of senescent cells in advanced atheromata and thus facilitate plaque rupture and life-threatening acute ischemic events. Remarkably, mice with short telomeres are protected from diet-induced atherosclerosis, possibly in part due to inhibition of neointimal cell proliferation. Clearly, more sophisticated mouse models using inducible and cell type-specific promoters are required to unravel the consequences of altering telomere dynamics in different phases of CVD and in distinct cell populations. These studies should examine the role of telomerase and other telomere-associated proteins that control telomere homeostasis.

Also note that most human studies investigating TL and CVD are cross-sectional, therefore information on the relation between CVD progression and the rate of telomere attrition over time at the individual level is very limited. Longitudinal studies in large cohorts are thus required to assess whether the rate of telomere attrition is a better prognostic marker of aging and associated CVD than mean TL, or whether these measures could be used in combination, such that accelerated CVD would occur in individuals with short LTL and higher rates of telomere attrition, whereas protection would be conferred to individuals with long LTL and lower rates of telomere attrition. Notably, accelerated telomere attrition in HGPS and Werner syndrome patients is associated with exaggerated atherosclerosis and premature aging and death. However, abnormally high telomere shortening in DKC is often associated with heart failure in the absence of atherosclerosis. Thus, short telomeres alone do not necessarily cause premature atherosclerosis, but might accelerate the process in HGPS and Werner Syndrome.

Another limitation in most human population studies is the use of mean TL of mixed populations of cells (mostly circulating leukocytes) with differing telomere dynamics. Moreover, there is evidence that the accumulation of a few critically short telomeres in a cell can trigger cellular senescence and there is large heterogeneity in TL across chromosomes and cells during aging (Abdallah et al. 2009, Bendix et al. 2010, Hemann et al. 2001, Vera et al. 2012, Baird et al. 2003, Kimura et al. 2007, Lin et al. 2014). Studies are thus warranted to address whether the amount of critically short telomeres, rather than mean TL, is a better predictor of aging and associated disease.

Acknowledgements

We apologize to colleagues whose primary work could not be cited due to space limitations. We thank Simon Bartlett for English editing. Work in the laboratory of VA is supported by grants from the Spanish Ministry of Economy and Competitiveness (MINECO), the FondoEuropeo de Desarrollo Regional (FEDER), the Instituto de Salud Carlos III (SAF2013-46663-R, RD12/0042/0028), and the Progeria Research Foundation (Established Investigator Award PRF 2014-52). IS is supported by grants from the British Heart Foundation. The Centro Nacional de Investigaciones Cardiovasculares Carlos III (CNIC) is supported by the MINECO and the Pro-CNIC Foundation, and is a Severo Ochoa Center of Excellence (MINECO award SEV-2015-0505).

Keywords: Telomere length, aging, epidemiology, cardiovascular disease, atherosclerosis, heart disease, hypertension, leukocyte

References

Aix, E., O. Gutiérrez-Gutiérrez, C. Sánchez-Ferrer, T. Aguado and I. Flores. 2016. Postnatal telomere dysfunction induces cardiomyocyte cell-cycle arrest through p21 activation. J Cell Biol 213(5): 571–83.

Abdallah, P., P. Luciano, K. W. Runge, M. Lisby, V. Geli, E. Gilson et al. 2009. A two-step model for senescence triggered by a single critically short telomere. Nat Cell Biol 11(8): 988–93. doi: 10.1038/ncb1911.

Akiyama, M., T. Hideshima, T. Hayashi, Y. T. Tai, C. S. Mitsiades, N. Mitsiades et al. 2003. Nuclear factor-kappaB p65 mediates tumor necrosis factor alpha-induced nuclear translocation of telomerase reverse transcriptase protein. Cancer Res 63(1): 18–21.

Andrés, V. and C. Castro. 2003. Antiproliferative strategies for the treatment of vascular proliferative disease. Curr Vasc Pharmacol 1(1): 85–98.

Andrés, V. and J. M. González. 2009. Role of A-type lamins in signaling, transcription, and chromatin organization. J Cell Biol 187(7): 945–57. doi: 10.1083/jcb.200904124.

Armanios, M. and E. H. Blackburn. 2012. The telomere syndromes. Nat Rev Genet 13(10): 693–704. doi: 10.1038/nrg3246.

Aviv, A. 2012. Genetics of leukocyte telomere length and its role in atherosclerosis. Mutat Res 730(1-2): 68–74. doi: 10.1016/j.mrfmmm.2011.05.001.

Aviv, A. and D. Levy. 2012. Telomeres, atherosclerosis, and the hemothelium: the longer view. Annu Rev Med 63: 293–301. doi: 10.1146/annurev-med-050311-104846.

Baird, D. M., J. Rowson, D. Wynford-Thomas and D. Kipling. 2003. Extensive allelic variation and ultrashort telomeres in senescent human cells. Nat Genet 33(2): 203–7. doi: 10.1038/ng1084.

Bar, C., B. Bernardes de Jesus, R. Serrano, A. Tejera, E. Ayuso, V. Jimenez et al. 2014. Telomerase expression confers cardioprotection in the adult mouse heart after acute myocardial infarction. Nat Commun 5: 5863. doi: 10.1038/ncomms6863.

Bendix, L., P. B. Horn, U. B. Jensen, I. Rubelj and S. Kolvraa. 2010. The load of short telomeres, estimated by a new method, Universal STELA, correlates with number of senescent cells. Aging Cell 9(3): 383–97. doi: 10.1111/j.1474-9726.2010.00568.x.

Broer, L., V. Codd, D. R. Nyholt, J. Deelen, M. Mangino, G. Willemsen et al. 2013. Meta-analysis of telomere length in 19,713 subjects reveals high heritability, stronger maternal inheritance and a paternal age effect. Eur J Hum Genet 21(10): 1163–8. doi: 10.1038/ejhg.2012.303.

Bu, D. X., M. E. Johansson, J. Ren, D. W. Xu, F. B. Johnson, K. Edfeldt et al. 2010. Nuclear factor kappa B-mediated transactivation of telomerase prevents intimal smooth muscle cell from replicative senescence during vascular repair. Arterioscler Thromb Vasc Biol 30(12): 2604–10. doi: 10.1161/ATVBAHA.110.213074.

Cao, K., C. D. Blair, D. A. Faddah, J. E. Kieckhaefer, M. Olive, M. R. Erdos et al. 2011. Progerin and telomere dysfunction collaborate to trigger cellular senescence in normal human fibroblasts. J Clin Invest 121(7): 2833–44. doi: 10.1172/JCI43578.

Cao, Y., H. Li, F. T. Mu, O. Ebisui, J. W. Funder and J. P. Liu. 2002. Telomerase activation causes vascular smooth muscle cell proliferation in genetic hypertension. FASEB J 16: 96–98.

Codd, V., C. P. Nelson, E. Albrecht, M. Mangino, J. Deelen, J. L. Buxton et al. 2013. Identification of seven loci affecting mean telomere length and their association with disease. Nat Genet 45(4): 422–7, 427e1-2. doi: 10.1038/ng.2528.

Cortés, M. J., A. Díez-Juan, P. Pérez, I. Pérez-Roger, R. Arroyo-Pellicer and V. Andrés. 2002. Increased early atherogenesis in young versus old hypercholesterolemic rabbits by a mechanism independent of arterial cell proliferation. FEBS Lett 522: 99–103.

Chang, E. and C. B. Harley. 1995. Telomere length and replicative aging in human vascular tissues. Proc Natl Acad Sci USA 92: 11190–11194.

Chimenti, C., J. Kajstura, D. Torella, K. Urbanek, H. Heleniak, C. Colussi et al. 2003. Senescence and death of primitive cells and myocytes lead to premature cardiac aging and heart failure. Circ Res 93: 604–613.

Chojnowski, A., P. F. Ong, E. S. Wong, J. S. Lim, R. A. Mutalif, R. Navasankari et al. 2015. Progerin reduces LAP2alpha-telomere association in Hutchinson-Gilford progeria. Elife 4. doi: 10.7554/eLife.07759.

De Meyer, T., E. R. Rietzschel, M. L. De Buyzere, M. R. Langlois, D. De Bacquer, P. Segers et al. 2009. Systemic telomere length and preclinical atherosclerosis: the Asklepios Study. Eur Heart J 30(24): 3074–81.

Fernández-Alvira, J. M., V. Fuster, B. Dorado, N. Soberón, I. Flores, M. Gallardo et al. 2016. Short telomere load, telomere length, and subclinical atherosclerosis: The PESA study. J Am Coll Cardiol 67(21): 2467–76.

Fuster, J. J., J. Díez and V. Andrés. 2007. Telomere dysfunction in hypertension. J Hypertens 25(11): 2185–92. doi: 10.1097/HJH.0b013e3282ef6196.

Fuster, J. J. and V. Andrés. 2006. Telomere biology and cardiovascular disease. Circ Res 99(11): 1167–1180. doi: 10.1161/01.res.0000251281.00845.18.

Gizard, F., E. B. Heywood, H. M. Findeisen, Y. Zhao, K. L. Jones, C. Cudejko et al. 2011. Telomerase activation in atherosclerosis and induction of telomerase reverse transcriptase expression by inflammatory stimuli in macrophages. Arterioscler Thromb Vasc Biol 31(2): 245–52. doi: 10.1161/ATVBAHA.110.219808.

Haycock, P. C., E. E. Heydon, S. Kaptoge, A. S. Butterworth, A. Thompson and P. Willeit. 2014. Leucocyte telomere length and risk of cardiovascular disease: systematic review and meta-analysis. BMJ 349: g4227. doi: 10.1136/bmj.g4227.

Hemann, M. T., M. A. Strong, L. Y. Hao and C. W. Greider. 2001. The shortest telomere, not average telomere length, is critical for cell viability and chromosome stability. Cell 107(1): 67–77.

Herrera, E., E. Samper, J. Martin-Caballero, J. M. Flores, H. W. Lee and M. A. Blasco. 1999. Disease states associated with telomerase deficiency appear earlier in mice with short telomeres. EMBO J 18: 2950–2960.

Imanishi, T., C. Moriwaki, T. Hano and I. Nishio. 2005. Endothelial progenitor cell senescence is accelerated in both experimental hypertensive rats and patients with essential hypertension. J Hypertens 23(10): 1831–7.

Kimura, M., M. Barbieri, J. P. Gardner, J. Skurnick, X. Cao, N. van Riel et al. 2007. Leukocytes of exceptionally old persons display ultra-short telomeres. Am J Physiol Regul Integr Comp Physiol 293(6): R2210–7. doi: 10.1152/ajpregu.00615.2007.

Kobayashi, K., T. Imanishi and T. Akasaka. 2006. Endothelial progenitor cell differentiation and senescence in an angiotensin II-infusion rat model. Hypertens Res 29: 449–455.

Koch, S., A. Larbi, E. Derhovanessian, D. Ozcelik, E. Naumova and G. Pawelec. 2008. Multiparameter flow cytometric analysis of CD4 and CD8 T cell subsets in young and old people. Immun Ageing 5: 6. doi: 10.1186/1742-4933-5-6.

Leri, A., S. Franco, A. Zacheo, L. Barlucchi, S. Chimenti, F. Limana et al. 2003. Ablation of telomerase and telomere loss leads to cardiac dilatation and heart failure associated with p53 upregulation. EMBO J 22: 131–139.

Lin, T. T., K. Norris, N. H. Heppel, G. Pratt, J. M. Allan, D. J. Allsup et al. 2014. Telomere dysfunction accurately predicts clinical outcome in chronic lymphocytic leukaemia, even in patients with early stage disease. Br J Haematol 167(2): 214–23. doi: 10.1111/bjh.13023.

Liu, Shih-Chi, Shoei-Shen Wang, Mu-Zon Wu, Deng-Chyang Wu, Fang-Jung Yu, Wen-Jone Chen et al. 2005. Activation of telomerase and expression of human telomerase reverse transcriptase in coronary atherosclerosis. Cardiovasc Pathol 14(5): 232–240.

Matthews, C., I. Gorenne, S. Scott, N. Figg, P. Kirkpatrick, A. Ritchie et al. 2006. Vascular smooth muscle cells undergo telomere-based senescence in human atherosclerosis: effects of telomerase and oxidative stress. Circ Res 99(2): 156–64.

Minamino, T., H. Miyauchi, T. Yoshida, Y. Ishida, H. Yoshida and I. Komuro. 2002. Endothelial cell senescence in human atherosclerosis. Role of telomere in endothelial dysfunction. Circulation 105: 1541–1544.

Moslehi, J., R. A. DePinho and E. Sahin. 2012. Telomeres and mitochondria in the aging heart. Circ Res 110(9): 1226–37. doi: 10.1161/CIRCRESAHA.111.246868.

Notaro, R., A. Cimmino, D. Tabarini, B. Rotoli and L. Luzzatto. 1997. *In vivo* telomere dynamics of human hematopoietic stem cells. Proc Natl Acad Sci USA 94(25): 13782–5.

Ogawa, Daisuke, Takashi Nomiyama, Takafumi Nakamachi, Elizabeth B. Heywood, Jeffrey F. Stone, Joel P. Berger et al. 2006. Activation of peroxisome proliferator-activated receptor gamma suppresses telomerase activity in vascular smooth muscle cells. Circ Res 98(7): e50–59. doi: 10.1161/01.RES.0000218271.93076.c3.

Oh, H., G. E. Taffet, K. A. Youker, M. L. Entman, P. A. Overbeek, L. H. Michael et al. 2001. Telomerase reverse transcriptase promotes cardiac muscle cell proliferation, hypertrophy, and survival. Proc Natl Acad Sci USA 98: 10308–10313.

Oh, Hidemasa, Sam C. Wang, Arun Prahash, Motoaki Sano, Christine S. Moravec, George E. Taffet et al. 2003. Telomere attrition and Chk2 activation in human heart failure. Proc Natl Acad Sci USA 100: 5378–5383.

Pérez-Rivero, G., M. P. Ruiz-Torres, J. V. Rivas-Elena, M. Jerkic, M. L. Díez-Marques, J. M. López-Novoa et al. 2006. Mice deficient in telomerase activity develop hypertension because of an excess of endothelin production. Circulation 114(4): 309–17.

Poch, E., P. Carbonell, S. Franco, A. Díez-Juan, M. A. Blasco and V. Andrés. 2004. Short telomeres protect from diet-induced atherosclerosis in apolipoprotein E-null mice. FASEB J 18(2): 418–20.

Robbins, C. S., I. Hilgendorf, G. F. Weber, I. Theurl, Y. Iwamoto, J. L. Figueiredo et al. 2013. Local proliferation dominates lesional macrophage accumulation in atherosclerosis. Nat Med 19(9): 1166–72. doi: 10.1038/nm.3258.

Rufer, N., T. H. Brummendorf, S. Kolvraa, C. Bischoff, K. Christensen, L. Wadsworth et al. 1999. Telomere fluorescence measurements in granulocytes and T lymphocyte subsets point to a high turnover of hematopoietic stem cells and memory T cells in early childhood. J Exp Med 190(2): 157–67.

Sahin, E. and R. A. Depinho. 2010. Linking functional decline of telomeres, mitochondria and stem cells during ageing. Nature 464(7288): 520–8. doi: 10.1038/nature08982.

Spyridopoulos, I., Y. Erben, T. H. Brummendorf, J. Haendeler, K. Dietz, F. Seeger et al. 2008. Telomere gap between granulocytes and lymphocytes is a determinant for hematopoetic progenitor cell impairment in patients with previous myocardial infarction. Arterioscler Thromb Vasc Biol 28(5): 968–74. doi: 10.1161/ATVBAHA.107.160846.

Spyridopoulos, I., J. Hoffmann, A. Aicher, T. H. Brummendorf, H. W. Doerr, A. M. Zeiher et al. 2009. Accelerated telomere shortening in leukocyte subpopulations of patients with coronary heart disease: role of cytomegalovirus seropositivity. Circulation 120(14): 1364–72. doi: 10.1161/CIRCULATIONAHA.109.854299.

Trigueros-Motos, L., J. M. González, J. Rivera and V. Andrés. 2011. Hutchinson-Gilford progeria syndrome, cardiovascular disease and oxidative stress. Front Biosci (Schol. Ed.) 3: 1285–97.

Vera, E., B. Bernardes de Jesus, M. Foronda, J. M. Flores and M. A. Blasco. 2012. The rate of increase of short telomeres predicts longevity in mammals. Cell Rep 2(4): 732–7. doi: 10.1016/j.celrep.2012.08.023.

Wong, K. K., R. S. Maser, R. M. Bachoo, J. Menon, D. R. Carrasco, Y. Gu et al. 2003. Telomere dysfunction and Atm deficiency compromises organ homeostasis and accelerates ageing. Nature 421: 643–648.

Wood, A. M., J. M. Rendtlew Danielsen, C. A. Lucas, E. L. Rice, D. Scalzo, T. Shimi et al. 2014. TRF2 and lamin A/C interact to facilitate the functional organization of chromosome ends. Nat Commun 5: 5467. doi: 10.1038/ncomms6467.

15

Telomeres in Cancer

João Vinagre,[1,2,a] *Ana Pestana,*[1,2,3,b] *Manuel Sobrinho Simões*[1,2,3,4,5,c] and *Paula Soares*[1,2,3,5,*]

INTRODUCTION

Development of cancer requires the orchestration of a series of multistep events in order to allow an incipient cancer cell gain growth advantage over the remaining normal cells. Such alterations in the genome of the cell may be the acquisition of oncogenic mutations, the loss of tumor suppressor genes or the epigenetic modulation of genes with a direct or indirect impact on cell death or proliferation. Overall, there is not an exact mathematical formula for the development of a cancer and many distinct genetic contributors may enter into the recipe. Additionally, to accomplish a state of replicative immortalization, the emerging cancer cell will still need to deceive the intrinsic telomere length-sensing control mechanisms that prevent an unrestrained division. Mechanistically, at each cell division, the telomeric DNA sequence is diminished, culminating into progressively shorter telomeres due to the lack of telomerase expression and the end replication limitation of DNA polymerase (Gunes and Rudolph 2013). In response to this critical shortening, telomeres lose the capping function, become dysfunctional and induce cell cycle arrest, senescence or apoptosis (Nandakumar and Cech 2013). Since telomere erosion and the subsequent

[1] Institute of Molecular Pathology and Immunology, University of Porto (IPATIMUP)/I3S, Rua Júlio Amaral de Carvalho, 45, 4200-135 Porto, Portugal.
[2] Institute for Research and Innovation in Health, University of Porto, Porto, Portugal (Instituto de Investigação e Inovação em Saúde, Universidade do Porto, Portugal).
[3] Medical Faculty, University of Porto, Porto, Portugal.
[4] Department of Pathology, Hospital S. João, Porto, Portugal.
[5] Department of Pathology, Medical Faculty, University of Porto, Porto, Portugal.
[a] Email: jvinagre@ipatimup.pt
[b] Email: apestana@ipatimup.pt
[c] Email: ssimoes@ipatimup.pt
* Corresponding author: psoares@ipatimup.pt

response limit the proliferative capacity of cancer cells, either as tumor mass growth or metastasis dissemination, they "need" to overcome this limitation that functions as a tumor suppressor mechanism. In other words, the neoplastic cells will only constitute a full-blown cancer if they have been able to overcome that limitation. In agreement with the previous premise is the fact that reactivation of telomerase is observed in human cancers and it is well established that proliferative cancer cells maintain their telomere length (Kim et al. 1994). Furthermore, there is a general stepwise increase in telomerase expression with tumor progression, whereas benign neoplasms and normal somatic cells apparently lack telomerase activity (Kim et al. 1994). Regarding normal cells, high levels of telomerase activity are only detected in germ cells and in stem cells of self-renewing tissues (Gunes and Rudolph 2013). The inevitable need of cancer cells to maintain telomere length in order to achieve replicative immortality has forced the appearance of mechanisms that could cope with telomere attrition. Again, from a Darwinist standpoint, such mechanisms are only found in a full blown cancer. So far there are two major pathways described that cells use for maintaining telomeres in a cancer setting: they either reactivate telomerase expression, a ribonucleoprotein polymerase, which elongates telomeres by adding hexameric tandem repeats to the chromosomal ends, or take advantage of a non-telomerase dependent alternative mechanism, ALT.

Telomerase Expression in Cancer

It has been known for a long time that high levels of telomerase activity can be detected in cancer cells, this finding contrasts with the scarcity of mutations affecting the telomerase-coding region (Aubert and Lansdorp 2008). A survey of telomerase activity in human cells and tumors coincided with the development of techniques that are more suitable for the evaluation of telomeres and telomerase, Telomeric Repeat Amplification Protocol (TRAP) is one of the most used assays since the initial studies (Kim et al. 1994). Using TRAP, Kim et al. demonstrated that in cultured cells representing 18 different human tissues, 98% of immortal cell lines were positive for telomerase. Similarly, presence of telomerase was also detected in 89% of 101 biopsies, representing 12 different human tumors, whereas normal somatic tissues were negative (Kim et al. 1994). Other studies have addressed how telomerase is expressed in neoplasias and affect the telomere size. A good example of a cancer model in which telomerase activity has been explored in depth is skin cancer. Among skin cancers, sporadic Basal Cell Carcinomas (BCC) were reported to present a variation between 20 to 100% of telomerase presence, both in tumor and/or tumor-free margins, presenting the "apparently" normal tissue lower expression levels of telomerase (Fabricius 2012). Worth mentioning is the fact that telomerase activity in tumor-free margins presents a higher prevalence in sun-exposed skin reflecting the association to the known ethiopatogenic factor, ultra-violet sun radiation (Ueda et al. 1997, Saleh et al. 2007). Less prevalent than BCC, melanoma is regarded as the most aggressive skin cancer and represents one of the best-studied models regarding telomerase. In cutaneous melanomas, the presence of telomerase follows a gradual increase with the aggressiveness of the lesions, i.e., less frequent in normal skin and benign nevi and increased frequency in dysplastic nevi and melanoma (Fabricius 2012).

In agreement with the previous stepwise increment, it has been described an association with high telomerase activity along with the presence of ulceration, vascular invasion, mitotic rate and thickness as well as with high proliferation rate and early metastization, all features linked to a worse prognosis in melanoma (Carvalho et al. 2006, Glaessl et al. 1999, Miracco et al. 2000). The link between higher telomerase expression and worse prognosis features is not exclusive to melanoma; a similar correlation can be detected in tumors such as those of the CNS arising from glial cells. Glioblastomas and oligodendrogliomas present higher levels of telomerase expression, 50 to 89% and 75 to 100%, respectively, when compared with less aggressive forms of tumors such as astrocytomas that range from 0 to 45% (Hiraga et al. 1998, Langford et al. 1995, Sano et al. 1998). In gliomas, telomerase activity correlates with tumor grading, worse prognosis, reduced survival, higher proliferation and recurrence (Falchetti et al. 2000, Langford et al. 1997, Simon et al. 2000). This correlation seems to be independent of the basal tissue proliferative capacity of the homing cancer tissue. Thyroid cancer arises from a low rate self-renewal tissue, that in the adult life rarely proliferates. Accordingly, telomerase activity is almost absent in normal thyroid tissue, being reported in less than 7% of the specimens (Capezzone et al. 2009, Szostak and Blackburn 1982). This expression seems to be restrained to a specific group of normal thyroid cells that represent embryonic remnants of the ultimobranchial body, the solid cell nests (Preto et al. 2004, Reis-Filho et al. 2003). Being the telomerase expression restrained to a small group of cells in thyroid, it was expected that thyroid malignancies would display less frequent telomerase activation than most human carcinomas. Despite this, it has been reported that two thirds of thyroid carcinomas displayed telomerase activation, and that was more frequent in undifferentiated than in differentiated carcinomas (Capezzone et al. 2009, Soares et al. 2011). Even though the differentiated thyroid carcinomas such as papillary and follicular carcinomas can display frequent telomerase expression, 48% and 71%, respectively (Capezzone et al. 2009) and some of the benign tumors, such as the follicular adenomas may display telomerase activity (Capezzone et al. 2011). The majority of the results suggest that telomerase expression is associated with a higher tumor grading and a more aggressive clinical behavior. High levels of telomerase expressions were also reported in cancers of the lung, ovarian, breast, colon, stomach, bladder, kidney and other organs. Overall, there is widespread increase of telomerase expression in human cancers, that in some models seem to be associated with worse prognostic features and decreased survival of the patients (Vinagre et al. 2014).

Telomerase Promoter Mutations in Cancer

Contrasting with the aforementioned knowledge of telomerase wide expression in multiple cancers, there was a lack of operative mechanisms that could explain all these findings. In the beginning of 2013, two seminal studies reported mutations in the TERTp gene in melanoma (Huang et al. 2013, Horn et al. 2013). The initial findings of these mutations prompted the study of such alterations in other human cancers. For the first time, our group and others reported the presence of recurrent somatic mutations in the TERTp in cancers of the CNS, hepatocellular carcinoma, thyroid (follicular cell-derived tumors) and tumors originated from tissues with relatively low rates of self-renewal (Killela et al. 2013, Liu et al. 2013b, Nault et al. 2013, Vinagre et al. 2013). However, TERTp mutations were not restricted to the above tumors and, soon, many other studies started reporting the association of TERTp mutations to other types of tumors, such as atypical fibroxantoma, pleomorphic dermal sarcoma (Griewank et al. 2013), bladder cancer (Allory et al. 2014, Hurst et al. 2014), basal cell carcinoma and squamous cell carcinoma of the skin (Scott et al. 2013), and clear cell carcinoma of the ovary (Wu et al. 2014). For a thorough review about TERT promoter mutations in human cancers, please refer to Vinagre et al. 2014.

These two seminal papers were groundbreaking not only by describing a novel mechanism for telomerase re-activation/re-expression but also because they brought to light the correspondence to an alteration in a noncoding region; taking different approaches, the authors reached the same results (Horn et al. 2013, Huang et al. 2013). In the study of Horn et al., a melanoma-prone family was investigated through linkage and by next generation sequencing in which a germ line disease-segregating mutation was identified in the TERTp (Horn et al. 2013). Huang et al., took a different approach: relying wholly on genome sequencing data publicly available, they detected the presence of TERTp mutations in several melanoma cases (Huang et al. 2013). Overall, the mutations clustered mostly, but not exclusively, in two hotspots that are located at -146 base pairs (bps) and −124 bps distance upstream of the start site (Huang et al. 2013). All the mutations correspond to cytidine to thymidine transitions at a dipyrimidine motif which indicates a putative signature of ultra-violet induced damage. These mutations create a novel binding consensus for ETS/TCFs transcription factors that respond to binding consensus with the sequence CCGGAA (Horn et al. 2013, Huang et al. 2013). Since the initial findings were made in melanoma in which BRAF V600E mutations are prevalent, it was proposed that BRAF mutations could cooperate with TERTp mutations; BRAF mutation leads to downstream activation of ETS/TCFs transcription factors, and ELK1 and ELK4 were pointed out as the main putative factors (Horn et al. 2013, Huang et al. 2013). *In vitro*, by luciferase reporter assay, it was demonstrated that the presence of these mutations lead to a two to four-fold increase of the TERTp activity (Horn et al. 2013, Huang et al. 2013). Recently, Bell et al. presented a study aiming to elucidate which transcription factors could be binding the novel created consensus by the TERTp mutations (Bell et al. 2015). Contrary to what was theoretically proposed at first, ELK1 and ELK4 were not the main functional transcriptions factors and Bell et al. discovered that the GABP-α was the critical ETS transcription factor activating telomerase expression in the presence of TERTp mutations (Bell et al. 2015). The authors reported that although many ETS transcription factors could bind to similar DNA sequence motifs in the TERTp, GABP-α was the only one able to recruit proximal ETS motifs and to assemble a heterotetrameric complex that leads to increased TERTp activation in a mutant-specific manner (Bell et al. 2015). Since the initial discovery of TERTp mutations, researchers have been trying to understand the biological meaning of such alterations and to elucidate how they contribute to human cancer. Chiba et al., in an attempt to understand TERTp mutations' contribution to cancer, used genome editing tools in pluripotent stem cells with basal physiological telomerase levels and inserted TERTp mutations; the engineered mutations resulted in a modest increase in telomerase transcription with no impact on telomerase activity (Chiba et al. 2015). However, upon differentiation into somatic cells, cells with TERTp mutations failed to silence telomerase expression, as expected, and presented increased telomerase activity as well as long telomeres (Chiba et al. 2015). Thus, the authors proposed that TERTp mutations were sufficient to overcome the proliferative barrier imposed by telomere shortening without additional tumor-selected mutations (Chiba et al. 2015).

The acquired knowledge, over the last 20 years, concerning the high expression of telomerase in human cancers, together with recent reports of the high frequency of TERTp mutations in several human cancers led to the hypothesis that such alterations might represent one of the missing links between telomerase gene regulation/reactivation. These novel findings triggered plenty of studies with large cohorts to determine if there were an association of telomerase alterations with clinicopathological features. The initial discovery was made in melanoma (Horn et al. 2013, Huang et al. 2013) and this discovery impelled us to determine whether or not there was an association of the mutations with clinicopathogical features. In melanoma, TERTp mutations were virtually restricted to intermittent sun-exposed areas and associated with nodular and superficial spreading subtypes, BRAF V600E mutation, and poor prognostic indicators (thickness, ulceration, mitotic rate) (Populo et al. 2014). Additionally, we observed that TERTp mutations potentially constitute a biological predictor of persistent disease and mortality in melanoma (Populo et al. 2014) is an independent prognostic factor in cutaneous melanoma (Griewank et al. 2014, Pópulo et al. 2015). Several studies have already reported TERTp mutations in follicular cell derived thyroid tumors (Landa et al. 2013, Liu et al. 2013a, Liu et al. 2013b, Vinagre et al. 2013), whereas no TERTp mutations were detected in normal thyroid tissue or medullary thyroid carcinomas (Vinagre et al. 2013, Liu et al. 2013c, Killela et al. 2013, de Biase et al. 2015). As previously described, for increased telomerase expression in more aggressive thyroid lesions, TERTp mutations also follow a stepwise increase from well to poorly differentiated and undifferentiated thyroid carcinomas (Landa et al. 2013, Liu et al. 2013a, Melo et al. 2014). When comparing the TERTp mutations with clinicopathological features and outcome, the mutations were significantly associated with older age at diagnosis, larger tumor size and higher stage (Melo et al. 2014, Liu et al. 2013a) and were also found to be an independent predictor of distant metastases and disease persistence at the end of follow-up in thyroid carcinomas (Melo et al. 2014, Liu et al. 2014, Xing et al. 2014, Gandolfi et al. 2015). Ultimately, patients with TERTp-mutated tumors were submitted to more radioiodine treatments with higher doses, as well as to other treatment modalities (Melo et al. 2014). Along with these two examples (melanoma and thyroid cancer), there are a fair number of studies done on other forms of cancer such as, CNS tumors and bladder cancer, with large cohorts, where TERTp mutations have been associated with a poorer prognosis of the patients (Lotsch et al. 2013, Heidenreich et al. 2015, Simon et al. 2015, Rachakonda et al. 2013). For the moment, TERTp mutations are recognized as one of the most common genetic events in a myriad of human cancers (Vinagre et al. 2014), that in some cancers significantly influence the patients survival.

Other mechanisms for telomere maintenance

Telomerase re-activation/re-expression is present in nearly 90% of human cancers and in the remaining 10% to 15%, in which no detectable telomerase activity is found, they still maintain telomere length (Cesare and Reddel 2010). The majority of these cases rely on a mechanism termed as ALT. The initial finding of a non-telomerase dependent mechanism was observed in telomerase-null mutant yeast. In this model, it was discovered that such cells maintained telomere length through a homologous recombination mechanism (Lundblad and Blackburn 1993). Heaphy et al. performed a survey on ALT phenotype in primary tumors from 94 different cancer subtypes and observed that in 3.7% of all tumors, ALT was present (Heaphy et al. 2011b). ALT is more prevalent in mesenchymal tumors, although not exclusively; glioblastoma and pancreatic endocrine tumors are examples of tumors in which ALT is frequent (Heaphy et al. 2011b, Heaphy et al. 2011a). Phenotypically, ALT cells present uncommon characteristics, the most striking being the abundance of telomeric DNA sequences separated from chromosomes. Such structures are commonly known as T-circles, C-circles or G-circles (Cesare and Griffith 2004). Other unique features include the co-localization with PML nuclear bodies and the highly heterogeneous sizes of telomeres within the same cell that is unbalanced (Cesare and Reddel 2010). CNS tumors are, so far, the most elucidated model in which ALT, TERTp mutations and hypermethylation of TERTp exist but do not overlap, reflecting that they might confer similar growth advantages. In low-grade CNS tumors, TERTp mutations are rare whereas ATRX mutations are very frequent (Jiao et al. 2012). It is known that ATRX mutations trigger ALT in low grade CNS tumors cells and it has been shown that this alternative mechanism is frequently activated in astrocytomas allowing telomere maintenance without the need for telomerase reactivation (Henson et al. 2005). In primary glioblastomas (high grade gliomas), TERTp mutations are very frequent, this being the main mechanism for telomere maintenance (Nonoguchi et al. 2013). TERTp mutations are rare in pediatric tumors of the CNS (Koelsche et al. 2013). The upregulation of telomerase expression in pediatric brain tumors is associated with hypermethylation of the TERTp, rather than with TERTp mutations (Castelo-Branco et al. 2013). This hypermethylation constitutes an additional and largely unknown mechanism for telomerase expression. These findings are consistent with the fact that the cells, from which pediatric CNS tumors are thought to originate, still have activated telomerase, thus not requiring the activation of telomerase through TERTp mutations. Finally, we can hypothesize that other mechanisms, at present not known, may lead to telomerase expression and subsequently telomere size maintenance. Some studies have already demonstrated an association between a single nucleotide polymorphism in the telomerase gene and an increased risk of glioma development (Shete et al. 2009, Zhao et al. 2012).

Conclusions

Telomere length and telomerase expression was soon perceived to be a fundamental (and limiting) characteristic for tumor cells maintenance but the underlying mechanisms remained unknown for long time. The recent discovery of recurrent TERTp mutations in several types of cancer and its association with prognostic parameters has shed new light and interest in telomerase regulation in cancer and in cancer cell immortalization mechanisms.

Telomerase arises as an important biomarker with strong potential for cancer diagnosis and prognosis, including its detection in tumor circulating DNA in body fluids. From a therapeutic perspective, small molecule inhibitors, gene therapy approaches and immunotherapy (DNA vaccines) are today being investigated actively, hoping to provide new tools to be applied in cancer treatment.

Acknowledgements

The authors are grateful to the Portuguese Foundation for Science and Technology for the support to JV (Ref. SFRH/BD/81940/2011) and to AP (SFRH/BD/110617/2015) and funding from the project "Cancer" from NORTE-01-0145-FEDER-000029, supported by Norte Portugal Regional Programme (NORTE 2020), under the PORTUGAL 2020 Partnership Agreement, through the European Regional Development Fund (ERDF). I3S is an Associate Laboratory of the Portuguese Ministry of Science, Technology and Higher Education that is partially supported by the FCT.

Keywords: Telomerase, cancer, telomere length, telomerase promoter mutations, ALT, telomerase prognostic association

References

Allory, Y., W. Beukers, A. Sagrera, M. Flandez, M. Marques, M. Marquez et al. 2014. Telomerase reverse transcriptase promoter mutations in bladder cancer: high frequency across stages, detection in urine, and lack of association with outcome. Eur Urol 65(2): 360–6. doi: 10.1016/j.eururo.2013.08.052.

Aubert, G. and P. M. Lansdorp. 2008. Telomeres and aging. Physiol Rev 88(2): 557–79. doi: 10.1152/physrev.00026.2007.

Bell, R. J., H. T. Rube, A. Kreig, A. Mancini, S. D. Fouse, R. P. Nagarajan et al. 2015. Cancer. The transcription factor GABP selectively binds and activates the mutant TERT promoter in cancer. Science 348(6238): 1036–9. doi: 10.1126/science.aab0015.

Capezzone, M., S. Marchisotta, S. Cantara and F. Pacini. 2009. Telomeres and thyroid cancer. Curr Genomics 10(8): 526–33. doi: 10.2174/138920209789503897.

Capezzone, M., S. Cantara, S. Marchisotta, G. Busonero, C. Formichi, M. Benigni et al. 2011. Telomere length in neoplastic and nonneoplastic tissues of patients with familial and sporadic papillary thyroid cancer. J Clin Endocrinol Metab 96(11): E1852–6. doi: 10.1210/jc.2011-1003.

Carvalho, L., M. Lipay, F. Belfort, I. Santos, J. Andrade, A. Haddad et al. 2006. Telomerase activity in prognostic histopathologic features of melanoma. J Plast Reconstr Aesthet Surg 59(9): 961–8. doi: 10.1016/j.bjps.2006.01.022.

Castelo-Branco, P., S. Choufani, S. Mack, D. Gallagher, C. Zhang, T. Lipman et al. 2013. Methylation of the TERT promoter and risk stratification of childhood brain tumours: an integrative genomic and molecular study. Lancet Oncol 14(6): 534–42. doi: 10.1016/S1470-2045(13)70110-4.

Cesare, A. J. and J. D. Griffith. 2004. Telomeric DNA in ALT cells is characterized by free telomeric circles and heterogeneous t-loops. Mol Cell Biol 24(22): 9948–57. doi: 10.1128/MCB.24.22.9948-9957.2004.

Cesare, A. J. and R. R. Reddel. 2010. Alternative lengthening of telomeres: models, mechanisms and implications. Nat Rev Genet 11(5): 319–30. doi: 10.1038/nrg2763.

Chiba, K., J. Z. Johnson, J. M. Vogan, T. Wagner, J. M. Boyle and D. Hockemeyer. 2015. Cancer-associated TERT promoter mutations abrogate telomerase silencing. Elife 4. doi: 10.7554/eLife.07918.

de Biase, D., G. Gandolfi, M. Ragazzi, M. Eszlinger, V. Sancisi, M. Gugnoni et al. 2015. TERT Promoter Mutations in Papillary Thyroid Microcarcinomas. Thyroid doi: 10.1089/thy.2015.0101.

Fabricius, E. V., B. Hoffmeister and J. R. Raguse. 2012. Molecularbiology of Basal Cell Carcinoma, Basal Cell Carcinoma, Dr. Vishal Madan (Ed.). Edited by InTech.

Falchetti, M. L., R. Pallini, E. D'Ambrosio, F. Pierconti, M. Martini, G. Cimino-Reale et al. 2000. *In situ* detection of telomerase catalytic subunit mRNA in glioblastoma multiforme. Int J Cancer 88(6): 895–901.

Gandolfi, G., M. Ragazzi, A. Frasoldati, S. Piana, A. Ciarrocchi and V. Sancisi. 2015. TERT promoter mutations are associated with distant metastases in papillary thyroid carcinoma. Eur J Endocrinol 172(4): 403–13. doi: 10.1530/EJE-14-0837.

Glaessl, A., A. K. Bosserhoff, R. Buettner, U. Hohenleutner, M. Landthaler and W. Stolz. 1999. Increase in telomerase activity during progression of melanocytic cells from melanocytic naevi to malignant melanomas. Arch Dermatol Res 291(2-3): 81–7.

Griewank, K. G., B. Schilling, R. Murali, N. Bielefeld, M. Schwamborn, A. Sucker et al. 2013. TERT promoter mutations are frequent in atypical fibroxanthomas and pleomorphic dermal sarcomas. Mod Pathol doi: 10.1038/modpathol.2013.168.

Griewank, K. G., R. Murali, Joan Anton Puig-Butille, B. Schilling, Elisabeth Livingstone, Miriam Potrony et al. 2014. TERT promoter mutation status as an independent prognostic factor in cutaneous melanoma. Journal of the National Cancer Institute 106(9). doi: 10.1093/jnci/dju246.

Gunes, C. and K. L. Rudolph. 2013. The role of telomeres in stem cells and cancer. Cell 152(3): 390–3. doi: 10.1016/j.cell.2013.01.010.

Heaphy, C. M., R. F. de Wilde, Y. Jiao, A. P. Klein, B. H. Edil, C. Shi et al. 2011a. Altered telomeres in tumors with ATRX and DAXX mutations. Science 333(6041): 425. doi: 10.1126/science.1207313.

Heaphy, C. M., A. P. Subhawong, S. M. Hong, M. G. Goggins, E. A. Montgomery, E. Gabrielson et al. 2011b. Prevalence of the alternative lengthening of telomeres telomere maintenance mechanism in human cancer subtypes. Am J Pathol 179(4): 1608–15. doi: 10.1016/j.ajpath.2011.06.018.

Heidenreich, B., P. S. Rachakonda, I. Hosen, F. Volz, K. Hemminki, A. Weyerbrock et al. 2015. TERT promoter mutations and telomere length in adult malignant gliomas and recurrences. Oncotarget 6(12): 10617–33.

Henson, J. D., J. A. Hannay, S. W. McCarthy, J. A. Royds, T. R. Yeager, R. A. Robinson et al. 2005. A robust assay for alternative lengthening of telomeres in tumors shows the significance of alternative lengthening of telomeres in sarcomas and astrocytomas. Clin Cancer Res 11(1): 217–25.

Hiraga, S., T. Ohnishi, S. Izumoto, E. Miyahara, Y. Kanemura, H. Matsumura et al. 1998. Telomerase activity and alterations in telomere length in human brain tumors. Cancer Res 58(10): 2117–25.

Horn, S., A. Figl, P. S. Rachakonda, C. Fischer, A. Sucker, A. Gast et al. 2013. TERT promoter mutations in familial and sporadic melanoma. Science 339(6122): 959–61. doi: 10.1126/science.1230062.

Huang, F. W., E. Hodis, M. J. Xu, G. V. Kryukov, L. Chin and L. A. Garraway. 2013. Highly recurrent TERT promoter mutations in human melanoma. Science 339(6122): 957–9. doi: 10.1126/science.1229259.

Hurst, C. D., F. M. Platt and M. A. Knowles. 2014. Comprehensive mutation analysis of the TERT promoter in bladder cancer and detection of mutations in voided urine. Eur Urol 5(2): 367–9. doi: 10.1016/j.eururo.2013.08.057.

Jiao, Y., P. J. Killela, Z. J. Reitman, A. B. Rasheed, C. M. Heaphy, R. F. de Wilde et al. 2012. Frequent ATRX, CIC, FUBP1 and IDH1 mutations refine the classification of malignant gliomas. Oncotarget 3(7): 709–22.

Killela, P. J., Z. J. Reitman, Y. Jiao, C. Bettegowda, N. Agrawal, L. A. Diaz, Jr. et al. 2013. TERT promoter mutations occur frequently in gliomas and a subset of tumors derived from cells with low rates of self-renewal. Proc Natl Acad Sci U S A 110(15): 6021–6. doi: 10.1073/pnas.1303607110.

Kim, N. W., M. A. Piatyszek, K. R. Prowse, C. B. Harley, M. D. West, P. L. Ho et al. 1994. Specific association of human telomerase activity with immortal cells and cancer. Science 266(5193): 2011–5.

Koelsche, C., F. Sahm, D. Capper, D. Reuss, D. Sturm, D. T. Jones et al. 2013. Distribution of TERT promoter mutations in pediatric and adult tumors of the nervous system. Acta Neuropathol 126(6): 907–15. doi: 10.1007/s00401-013-1195-5.

Landa, I., I. Ganly, T. A. Chan, N. Mitsutake, M. Matsuse, T. Ibrahimpasic et al. 2013. Frequent somatic TERT promoter mutations in thyroid cancer: higher prevalence in advanced forms of the disease. J Clin Endocrinol Metab 98(9): E1562–6. doi: 10.1210/jc.2013-2383.

Langford, L. A., M. A. Piatyszek, R. Xu, S. C. Schold, Jr. and J. W. Shay. 1995. Telomerase activity in human brain tumours. Lancet 346(8985): 1267–8.

Langford, L. A., M. A. Piatyszek, R. Xu, S. C. Schold, Jr., W. E. Wright and J. W. Shay. 1997. Telomerase activity in ordinary meningiomas predicts poor outcome. Hum Pathol 28(4): 416–20.

Liu, T., N. Wang, J. Cao, A. Sofiadis, A. Dinets, J. Zedenius et al. 2013a. The age- and shorter telomere-dependent TERT promoter mutation in follicular thyroid cell-derived carcinomas. Oncogene doi: 10.1038/onc.2013.446.

Liu, X., J. Bishop, Y. Shan, S. Pai, D. Liu, A. K. Murugan et al. 2013b. Highly prevalent TERT promoter mutations in aggressive thyroid cancers. Endocr Relat Cancer 20(4): 603–10. doi: 10.1530/ERC-13-0210.

Liu, X., G. Wu, Y. Shan, C. Hartmann, A. von Deimling and M. Xing. 2013c. Highly prevalent TERT promoter mutations in bladder cancer and glioblastoma. Cell Cycle 12(10): 1637–8. doi: 10.4161/cc.24662.

Liu, X., S. Qu, R. Liu, C. Sheng, X. Shi, G. Zhu et al. 2014. TERT promoter mutations and their association with BRAF V600E mutation and aggressive clinicopathological characteristics of thyroid cancer. J Clin Endocrinol Metab jc20134048. doi: 10.1210/jc.2013-4048.

Lotsch, D., B. Ghanim, M. Laaber, G. Wurm, S. Weis, S. Lenz et al. 2013. Prognostic significance of telomerase-associated parameters in glioblastoma: effect of patient age. Neuro Oncol 15(4): 423–32. doi: 10.1093/neuonc/nos329.

Lundblad, V. and E. H. Blackburn. 1993. An alternative pathway for yeast telomere maintenance rescues est1—senescence. Cell 73(2): 347–60.

Melo, M., A. G. da Rocha, J. Vinagre, R. Batista, J. Peixoto, C. Tavares et al. 2014. TERT promoter mutations are a major indicator of poor outcome in differentiated thyroid carcinomas. J Clin Endocrinol Metab 99(5): E754–65. doi: 10.1210/jc.2013-3734.

Miracco, C., L. Pacenti, R. Santopietro, L. Laurini, M. Biagioli and P. Luzi. 2000. Evaluation of telomerase activity in cutaneous melanocytic proliferations. Hum Pathol 31(9): 1018–21. doi: 10.1053/hupa.2000.9779.

Nandakumar, J. and T. R. Cech. 2013. Finding the end: recruitment of telomerase to telomeres. Nat Rev Mol Cell Biol 14(2): 69–82. doi: 10.1038/nrm3505.

Nault, J. C., M. Mallet, C. Pilati, J. Calderaro, P. Bioulac-Sage, C. Laurent et al. 2013. High frequency of telomerase reverse-transcriptase promoter somatic mutations in hepatocellular carcinoma and preneoplastic lesions. Nat Commun 4: 2218. doi: 10.1038/ncomms3218.

Nonoguchi, N., T. Ohta, J. E. Oh, Y. H. Kim, P. Kleihues and H. Ohgaki. 2013. TERT promoter mutations in primary and secondary glioblastomas. Acta Neuropathol 126(6): 931–7. doi: 10.1007/s00401-013-1163-0.

Populo, H., P. Boaventura, J. Vinagre, R. Batista, A. Mendes, R. Caldas et al. 2014. TERT promoter mutations in skin cancer: the effects of sun exposure and X-irradiation. J Invest Dermatol 134(8): 2251–7. doi: 10.1038/jid.2014.163.

Pópulo, H., J. M. Lopes, M. Sobrinho-Simões and P. Soares. 2015. RE: TERT promoter mutation status as an independent prognostic factor in cutaneous melanoma. Journal of the National Cancer Institute 107(4). doi: 10.1093/jnci/djv049.

Preto, A., J. Cameselle-Teijeiro, J. Moldes-Boullosa, P. Soares, J. F. Cameselle-Teijeiro, P. Silva et al. 2004. Telomerase expression and proliferative activity suggest a stem cell role for thyroid solid cell nests. Mod Pathol 17(7): 819–26. doi: 10.1038/modpathol.3800124.

Rachakonda, P. S., I. Hosen, P. J. de Verdier, M. Fallah, B. Heidenreich, C. Ryk et al. 2013. TERT promoter mutations in bladder cancer affect patient survival and disease recurrence through modification by a common polymorphism. Proc Natl Acad Sci U S A 110(43): 17426–31. doi: 10.1073/pnas.1310522110.

Reis-Filho, J. S., A. Preto, P. Soares, S. Ricardo, J. Cameselle-Teijeiro and M. Sobrinho-Simoes. 2003. p63 expression in solid cell nests of the thyroid: further evidence for a stem cell origin. Mod Pathol 16(1): 43–8. doi: 10.1097/01.MP.0000047306.72278.39.

Saleh, S., A. King-Yin Lam, P. Gertraud Buettner, M. Glasby, B. Raasch and Y. H. Ho. 2007. Telomerase activity of basal cell carcinoma in patients living in North Queensland, Australia. Hum Pathol 38(7): 1023–9. doi: 10.1016/j.humpath.2006.12.006.

Sano, T., A. Asai, K. Mishima, T. Fujimaki and T. Kirino. 1998. Telomerase activity in 144 brain tumours. Br J Cancer 77(10): 1633–7.

Scott, G. A., T. S. Laughlin and P. G. Rothberg. 2013. Mutations of the TERT promoter are common in basal cell carcinoma and squamous cell carcinoma. Mod Pathol doi: 10.1038/modpathol.2013.167.

Shete, S., F. J. Hosking, L. B. Robertson, S. E. Dobbins, M. Sanson, B. Malmer et al. 2009. Genome-wide association study identifies five susceptibility loci for glioma. Nat Genet 41(8): 899–904. doi: 10.1038/ng.407.

Simon, M., T. W. Park, S. Leuenroth, V. H. Hans, T. Loning and J. Schramm. 2000. Telomerase activity and expression of the telomerase catalytic subunit, hTERT, in meningioma progression. J Neurosurg 92(5): 832–40. doi: 10.3171/jns.2000.92.5.0832.

Simon, M., I. Hosen, K. Gousias, S. Rachakonda, B. Heidenreich, M. Gessi et al. 2015. TERT promoter mutations: a novel independent prognostic factor in primary glioblastomas. Neuro Oncol 17(1): 45–52. doi: 10.1093/neuonc/nou158.

Soares, P., J. Lima, A. Preto, P. Castro, J. Vinagre, R. Celestino et al. 2011. Genetic alterations in poorly differentiated and undifferentiated thyroid carcinomas. Curr Genomics 12(8): 609–17. doi: 10.2174/138920211798120853.

Szostak, J. W. and E. H. Blackburn. 1982. Cloning yeast telomeres on linear plasmid vectors. Cell 29(1): 245–55.

Ueda, M., A. Ouhtit, T. Bito, K. Nakazawa, J. Lubbe, M. Ichihashi et al. 1997. Evidence for UV-associated activation of telomerase in human skin. Cancer Res 57(3): 370–4.

Vinagre, J., A. Almeida, H. Populo, R. Batista, J. Lyra, V. Pinto et al. 2013. Frequency of TERT promoter mutations in human cancers. Nat Commun 4: 2185. doi: 10.1038/ncomms3185.

Vinagre, J., V. Pinto, R. Celestino, M. Reis, H. Populo, P. Boaventura et al. 2014. Telomerase promoter mutations in cancer: an emerging molecular biomarker? Virchows Arch doi: 10.1007/s00428-014-1608-4.

Wu, R. C., A. Ayhan, D. Maeda, K. R. Kim, B. A. Clarke, P. Shaw et al. 2014. Frequent somatic mutations of the telomerase reverse transcriptase promoter in ovarian clear cell carcinoma but not in other major types of gynaecological malignancy. J Pathol 232(4): 473–81. doi: 10.1002/path.4315.

Xing, M., R. Liu, X. Liu, A. K. Murugan, G. Zhu, M. A. Zeiger et al. 2014. BRAF V600E and TERT promoter mutations cooperatively identify the most aggressive papillary thyroid cancer with highest recurrence. J Clin Oncol 32(25): 2718–26. doi: 10.1200/JCO.2014.55.5094.

Zhao, Y., G. Chen, Y. Zhao, X. Song, H. Chen, Y. Mao et al. 2012. Fine-mapping of a region of chromosome 5p15.33 (TERT-CLPTM1L) suggests a novel locus in TERT and a CLPTM1L haplotype are associated with glioma susceptibility in a Chinese population. Int J Cancer 131(7): 1569–76. doi: 10.1002/ijc.27417.

16

Forensic Application of Telomere Shortening in Age-at-Death Estimation

Joe Adserias-Garriga

INTRODUCTION TO FORENSIC IDENTIFICATION

Identification of human remains is a very important issue for both legal and humanitarian reasons. The establishment of an identity is vital to the grieving process and for providing closure. Also, a death certificate after identification is necessary to settle personal and business affairs such as insurance proceeds, remarriage or child custody. Moreover, criminal investigations in a homicide case may not be able to proceed without identification of the victim (Senn and Weems 2013). The requirements to establish the deceased's identity generally falls into criminal investigation, accidents and mass disaster incidents or war crimes and genocide.

In criminal investigations, context identification is a key factor since it is so difficult to investigate a case where the corpse has not been identified. Also, in accidents and mass disaster incidents, the ultimate aim of Disaster Victim Identification (DVI) operations is the establishment of the identity of each victim, as well as in war crimes and genocide, where the victims must be identified when possible (Thompson and Black 2006).

A positive or confirmed identity occurs when two data sets of information (ante mortem and postmortem) are compared and enough specific data markers match to conclude that the records were created from the same individual (Thompson and Black 2006). However, most of the time, before achieving a positive identity comparing the ante mortem and postmortem data, forensic experts must reconstruct a biological profile, determining the most probable sex, ancestry, age, stature and possible pathological conditions of the individual. Thus, age estimation is one of the most relevant pieces of data in the biological profile.

University of Girona, Postgraduate in Forensic Anthropology, C/Emili Grahit 77 Girona, Spain.
  Email: mjadserias@ub.edu

Age Estimation and Forensic Identification

Age estimation involves first assessing physiological age and then attempting to correlate it with chronological age. The estimation of age can be affected by sex and ancestry based variations. Thus, standards for age estimation should be used while considering the population of origin when it is possible.

Age estimation from the observation of bone and teeth traits becomes less precise as a person gets older (Adams 2007). One of the challenges in forensic science is to estimate the age accurately in aged individuals. Growth and development normally occur at a predictable rate, which allows age-at-death estimates to be quite precise in subadults (infants, children and adolescents) (Adams 2007). But when growth has ceased, age estimation in adults is basically based on the degenerative changes of bone and teeth, which can be affected by pathology. Therefore, the estimation of age in adult individuals is generally less precise than in subadults (Adams 2007). Because of that reason, most researches suggest that when determining age in adult individuals, assessing multiple age indicators provides more accurate results than using a single indicator (SGWATH 2010).

Methods for age assessment

Age-at-death can be estimated using different methods. Selection of the methods to be employed in age estimation depends on the materials available for examination, their condition and the age category of the individual (SGWATH 2013). Traditionally, forensic anthropologists and forensic odontologists estimated the individual's age-at-death by the macroscopic changes in bones and teeth due to growth and development in subadults and degenerative changes in adults.

The most reliable age indicator in fetuses is dentition development, rather than skeletal development (Schour and Masler 1941, Ubelaker 1987, AlQahtani et al. 2010). Presence of ossification nuclei (Kosa 1989, Fazekas and Kosa 1978) and long bone development (Scheuer and Black 2000) can also be used for assessing age.

In the case of newborns, different morphological methods can be used for age assessment, such as dental development (Schour and Masler 1941, Ubelaker 1987, AlQahtani et al. 2010), diaphyseal length of long bones (Scheuer and Black 2000), and presence of ossification nuclei (Kosa 1989).

Dental mineralization and eruption (Demirjian et al. 1973, Gustafson and Koch 1974, Hagg and Matsson 1985, Liliequist and Lundberg 1971, Melsen et al. 1986, AlQahtani et al. 2010), presence of ossification centers and epiphyseal fusion (Albert and Maples 1995, Buken et al. 2007, Haavikko and Kilpinen 1973, Kullman 1995, Lynnerup et al. 2008, Melsen et al. 1986, Ubelaker 1987), bone dimensions (Pfau and Sciulli 1994) and development of hand and wrist bones (Greulich and Pyle 1959) can be used for estimating age in infants and children.

With regards to adolescents, dental development (Demirjian et al. 1973, Mincer et al. 1993), long bone development (Scheuer and Black 2000, Albert and Maples 1995) and development of hand and wrist bones (Greulich and Pyle 1959, Tanner and Whitehouse 1975) can be used for age estimation.

From adolescence to early adulthood (20 to 25 year old individuals), third molar development (Mincer et al. 1993), development of hand and wrist bones (Greulich and Pyle 1959, Tanner and Whitehouse 1975), spheno-occipital or basilar synchondrosis fusion (Scheuer and Black 2000, Madeline and Elster 1995), and clavicle sternal end fusion (Schulz et al. 2005, Schmidt et al. 2007) are good age indicators. Likewise, it is always recommended to use as many indicators as possible in order to achieve the most accurate results.

Age estimation in adult individuals can be based on different indicators such as coxae's pubic symphysis changes (Brooks 1955, Todd 1921) and auricular region changes (Lovejoy et al. 1985), acetabulum surface changes (Rissech et al. 2006, Rissech and Malgosa 2007, Rissech et al. 2007, Rouge-Maillart et al. 2007), sternal end of the fourth ribs (Iscan et al. 1984, 1985, 1987), dental changes (Cameriere et al. 2004a,b, Lamendin et al. 1992) and histological changes in bone by microscopic osteon counting (Ahlqvist and Damsten 1969, Cool et al. 1995, Kerley 1965, Kerley and Ubelaker 1978).

All those methods for estimating age in adults have some limitations to take into account when applying them in casework. The Suchey-Brooks method (Brooks 1955) as well as Lamendin's method (Lamendin et al. 1992) show higher reliability from 20 to 40 year old individuals; while Iscan is more reliable after 60 years. Lovejoy's method of observing changes in the auricular surface or cranial suture closure methods (Meindl and Lovejoy 1985), show inter-individual and intra-individual variation. Also, the microscopic osteon counting method (Ahlqvist and Damsten 1969, Cool

Table 1. Morphological methods for age estimation for individuals of different age ranges.

Age range	Method
FETUSES	Dental development Presence of ossification nuclei Long bone development
NEWBORNS	Dental development Presence of ossification nuclei Diaphyseal fusion
INFANT AND CHILDREN	Dental development Presence of ossification nuclei Epiphyseal fusion Bone dimensions Hand and wrist bones development
ADOLESCENTS	Dental development Epiphyseal fusion Hand and wrist bones development
TRANSITION	Third molar development Epiphyseal fusion Hand and wrist bones developments Spheno-occipital-basilar synchondrosis fusion Clavicle sternal end fusion
ADULTS	Pubic symphysis changes Acetabulum surface changes Sternal end of the fourth ribs Bone histological changes

et al. 1995, Kerley 1965, Kerley and Ubelaker 1978) is more time consuming compared with the other observational methods mentioned.

As mentioned above, the precision of the age estimation decreases as the individual grows older. Due to this problem, different methods have been developed to estimate age-at-death in adults. Those new methods to estimate age-at-death are based on the natural process of aging, which leads to alterations of tissues and organs on different biochemical levels (Balin and Allen 1989). Biological aging can be thought of as a progressive decline in the function of the cells of organisms, so that aging is associated with the accumulation of cellular damage (Kaeberlein 2007).

One of these methodologies is telomere shortening. Telomeres form the ends of human chromosomes. Telomeres shorten with each round of cell division and this mechanism limits proliferation of human cells to a finite number of cell divisions by inducing replicative senescence, differentiation or apoptosis (Zakian 1995, Jiang et al. 2007). Based on that concept, telomere shortening has been studied as a possible molecular age indicator. Towards this purpose, several methodologies have been used to estimate telomere length.

Methodologies to estimate telomere length

Different methodologies to measure telomere length have been developed over the last 25 years. All of them rely on the primer's binding to telomere specific DNA repeats. The different methodologies available nowadays to estimate telomere length are the following:

Telomere Restriction Fragment (TRF) analysis: TRF analysis was the first method devised to estimate the average telomere length of a cell population (Harley et al. 1990, Allsopp et al. 1992). This method has been used as the point of reference for all other methods developed since (Aubert and Lansdorp 2008). TRF analysis consists of digesting genomic DNA with common restriction enzymes (like RsaI and HinfI) that cannot use telomere repeats as a substrate. The resulting product is an intact TRF including subtelomeric tracts (Bekaert et al. 2005a). DNA fragments produced in the digestion are resolved by gel electrophoresis, Southern blotted and probed with a labelled telomere oligonucleotide. The telomere length is estimated by comparison with known size markers. This method requires a large amount of DNA (2 to 10 µg) and has a low resolution. On the other hand, it is simple to implement, and does not require specialized lab equipment.

Hybridization Protection Assay (HPA): DNA solution or cell lysate is hybridized with an acridinium ester labelled telomere-specific probe. Quantification is performed by chemiluminescence. The value is normalised for the total DNA amount by also measuring an AE-labelled Alu probe and calculating the ratio of both (Nakamura et al. 1999).

Direct lysate method: This technique uses a biotin-streptavidin-based acetylcholinesterase hybridization assay for quantification of the telomeric sequence and a DNA SYBR Green assay to correct for the total amount of DNA. This method permits rapid measuring of telomere length on microtitre plates without any DNA purification (Freulet-Marriere et al. 2004).

Telomeric-Oligonucleotide Ligation Assay (T-OLA): This technique consists of [γ-32P]-labelled telomeric oligonucleotides, hybridised to the undenaturated DNA and ligated. The concatenated products are released by heat denaturation, resolved by denaturating PAGE and visualised by autoradiography (Cimino-Reale et al. 2001). This technique is capable of measuring the length of the G-rich strand 3' overhang (Cimino-Reale et al. 2001), but also gives information on the general telomere size as telomere shortening is found to be directly proportional to the length of the overhang (Huffman et al. 2000).

Quantitative-Fluorescence *In Situ* Hybridization (Q-FISH): uses directly fluorescently labeled (CCCTAA) 3 Peptide Nucleic Acid (PNA) probes as a high-affinity alternative to DNA oligonucleotide probes that specifically hybridize to denature telomere DNA repeat arrays. Then, the fluorescent signal can be detected and measured relative to standards of known telomere length in metaphase spreads with specific software for Q-FISH image analysis (Poon and Lansdorp 2001). Q-FISH is the method of choice for high-resolution telomere length measurements at specific chromosome ends. Q-FISH has also been used to detect ends without detectable repeats (0.5 kb) as well as chromosome fusion events.

Flow FISH: The combination of Q-FISH with flow cytometry (Flow-FISH) provides more sensitivity, accuracy and speed but has a few major disadvantages, namely complexity and cost (Baerlocher and Lansdorp 2004). Flow FISH has become the method of choice for the measurement of telomere length in peripheral blood cells from human samples and has allowed the determination of the normal range of telomere lengths for specific cell subsets (Baerlocher and Lansdorp 2004, Aubert and Lansdorp 2008) that are being used both for research and for clinical investigations (Alter et al. 2007, Brummendorf et al. 2001, Yamaguchi et al. 2005).

Q-PCR: as an alternative to the hybridization based techniques, Q-PCR based techniques have been developed to estimate telomere length. The difficulty with using PCR is the lack of suitable primer binding sites. A possible solution is the use of primers, which are not perfectly complementary to the telomeric sequence, still allowing the DNA-polymerase to amplify telomeric fragments (Bekaert et al. 2005a).

Table 2. Different methods available to estimate telomere length (modified from Bekaert et al. 2005).

Method	Methodology base	Study material	Amount required
Telomere restriction fragment (TRF) analysis	Hybridization	DNA	2–10 µg
Hybridization protection assay (HPA)	Hybridization	DNA or cell lysate	10 ng/1000 cells
Direct lysate method	Hybridization	cell lysate	> 2 x 105 cells
Telomeric-oligonucleotide ligation assay (T-OLA)	Hybridization	DNA	5 µg
Quantitative fluorescence *in situ* hybridization (Q-FISH)	Hybridization	Metaphase preparations	> 30 nuclei
Flow FISH	Hybridization	cells	> 1000 cells
Quantitative polymerase Chain Reaction (Q-PCR)	PCR	DNA	35 ng
Single telomere length analysis (STELA)	PCR	DNA	10–100 ng

Single Telomere Length Analysis (STELA): While PCR products are directly quantified in Q-PCR, STELA requires gel electrophoresis to separate the amplified products that are characterized by Southern blot hybridization (Aubert and Lansdorp 2008). Even though STELA is very labor intensive, it offers the most precise measurement of telomere length currently available. In addition, it can be used with very few and even single cells and has the advantage that it can detect short outlier telomeres in a sample (Baird et al. 2003).

Telomere Shortening and Age Estimation in Forensic Science

Terminal Restriction Fragment length of telomeres and age are inversely correlated, which can be used to determine the age of an individual (Ren et al. 2009). Telomere shortening during the aging process occurs and has been tested in different populations and in many cells and tissues, such as fibroblasts (Harley et al. 1990), peripheral blood cells (Iwama et al. 1998), colonic mucosa (Hastie et al. 1990), kidneys (Melk et al. 2000), dental pulp (Takasaki et al. 2003) and buccal cells (Hewakapuge et al. 2008).

Further studies developed in accessible tissues like blood, skin or teeth pointed out this inverse relationship between telomere length and age. Butler et al. (Butler et al. 1998) demonstrated this correlation in blood and skin, finding the shortest telomere length in a 72 year-old man. Lately, through Southern blot assay in blood samples, Tsuji et al. (Tsuji et al. 2002) confirmed this relationship, finding a standard error between chronological and estimated age of 7.037 years. Using the same methodology and pulp samples, Takasaki et al. (Takasaki et al. 2003) found a high correlation between telomere shortening and age and approximately the same standard error between chronological and estimated age (7.52 years). Furthermore, they applied this methodology to forensic cases; although in the majority the difference between chronological and estimated age was inside the range of the technique, in few cases, the error reached to 10 years or more. In the same way, the study of Ren et al. (Ren et al. 2009) demonstrated the same correlation in blood samples from a Tibetan population using a Southern blot assay. The error was high, 9.832 years, however, they compared their results between populations and they did not find any differences.

Although these previous studies demonstrated an inverse correlation between telomere length and age, it seems to be affected by different factors, like cause of death and postmortem interval. In fact, studies of Tsuji et al. (Tsuji et al. 2002) found that telomere length was shorter from bloodstains stored at room temperature for five months than of blood samples taken recently from the same person. In the same line of research, but using pulp samples, Takasaki et al. (Takasaki et al. 2003), found a shortened length of telomeres after one year of storage at room temperature and their studies in forensic cases pointed out that the cause of death can influence telomere shortening. Apart from these studies developed with forensic purposes, other works indicated that the telomere length can be affected by a combination of genetic, epigenetic and environmental factors including sexual dimorphism and diseases (Bekaert et al. 2005a, Bekaert et al. 2005b, Ren et al. 2009, Shammas 2011).

In spite of these previous studies, the major concern in the use of telomere length to estimate age in forensics is the great variability that is found in different individuals (Hewakapuge et al. 2008). Due to the presence of large inter-individual variations in telomere length, there are some authors that reject its use for forensic purposes (Hewakapuge et al. 2008). In contrast, others find in telomere shortening a tool of great value to forensics, due to the possibility of giving a rough estimation of age of subjects in forensic samples that carry no morphological information (Tsuji et al. 2002, Ren et al. 2009).

Further studies are needed in order to set the standards for the proper use of telomere length for age assessment. This would permit research to go forward in the development of new techniques robust enough to determine age across a broad spectrum of age ranges and the effect of other variables such as gender, ancestry, disease, and others.

Comparison of age estimation using telomere shortening and other methods

In estimating age-at-death, it is important to recognize the limitations of the methods used. Some specific methods are more prone to produce biased age estimates than others, for statistical reasons and the nature of the original reference sample. Proper documentation of the variation/error rate must be provided when reporting age estimates (SWGATH 2013).

The different studies on telomere shortening used as an age indicator showed an error rate of around 10 years (Ren et al. 2009, Tsuji et al. 2002, Takasaki et al. 2003). This rate is lower than the error rate of other methods based on biochemical approaches like Collagen crosslinks, showing a standard error of 14.9 years, just to achieve the 65% of confidence; AGEs method showing an error rate of 13.7 years; and sjTRECs rearrangements, showing an error rate of 10.47 years (Zapico and Ubelaker 2013) (Table 3).

Table 3. The standard error presented in biochemical methods to estimate age.

Type of Method	Method	Standard error
CHEMICAL METHODS	Aspartic acid racemization Lead accumulation Collagen crosslinks Chemical composition of teeth Advanced glycation end products (AGEs)	± 3 years ± 4.8 years ± 14.9 years (65% confidence) ± 5 years ± 13.7 years
MOLECULAR METHODS	Mitochondrial mutation sjSTREC rearrangements Telomere shortening	± 5 years ± 10.47 years ± 9.8 years

Although there are other biochemical methods that present higher accuracy than the telomere shortening method, such as chemical composition of teeth analysis and lead accumulation, showing similar error rates of around five years; the most accurate method is the racemization of aspartic acid with an error rate of three years (Zapico and Ubelaker 2013) (Table 3).

Error rates of around 10 years are also shown in morphological methods such as the Lamendin technique. Likewise, the Meindl and Lovejoy methods for aging adults by the lateral-anterior system provide nine possible scores with a standard deviation from 6.2 to 10.5 years.

Conclusions

When estimating the age-at-death of an individual, forensic scientists first assess chronological age with different methods. Since the estimate of age can be affected by sex and ancestry based variations, it is imperative to choose the standards for age estimation appropriate for the population of origin of the remains and a standard deviation of the method used must always be provided.

Morphological age estimation methods in subadults are based on skeletal and dental development, as growth and development normally occurs at a predictable rate. Thus, age estimation methods in subadults, especially dental development methods, are highly reliable. However, when growth has ceased, developmental methods cannot be applied anymore and the age estimation in adults is primarily based on skeletal and dental degenerative changes. Since those degenerative changes can be affected by pathology, morphological methods to estimate age-at-death in adults used to be less precise than those in subadults. Because of that, the biggest concern of age estimation is aging individuals over 60 years of age.

To address the limitation of aging adults, different biochemical methods have been developed, such as lead accumulation, collagen crosslinks, chemical composition of teeth, AGEs and aspartic acid racemization, with the last one as the most accurate.

Apart from these methods, telomere shortening has been proposed as an age indicator. Telomere shortening during the aging process occurs and has been tested in different populations and in many cells and tissues. This wide range of application in different tissues and different populations is indeed a great advantage for its application for forensic purposes.

In spite of the promising results, research efforts must be directed to reduce the standard error of age estimation using the telomere shortening method.

Keywords: Forensic identification, age-at-death estimation, telomere shortening, Southern blot, postmortem interval, cause of death, environmental factors, forensic cases

References

Adams, B. 2007. Forensic Anthropology. Chelsea House Publications. New York.

Ahlqvist, J. and O. Damsten. 1969. A modification of Kerley's method for the microscopic determination of age in human bone. J Forensic Sci 14(2): 205–12.

Albert, A. M. and W. R. Maples. 1995. Stages of epiphyseal union for thoracic and lumbar vertebral centra as a method of age determination for teenage and young adult skeletons. J Forensic Sci 40(4): 623–33.

Allsopp, R. C., H. Vaziri, C. Patterson, S. Goldstein, E. V. Younglai, A. B. Futcher et al. 1992. Telomere length predicts replicative capacity of human fibroblasts. Proc Natl Acad Sci U S A 89(21): 10114–8.

AlQahtani, S. J., M. P. Hector and H. M. Liversidge. 2010. Brief communication: The London atlas of human tooth development and eruption. Am J Phys Anthropol 142(3): 481–90. doi: 10.1002/ajpa.21258.

Alter, B. P., G. M. Baerlocher, S. A. Savage, S. J. Chanock, B. B. Weksler, J. P. Willner et al. 2007. Very short telomere length by flow fluorescence in situ hybridization identifies patients with dyskeratosis congenita. Blood 110(5): 1439–47. doi: 10.1182/blood-2007-02-075598.

Aubert, G. and P. M. Lansdorp. 2008. Telomeres and aging. Physiol Rev 88(2): 557–79. doi: 10.1152/physrev.00026.2007.

Baerlocher, G. M. and P. M. Lansdorp. 2004. Telomere length measurements using fluorescence *in situ* hybridization and flow cytometry. Methods Cell Biol 75: 719–50.

Baird, D. M., J. Rowson, D. Wynford-Thomas and D. Kipling. 2003. Extensive allelic variation and ultrashort telomeres in senescent human cells. Nat Genet 33(2): 203–7. doi: 10.1038/ng1084.

Balin, A. K. and R. G. Allen. 1989. Molecular bases of biologic aging. Clin Geriatr Med 5(1): 1–21.

Bekaert, S., T. De Meyer and P. Van Oostveldt. 2005a. Telomere attrition as ageing biomarker. Anticancer Res 25(4): 3011–21.

Bekaert, S., I. Van Pottelbergh, T. De Meyer, H. Zmierczak, J. M. Kaufman, P. Van Oostveldt et al. 2005b. Telomere length versus hormonal and bone mineral status in healthy elderly men. Mech Ageing Dev 126(10): 1115–22. doi: 10.1016/j.mad.2005.04.007.

Brummendorf, T. H., J. P. Maciejewski, J. Mak, N. S. Young and P. M. Lansdorp. 2001. Telomere length in leukocyte subpopulations of patients with aplastic anemia. Blood 97(4): 895–900.

Brooks, S. T. 1955. Skeletal age at death: the reliability of cranial and pubic age indicators. American Journal of Physical Anthropology 13: 567–597.

Buken, B., A. A. Safak, B. Yazici, E. Buken and A. S. Mayda. 2007. Is the assessment of bone age by the Greulich-Pyle method reliable at forensic age estimation for Turkish children? Forensic Sci Int 173(2-3): 146–53. doi: 10.1016/j.forsciint.2007.02.023.

Butler, M. G., J. Tilburt, A. DeVries, B. Muralidhar, G. Aue, L. Hedges et al. 1998. Comparison of chromosome telomere integrity in multiple tissues from subjects at different ages. Cancer Genet Cytogenet 105(2): 138–44.

Cameriere, R., L. Ferrante and M. Cingolani. 2004a. Precision and reliability of pulp/tooth area ratio (RA) of second molar as indicator of adult age. J Forensic Sci 49(6): 1319–23.

Cameriere, R., L. Ferrante and M. Cingolani. 2004b. Variations in pulp/tooth area ratio as an indicator of age: a preliminary study. J Forensic Sci 49(2): 317–9.

Cimino-Reale, G., E. Pascale, E. Battiloro, G. Starace, R. Verna and E. D'Ambrosio. 2001. The length of telomeric G-rich strand 3'-overhang measured by oligonucleotide ligation assay. Nucleic Acids Res 29(7): E35.

Cool, S. M., J. K. Hendrikz and W. B. Wood. 1995. Microscopic age changes in the human occipital bone. J Forensic Sci 40(5): 789–96.

Demirjian, A., H. Goldstein and J. M. Tanner. 1973. A new system of dental age assessment. Hum Biol 45(2): 211–27.

Freulet-Marriere, M. A., G. Potocki-Veronese, J. R. Deverre and L. Sabatier. 2004. Rapid method for mean telomere length measurement directly from cell lysates. Biochem Biophys Res Commun 314(4): 950–6.

Fazekas, I. G. Y. and F. Kosa. 1978. Forensic Fetal Osteology. Budapest: Akademiai Kaido.

Gustafson, G. and G. Koch. 1974. Age estimation up to 16 years of age based on dental development. Odontol Revy 25(3): 297–306.

Greulich, W. W., Pyle SI. 2nd ed. Stanford, California, USA: Stanford University Press; 1959. Radiograph Atlas of Skeletal Development of the Hand and Wrist.

Haavikko, K. and E. Kilpinen. 1973. Skeletal development of Finnish children in the light of hand-wrist roentgenograms. Proc Finn Dent Soc 69(5): 182–90.

Hagg, U. and L. Matsson. 1985. Dental maturity as an indicator of chronological age: the accuracy and precision of three methods. Eur J Orthod 7(1): 25–34.

Harley, C. B., A. B. Futcher and C. W. Greider. 1990. Telomeres shorten during ageing of human fibroblasts. Nature 345(6274): 458–60. doi: 10.1038/345458a0.

Hastie, N. D., M. Dempster, M. G. Dunlop, A. M. Thompson, D. K. Green and R. C. Allshire. 1990. Telomere reduction in human colorectal carcinoma and with ageing. Nature 346(6287): 866–8. doi: 10.1038/346866a0.

Hewakapuge, S., R. A. van Oorschot, P. Lewandowski and S. Baindur-Hudson. 2008. Investigation of telomere lengths measurement by quantitative real-time PCR to predict age. Leg Med (Tokyo) 10(5): 236–42. doi: 10.1016/j.legalmed.2008.01.007.

Huffman, K. E., S. D. Levene, V. M. Tesmer, J. W. Shay and W. E. Wright. 2000. Telomere shortening is proportional to the size of the G-rich telomeric 3'-overhang. J Biol Chem 275(26): 19719–22. doi: 10.1074/jbc.M002843200.

Iscan, M. Y., S. R. Loth and R. K. Wright. 1984. Age estimation from the rib by phase analysis: white males. J Forensic Sci 29(4): 1094–104.

Iscan, M. Y., S. R. Loth and R. K. Wright. 1985. Age estimation from the rib by phase analysis: white females. J Forensic Sci 30(3): 853–63.

Iscan, M. Y., S. R. Loth and R. K. Wright. 1987. Racial variation in the sternal extremity of the rib and its effect on age determination. J Forensic Sci 32(2): 452–66.

Iwama, H., K. Ohyashiki, J. H. Ohyashiki, S. Hayashi, N. Yahata, K. Ando et al. 1998. Telomeric length and telomerase activity vary with age in peripheral blood cells obtained from normal individuals. Hum Genet 102(4): 397–402.

Jiang, H., Z. Ju and K. L. Rudolph. 2007. Telomere shortening and ageing. Z Gerontol Geriatr 40(5): 314–24. doi: 10.1007/s00391-007-0480-0.

Kaeberlein, M. 2007. Molecular basis of ageing. EMBO Rep 8(10): 907–11. doi: 10.1038/sj.embor.7401066.

Kerley, E. R. 1965. The microscopic determination of age in human bone. Am J Phys Anthropol 23(2): 149–63.

Kerley, E. R. and D. H. Ubelaker. 1978. Revisions in the microscopic method of estimating age at death in human cortical bone. Am J Phys Anthropol 49(4): 545–6. doi: 10.1002/ajpa.1330490414.

Kosa, F. 1989. Age estimation from the fetal skeleton. pp. 21–54. In: I N. İşçan (ed.) Age Markers in the Human Skeleton. Charles C. Thomas.

Kullman, L. 1995. Accuracy of two dental and one skeletal age estimation method in Swedish adolescents. Forensic Sci Int 75(2-3): 225–36.

Lamendin, H., E. Baccino, J. F. Humbert, J. C. Tavernier, R. M. Nossintchouk and A. Zerilli. 1992. A simple technique for age estimation in adult corpses: the two criteria dental method. J Forensic Sci 37(5): 1373–9.

Liliequist, B. and M. Lundberg. 1971. Skeletal and tooth development. A methodologic investigation. Acta Radiol Diagn (Stockh) 11(2): 97–112.

Lovejoy, C. O., R. S. Meindl, T. R. Pryzbeck and R. P. Mensforth. 1985. Chronological metamorphosis of the auricular surface of the ilium: a new method for the determination of adult skeletal age at death. Am J Phys Anthropol 68(1): 15–28. doi: 10.1002/ajpa.1330680103.

Lynnerup, N., E. Belard, K. Buch-Olsen, B. Sejrsen and K. Damgaard-Pedersen. 2008. Intra- and interobserver error of the Greulich-Pyle method as used on a Danish forensic sample. Forensic Sci Int 179(2-3): 242 e1–6. doi: 10.1016/j.forsciint.2008.05.005.

Madeline, L. A. and A. D. Elster. 1995. Suture closure in the human chondrocranium: CT assessment. Radiology 196(3): 747–56. doi: 10.1148/radiology.196.3.7644639.

Meindl, R. S. and C. O. Lovejoy. 1985. Ectocranial suture closure: a revised method for the determination of skeletal age at death based on the lateral-anterior sutures. Am J Phys Anthropol 68(1): 57–66. doi: 10.1002/ajpa.1330680106.

Melk, A., V. Ramassar, L. M. Helms, R. Moore, D. Rayner, K. Solez et al. 2000. Telomere shortening in kidneys with age. J Am Soc Nephrol 11(3): 444–53.

Melsen, B., A. Wenzel, T. Miletic, J. Andreasen, P. L. Vagn-Hansen and S. Terp. 1986. Dental and skeletal maturity in adoptive children: assessments at arrival and after one year in the admitting country. Ann Hum Biol 13(2): 153–9.

Mincer, H. H., E. F. Harris and H. E. Berryman. 1993. The A.B.F.O. study of third molar development and its use as an estimator of chronological age. J Forensic Sci 38(2): 379–90.

Nakamura, Y., M. Hirose, H. Matsuo, N. Tsuyama, K. Kamisango and T. Ide. 1999. Simple, rapid, quantitative, and sensitive detection of telomere repeats in cell lysate by a hybridization protection assay. Clin Chem 45(10): 1718–24.

Pfau, R. O. and P. W. Sciulli. 1994. A method for establishing the age of subadults. J Forensic Sci 39(1): 165–76.

Poon, S. S. and P. M. Lansdorp. 2001. Measurements of telomere length on individual chromosomes by image cytometry. Methods Cell Biol 64: 69–96.

Ren, F., C. Li, H. Xi, Y. Wen and K. Huang. 2009. Estimation of human age according to telomere shortening in peripheral blood leukocytes of Tibetan. Am J Forensic Med Pathol 30(3): 252–5. doi: 10.1097/PAF.0b013e318187df8e.

Rissech, C., G. F. Estabrook, E. Cunha and A. Malgosa. 2006. Using the acetabulum to estimate age at death of adult males. J Forensic Sci 51(2): 213–29. doi: 10.1111/j.1556-4029.2006.00060.x.

Rissech, C., G. F. Estabrook, E. Cunha and A. Malgosa. 2007. Estimation of age-at-death for adult males using the acetabulum, applied to four Western European populations. J Forensic Sci 52(4): 774–8. doi: 10.1111/j.1556-4029.2007.00486.x.

Rissech, C. and A. Malgosa. 2007. Pubis growth study: applicability in sexual and age diagnostic. Forensic Sci Int 173(2-3): 137–45. doi: 10.1016/j.forsciint.2007.02.022.

Rouge-Maillart, C., N. Jousset, B. Vielle, A. Gaudin and N. Telmon. 2007. Contribution of the study of acetabulum for the estimation of adult subjects. Forensic Sci Int 171(2-3): 103–10. doi: 10.1016/j. forsciint.2006.10.007.

Scheuer, L. and S. Black. 2000. Developmental Juvenile Osteology Academic Press.

Schour, I. and M. Massler. 1941. The development of the human dentition. Journal of the American Dental Association 28: 1153–1160.

Schmidt, S., M. Muhler, A. Schmeling, W. Reisinger and R. Schulz. 2007. Magnetic resonance imaging of the clavicular ossification. Int J Legal Med 121(4): 321–4. doi: 10.1007/s00414-007-0160-z.

Schulz, R., M. Muhler, S. Mutze, S. Schmidt, W. Reisinger and A. Schmeling. 2005. Studies on the time frame for ossification of the medial epiphysis of the clavicle as revealed by CT scans. Int J Legal Med 119(3): 142–5. doi: 10.1007/s00414-005-0529-9.

Senn, D. and R. Weems. 2013. Manual of Forensic Odontology. Fifth edition. CRC Press.

Shammas, M. A. 2011. Telomeres, lifestyle, cancer, and aging. Curr Opin Clin Nutr Metab Care 14(1): 28–34. doi: 10.1097/MCO.0b013e32834121b1.

Scientific Working Group for Forensic Anthropology (SGWATH) Age Estimation, 2010.

Scientific Working Group for Forensic Anthropology (SWGANTH) Age Estimation Issue (01/22/2013) http://www.swganth.org/.

Takasaki, T., A. Tsuji, N. Ikeda and M. Ohishi. 2003. Age estimation in dental pulp DNA based on human telomere shortening. Int J Legal Med 117(4): 232–4. doi: 10.1007/s00414-003-0376-5.

Tanner, J. M. and R. H. Whitehouse. 1975. A note on the bone age at which patients with true isolated growth hormone deficiency enter puberty. J Clin Endocrinol Metab 41(4): 788–90. doi: 10.1210/jcem-41-4-788.

Thompson, T. and S. Black. 2006. Forensic Human Identification: An Introduction. CRC Press.

Todd, T. W. 1921. Age changes in the pubic bone II. The pubis of the male Negro-White hybrid. III The pubis of the White female. IV. The pubis of the female Negro-White hybrid. Am J Phys Anthropol 4: 1–70.

Tsuji, A., A. Ishiko, T. Takasaki and N. Ikeda. 2002. Estimating age of humans based on telomere shortening. Forensic Sci Int 126(3): 197–9.

Ubelaker, D. H. 1987. Estimating age at death from immature human skeletons: an overview. J Forensic Sci 32(5): 1254–63.

Yamaguchi, H., R. T. Calado, H. Ly, S. Kajigaya, G. M. Baerlocher, S. J. Chanock et al. 2005. Mutations in TERT, the gene for telomerase reverse transcriptase, in aplastic anemia. N Engl J Med 352(14): 1413–24. doi: 10.1056/NEJMoa042980.

Zakian, V. A. 1995. Telomeres: beginning to understand the end. Science 270(5242): 1601–7.

Zapico, S. C. and D. H. Ubelaker. 2013. Applications of physiological bases of ageing to forensic sciences. Estimation of age-at-death. Ageing Res Rev 12(2): 605–17. doi: 10.1016/j.arr.2013.02.002.

Section IV
Mitochondrial DNA Mutations

17

Introduction to Mitochondria Biology

Christian Thomas[a] and *Sara C. Zapico**

INTRODUCTION

Mitochondria are intracellular organelles that evolved from an ancient endosymbiotic α-purple bacteria ingested by an eukaryotic ancestor approximately 1.5 billion years ago (Gray 2001). As a consequence of their origin, mitochondria have a double-membrane structure, composed by an outer membrane, highly permeable with many pores surrounding the inter-membrane space and an impermeable inner membrane, delimiting the internal matrix (Frey and Mannella 2000) (Fig. 1).

Mitochondria generate approximately 90% of the energy that cells need to survive. Because of that, the number of mitochondria found in a human cell depends upon its energy needs. Apart from this main function, mitochondria play a role in intracellular

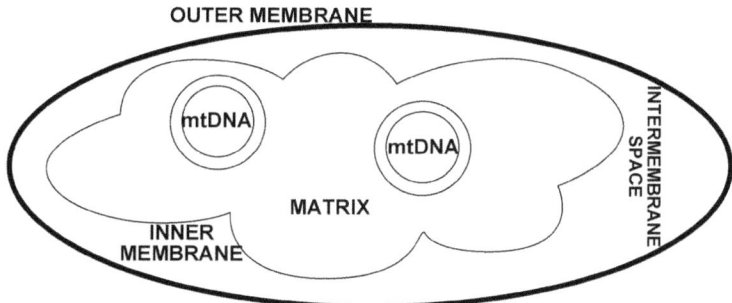

Figure 1. Mitochondria structure.

Smithsonian Institution, National Museum of Natural History, Anthropology Department 10th and Constitution Ave, NW, PO Box 37012 Washington, DC 20560 USA.
[a] Email: crf.thomas@gmail.com
* Corresponding author: saiczapico@gmail.com

signaling and apoptosis, intermediary metabolism and in the metabolism of amino acids, lipids, cholesterol, steroids, and nucleotides (Druzhyna et al. 2008).

Oxidative Phosphorylation (OXPHOS)

OXPHOS is the process of ATP synthesis. It occurs through the Electron Transport Chain (ETC), located at the inner mitochondrial membrane and consists of five protein complexes (from I to V) (Druzhyna et al. 2008). Various substrates can be metabolized to produce ATP; reduced cofactors (NADPH and $FADH_2$), generated from the intermediary metabolism of carbohydrates (within TCA cycle, citric acid/tricarboxylic) and proteins and fats (β-oxidation) donate electrons to complex I and complex II. These electrons pass sequentially to ubiquinone (coenzyme Q or CoQ) to form CoQH and then $CoQH_2$. Ubiquinol transfers its electrons to complex III, which transfers them to cytochrome c. From cytochrome c, the electrons go to complex IV, which donates an electron to oxygen to produce water. Each of these complexes incorporates multiple electron carriers. Complexes I, II and III comprise of several ion-sulfur (Fe-S) centers and complexes III and IV include the b + c_1 and a + a_3 cytochromes. The energy released by the flow of electrons is used by complexes I, III and IV to send protons (H^+) out of the mitochondrial inner membrane into the intermembrane space. This proton gradient produces the mitochondrial membrane potential that is coupled with ATP synthesis by complex V from ADP and Pi. ATP is released from the mitochondria in exchange for cytosolic ADP using an ANT (Chinnery and Schon 2003, Wallace 2005) (Fig. 2).

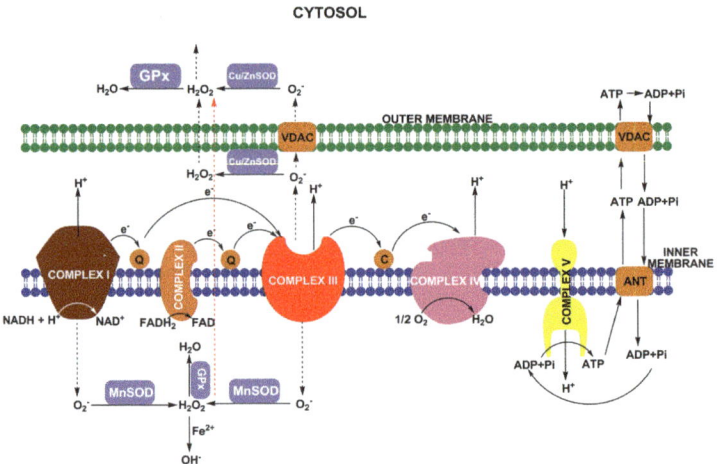

Figure 2. Oxidative Phosphorilation (OXPHOS) and generation of mitochondrial ROS.

As a by-product of OXPHOS, the mitochondria produce endogenous ROS (Fig. 2). Excess electrons from complexes I–III can be transferred directly to O_2 to generate superoxide anion (O_2^-). It is converted into H_2O_2 by the matrix enzyme MnSOD or SOD2 or by the mitochondrial intermembrane space and cytosol enzyme Cu/ZnSOD or SOD1. H_2O_2 is more stable than O_2^- and can diffuse into the cytosol and nucleus to activate redox-sensitive signaling. GPx in mitochondria and CAT in peroxisomes detoxify hydrogen peroxide into water. In the presence of reduced transition metals (like Fe^{2+}), hydrogen peroxide is converted to OH· through the Fenton reaction. OH· is the most reactive ROS (Martin 2012, Dai and Rabinovitch 2009, Fridovich 1995). ROS are highly reactive molecules and as oxidants, can extract electrons from DNA, proteins, lipids and other molecules. As a result, they induce oxidative damage, which ultimately leads to inactivation of proteins, injury to the integrity of biological membranes and genotoxicity. High levels of ROS induce cell death by apoptotic and/ or necrotic mechanisms. In contrast, low levels of ROS can act as signaling molecules in the cell (Ott et al. 2007, D'Autreaux and Toledano 2007, Fogg 2011).

Mitochondrial DNA

Mitochondria contain their own genome mitochondrial DNA (mtDNA), which is inherited through the maternal side. During the fertilization of an ovum, the sperm head is the only bit of a spermatozoon that enters the egg. The spermatozoon's mid piece contains mitochondria, but shears off as the head enters the egg's perivitelline space. Although some paternal mtDNA enters the ovum, it is actively removed (Manfredi et al. 1997, Thompson et al. 2003). Consequently, the only mtDNA present in the developing embryo is that derived from the egg.

Each mitochondrion contains between two to ten copies of the mtDNA. Human mtDNA is a double-stranded negatively supercoiled circular molecule (Fig. 3). It is organized into protein-DNA complexes, nucleoids, within the mitochondrial matrix

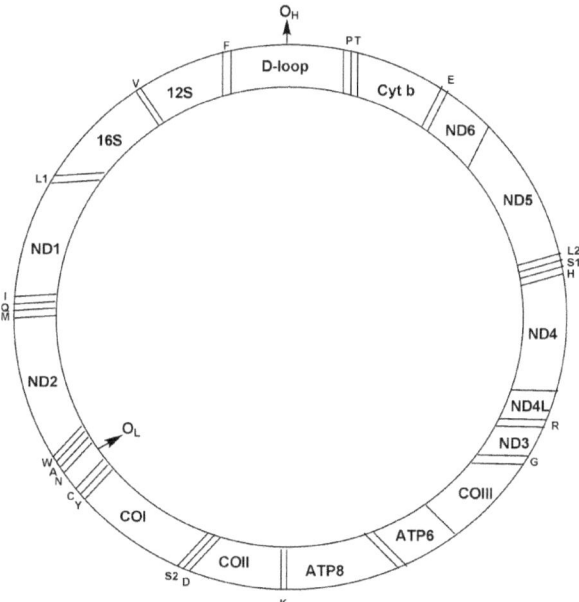

Figure 3. Mitochondrial DNA. The origins of Heavy-strand replication (OH) and Light-strand replication (LH) are shown. Complex I genes (ND1-ND6 and ND4L); cytochrome c oxidase (COI-COIII); cyt b, cytochrome b; ATP8 and ATP6, ATPase 8 and 6; Genes for two rRNAs (12S and 16S) and 22 tRNAs (F, V, L1, I, M, W, D, K, G, R, H, S1, L2, T, P, E, S2, Y, C, N, A).

(Gilkerson 2009). There is evidence that some of the genes from the ancestral bacteria have been transferred to the nuclear DNA; however, 16,569 base pairs remain within the mitochondrion. It contains 37 genes, which in turn code for the synthesis of 12S and 16S rRNAs, 22 tRNAs and 13 proteins that are structural subunits of OXPHOS enzyme complexes, seven of Complex I; three of Complex IV; two of Complex V and one of Complex III (Andrews et al. 1999, Anderson et al. 1981). Apart from that, nuclear genes encode approximately 1,500 mitochondrial proteins, the majority of mitochondrial respiratory chain polypeptides, including all four subunits of complex II, replication proteins such as the mitochondrial DNA polymerase γ (DNA pol γ), the mitochondria RNA polymerase components, the mtTFA, the mitochondrial ribosomal proteins and elongation factors and the mitochondrial metabolic enzymes (Chinnery and Schon 2003, Wallace 2005, Schatz 1996, Shoubridge 2001).

Unlike nuclear DNA, the mitochondrial genome is extremely compact, does not contain introns, has no space between genes and lacks 5' or 3' non-coding sequences (Singh and Kulawiec 2009), and about 93% of the DNA represents coding sequences. The remaining, non-coding region consists of about 1,122 base pairs and is called the control region or displacement loop (D-loop) since it contains the origin of replication for one of the mtDNA strands. There is an asymmetric distribution of nucleotides in the mtDNA that gives rise to the "heavy" (purine-rich) strand, in which most of the information is encoded, and the "light" (pyrimidine-rich) strand, which contains the information for only one polypeptide and 8 tRNAs (Kasamatsu and Vinograd 1974). Replication of mtDNA is developed by the concerted action of DNA pol γ, its accessory subunit p55 and replication factors such as the mitochondrial single-stranded DNA helicase (Twinkle). The mode of mtDNA replication is controversial, as two competing models have been advocated: strand-coupled replication (bi-directional) and strand-asymmetric. The bi-directional replication is initiated at two spatially and temporally distinct origins of replication OL and OH, for the light and heavy strands (in the D-loop, in this case) respectively (Taanman 1999, Yasukawa et al. 2006). According to the strand-asymmetric model, the leading strand replication is two thirds complete before lagging strand replication is initiated (Yang et al. 2002, Brown et al. 2005).

mtDNA mutations

mtDNA is exposed to more damage than nuclear DNA, accumulating mutations relatively faster than nuclear DNA. This is partly due to the exposure of the mtDNA to ROS, produced as by-products in OXPHOS (Brown et al. 1979, Hashiguchi et al. 2004). However, it has repair mechanisms. The major DNA repair mechanism is BER, encoded by nuclear DNA, although it is present at lower levels in mitochondria (Driggers et al. 1993). BER starts with the recognition of damage, followed by removal of the lesion through enzymatic processes, restoring genomic integrity (Gredilla 2010). There are DNA repair enzymes in the mitochondria such as damage-specific DNA glycosylases, like OGG1, an NTH1, APE and DNA ligase IIIβ. DNA pol γ is implicated in mtDNA BER (Larsen et al. 2005, Szczesny et al. 2003, Lakshmipathy and Campbell 1999). Apart from BER, there is some evidence that mitochondria possess mismatch repair activities, homologous recombination and non-homologous end joining (Mason et al. 2003, Thyagarajan et al. 1996, Coffey et al. 1999).

In spite of these potential repair mechanisms, the D-loop region exhibits a higher mutation rate, mainly substitutions, than the coding region because the sequences do not code for any substances necessary for the cell's function. Following the sequencing of the human mtDNA genome, it was apparent that the D-loop was not under the same functional constraints as the rest of the genome. Some blocks within the control region are highly conserved but large parts are not. The D-loop is divided into two regions, HV1 (from 16,024 to 16,365 bp) and HV2 (from 73 to 268 bp) and they contain the highest levels of variation within the mtDNA. There is a third region, less studied, HV3 (from 438 to 574 bp). The rate of mutation in these regions is not constant and some sites are hotspots for mutations, while others show much lower rates of change (Excoffier and Yang 1999, Tamura and Nei 1993, Meyer et al. 1999). It has been estimated that mtDNA may vary approximately 1–2.3% between unrelated individuals. Direct analysis of mother-to-child transmissions has estimated that a mutation in the hypervariable regions is passed from mother to child approximately once in every 30 to 40 events. In the vast majority of cases where a mutation is detected, there is only one base change between the mother and child (Parsons et al. 1997, Forster et al. 2002).

Homoplasmy and Heteroplasmy

The mtDNA sequence of all the mitochondria in any one individual is usually identical; this condition is referred to as "homoplasmy". However, mutations are bound to occur within some of the thousands of copies of mtDNA, creating more than one mtDNA type in some people between cells, within a single cell, or within a single mitochondrion. This condition is called "heteroplasmy" (Lo et al. 2005). Heteroplasmy used to be considered relatively rare, but it is now believed that all individuals are heteroplasmic at some level (many below the limits of detection in DNA sequence analysis).

It is apparent that heteroplasmy is not necessarily expressed to the same extent in all the tissues of the body, since it can be observed in several ways: (a) individuals may have more than one mtDNA type in a single tissue; (b) individuals may exhibit one mtDNA type in one tissue and a different type in another tissue; (c) individuals may be heteroplasmic in one tissue sample and homoplasmic in another tissue sample. Heteroplasmy may result from the high mutation rate or from either inheritance at the germline level or the level of somatic cell mitosis and mtDNA replication.

Once a mutation has led to significant heteroplasmy in the germline, the offspring will either be heteroplasmic or one mitotype will be fixed, so that heteroplasmy can no longer be detected. At some point a genetic bottleneck occurs before the formation of a mature oocyte. The bottleneck allows only a few mtDNA molecules to pass into the oocyte during its formation, altering the ratio of heteroplasmic variants or to fix on one mitotype (Marchington et al. 1998, Sekiguchi et al. 2003, Chinnery et al. 2000).

Likewise it seems like mutations accumulate during an individual's lifetime. Calloway et al. (Calloway et al. 2000) found the highest levels of heteroplasmy in muscle, increasing with age. Other studies supported these hypotheses in several tissues (Wei et al. 2009, Lacan et al. 2011, Zapico and Ubelaker 2015).

Keywords: Mitochondria, bacteria, energy, ATP, oxidative phosphorylation, reactive oxygen species, mtDNA, maternal inheritage, D-loop, hypervariable region 1, hypervariable region 2, homoplasmy, heteroplasmy

References

Anderson, S., A. T. Bankier, B. G. Barrell, M. H. de Bruijn, A. R. Coulson, J. Drouin et al. 1981. Sequence and organization of the human mitochondrial genome. Nature 290(5806): 457–65.

Andrews, R. M., I. Kubacka, P. F. Chinnery, R. N. Lightowlers, D. M. Turnbull and N. Howell. 1999. Reanalysis and revision of the Cambridge reference sequence for human mitochondrial DNA. Nat Genet 23(2): 147. doi: 10.1038/13779.

Brown, T. A., C. Cecconi, A. N. Tkachuk, C. Bustamante and D. A. Clayton. 2005. Replication of mitochondrial DNA occurs by strand displacement with alternative light-strand origins, not via a strand-coupled mechanism. Genes Dev 19(20): 2466–76. doi: 10.1101/gad.1352105.

Brown, W. M., M. George, Jr. and A. C. Wilson. 1979. Rapid evolution of animal mitochondrial DNA. Proc Natl Acad Sci U S A 76(4): 1967–71.

Calloway, C. D., R. L. Reynolds, G. L. Herrin, Jr. and W. W. Anderson. 2000. The frequency of heteroplasmy in the HVII region of mtDNA differs across tissue types and increases with age. Am J Hum Genet 66(4): 1384–97. doi: 10.1086/302844.

Chinnery, P. F. and E. A. Schon. 2003. Mitochondria. J Neurol Neurosurg Psychiatry 74(9): 1188–99.

Chinnery, P. F., D. R. Thorburn, D. C. Samuels, S. L. White, H. M. Dahl, D. M. Turnbull et al. 2000. The inheritance of mitochondrial DNA heteroplasmy: random drift, selection or both? Trends Genet 16(11): 500–5.

Coffey, G., U. Lakshmipathy and C. Campbell. 1999. Mammalian mitochondrial extracts possess DNA end-binding activity. Nucleic Acids Res 27(16): 3348–54.

D'Autreaux, B. and M. B. Toledano. 2007. ROS as signalling molecules: mechanisms that generate specificity in ROS homeostasis. Nat Rev Mol Cell Biol 8(10): 813–24. doi: 10.1038/nrm2256.

Dai, D. F. and P. S. Rabinovitch. 2009. Cardiac aging in mice and humans: the role of mitochondrial oxidative stress. Trends Cardiovasc Med 19(7): 213–20. doi: 10.1016/j.tcm.2009.12.004.

Driggers, W. J., S. P. LeDoux and G. L. Wilson. 1993. Repair of oxidative damage within the mitochondrial DNA of RINr 38 cells. J Biol Chem 268(29): 22042–5.

Druzhyna, N. M., G. L. Wilson and S. P. LeDoux. 2008. Mitochondrial DNA repair in aging and disease. Mech Ageing Dev 129(7-8): 383–90. doi: 10.1016/j.mad.2008.03.002.

Excoffier, L. and Z. Yang. 1999. Substitution rate variation among sites in mitochondrial hypervariable region I of humans and chimpanzees. Mol Biol Evol 16(10): 1357–68.

Fogg, V. C., N. J. Lanning and J. P. Mackeigan. 2011. Mitochondria in cancer: at the crossroads of life and death. Chin J Cancer 30(8): 526–39. doi: 10.5732/cjc.011.10018.

Forster, L., P. Forster, S. Lutz-Bonengel, H. Willkomm and B. Brinkmann. 2002. Natural radioactivity and human mitochondrial DNA mutations. Proc Natl Acad Sci U S A 99(21): 13950–4. doi: 10.1073/pnas.202400499.

Frey, T. G. and C. A. Mannella. 2000. The internal structure of mitochondria. Trends Biochem Sci 25(7): 319–24.

Fridovich, I. 1995. Superoxide radical and superoxide dismutases. Annu Rev Biochem 64: 97–112. doi: 10.1146/annurev.bi.64.070195.000525.

Gilkerson, R. W. 2009. Mitochondrial DNA nucleoids determine mitochondrial genetics and dysfunction. Int J Biochem Cell Biol 41(10): 1899–906. doi: 10.1016/j.biocel.2009.03.016.

Gray, M. W., G. Burger and B. F. Lang. 2001. The origin and early evolution of mitochondria. Genome Biol 2(6): REVIEWS1018.

Gredilla, R. 2010. DNA damage and base excision repair in mitochondria and their role in aging. J Aging Res 2011: 257093. doi: 10.4061/2011/257093.

Hashiguchi, K., V. A. Bohr and N. C. de Souza-Pinto. 2004. Oxidative stress and mitochondrial DNA repair: implications for NRTIs induced DNA damage. Mitochondrion 4(2-3): 215–22. doi: 10.1016/j.mito.2004.05.014.

Kasamatsu, H. and J. Vinograd. 1974. Replication of circular DNA in eukaryotic cells. Annu Rev Biochem 43(0): 695–719. doi: 10.1146/annurev.bi.43.070174.003403.

Lacan, M., C. Theves, C. Keyser, A. Farrugia, J. P. Baraybar, E. Crubezy et al. 2011. Detection of age-related duplications in mtDNA from human muscles and bones. Int J Legal Med 125(2): 293–300. doi: 10.1007/s00414-010-0440-x.

Lakshmipathy, U. and C. Campbell. 1999. The human DNA ligase III gene encodes nuclear and mitochondrial proteins. Mol Cell Biol 19(5): 3869–76.

Larsen, N. B., M. Rasmussen and L. J. Rasmussen. 2005. Nuclear and mitochondrial DNA repair: similar pathways? Mitochondrion 5(2): 89–108. doi: 10.1016/j.mito.2005.02.002.

Lo, M. C., H. M. Lee, M. W. Lin and C. Y. Tzen. 2005. Analysis of heteroplasmy in hypervariable region II of mitochondrial DNA in maternally related individuals. Ann N Y Acad Sci 1042: 130–5. doi: 10.1196/annals.1338.013.

Manfredi, G., D. Thyagarajan, L. C. Papadopoulou, F. Pallotti and E. A. Schon. 1997. The fate of human sperm-derived mtDNA in somatic cells. Am J Hum Genet 61(4): 953–60. doi: 10.1086/514887.

Marchington, D. R., V. Macaulay, G. M. Hartshorne, D. Barlow and J. Poulton. 1998. Evidence from human oocytes for a genetic bottleneck in an mtDNA disease. Am J Hum Genet 63(3): 769–75. doi: 10.1086/302009.

Martin, L. J. 2012. Biology of mitochondria in neurodegenerative diseases. Prog Mol Biol Transl Sci 107: 355–415. doi: 10.1016/B978-0-12-385883-2.00005-9.

Mason, P. A., E. C. Matheson, A. G. Hall and R. N. Lightowlers. 2003. Mismatch repair activity in mammalian mitochondria. Nucleic Acids Res 31(3): 1052–8.

Meyer, S., G. Weiss and A. von Haeseler. 1999. Pattern of nucleotide substitution and rate heterogeneity in the hypervariable regions I and II of human mtDNA. Genetics 152(3): 1103–10.

Ott, M., V. Gogvadze, S. Orrenius and B. Zhivotovsky. 2007. Mitochondria, oxidative stress and cell death. Apoptosis 12(5): 913–22. doi: 10.1007/s10495-007-0756-2.

Parsons, T. J., D. S. Muniec, K. Sullivan, N. Woodyatt, R. Alliston-Greiner, M. R. Wilson et al. 1997. A high observed substitution rate in the human mitochondrial DNA control region. Nat Genet 15(4): 363–8. doi: 10.1038/ng0497-363.

Schatz, G. 1996. The protein import system of mitochondria. J Biol Chem 271(50): 31763–6.

Sekiguchi, K., K. Kasai and B. C. Levin. 2003. Inter- and intragenerational transmission of a human mitochondrial DNA heteroplasmy among 13 maternally-related individuals and differences between and within tissues in two family members. Mitochondrion 2(6): 401–14. doi: 10.1016/S1567-7249(03)00028-X.

Shoubridge, E. A. 2001. Nuclear genetic defects of oxidative phosphorylation. Hum Mol Genet 10(20): 2277–84.

Singh, K. K. and M. Kulawiec. 2009. Mitochondrial DNA polymorphism and risk of cancer. Methods Mol Biol 471: 291–03. doi: 10.1007/978-1-59745-416-2_15.

Szczesny, B., T. K. Hazra, J. Papaconstantinou, S. Mitra and I. Boldogh. 2003. Age-dependent deficiency in import of mitochondrial DNA glycosylases required for repair of oxidatively damaged bases. Proc Natl Acad Sci U S A 100(19): 10670–5. doi: 10.1073/pnas.1932854100.

Taanman, J. W. 1999. The mitochondrial genome: structure, transcription, translation and replication. Biochim Biophys Acta 1410(2): 103–23.

Tamura, K. and M. Nei. 1993. Estimation of the number of nucleotide substitutions in the control region of mitochondrial DNA in humans and chimpanzees. Mol Biol Evol 10(3): 512–26.

Thompson, W. E., J. Ramalho-Santos and P. Sutovsky. 2003. Ubiquitination of prohibitin in mammalian sperm mitochondria: possible roles in the regulation of mitochondrial inheritance and sperm quality control. Biol Reprod 69(1): 254–60. doi: 10.1095/biolreprod.102.010975.

Thyagarajan, B., R. A. Padua and C. Campbell. 1996. Mammalian mitochondria possess homologous DNA recombination activity. J Biol Chem 271(44): 27536–43.

Wallace, D. C. 2005. A mitochondrial paradigm of metabolic and degenerative diseases, aging, and cancer: a dawn for evolutionary medicine. Annu Rev Genet 39: 359–407. doi: 10.1146/annurev.genet.39.110304.095751.

Wei, Y. H., S. B. Wu, Y. S. Ma and H. C. Lee. 2009. Respiratory function decline and DNA mutation in mitochondria, oxidative stress and altered gene expression during aging. Chang Gung Med J 32(2): 113–32.

Yang, M. Y., M. Bowmaker, A. Reyes, L. Vergani, P. Angeli, E. Gringeri et al. 2002. Biased incorporation of ribonucleotides on the mitochondrial L-strand accounts for apparent strand-asymmetric DNA replication. Cell 111(4): 495–505.

Yasukawa, T., A. Reyes, T. J. Cluett, M. Y. Yang, M. Bowmaker, H. T. Jacobs et al. 2006. Replication of vertebrate mitochondrial DNA entails transient ribonucleotide incorporation throughout the lagging strand. EMBO J 25(22): 5358–71. doi: 10.1038/sj.emboj.7601392.

Zapico, S. C. and D. H. Ubelaker. 2015. Relationship between mitochondrial DNA mutations and aging. Estimation of age-at-death. J Gerontol A Biol Sci Med Sci doi: 10.1093/gerona/glv115.

18

Mitochondrial DNA Mutations and Aging

Rebecca Gordon[1] and *Sara C. Zapico*[2,*]

INTRODUCTION

Bratic and Larsson (2013) proposed the "mitochondrial theory of aging", which is based on the following observations: "(1) mitochondrial ROS production increases with age because of a decline in mitochondrial function; (2) activity of several free radical-scavenging enzymes declines with age; (3) mutations of mitochondrial DNA (mtDNA) accumulate during aging; (4) a vicious cycle occurs because mtDNA mutations impair Electron Transport Chain function, increasing ROS production and accumulating oxidative damage to proteins, lipids and DNA." This theory builds on earlier studies which focused on the role of mitochondrial mutations and aging.

In 1956, Harman proposed that "aging and the degenerative diseases associated with it are attributed basically to the deleterious side attacks of free radicals on cell constituents and on the connected tissues" (Harman 1956). This "free radical theory of aging" was modified over time, leading to the "oxidative stress theory." Sohal and Weindruch (1996) postulated "a chronic state of oxidative stress exists in cells of aerobic organisms even under normal physiological conditions because of an imbalance of prooxidants and antioxidants. This imbalance results in a steady-state accumulation of oxidative damage in a variety of macromolecules. Oxidative damage increases during aging, which results in a progressive loss in the functional efficiency of various cellular processes". Recently, Sohal and Orr (2012) proposed a new modification, "the redox stress theory of aging". This theoretical modification indicated that "if free radicals cause a stress that cells can manage with, then damage will not occur

---

[1] American University, Anthropology Department, Hamilton 100, 4400 Massachusetts Avenue NW, Washington, DC 20016.
  Email: rebeccagordon@gmail.com
[2] Smithsonian Institution, National Museum of Natural History, Anthropology Department 10th and Constitution Ave, NW, PO Box 37012, Washington, DC 20560, USA.
* Corresponding author: saiczapico@gmail.com

because antioxidant defenses will overwhelm such stress. Only if the stress is of such magnitude that it disrupts cellular signaling mechanisms, age-associated damage will take place". Based on this theory and their previous studies, Viña et al. (Viña et al. 2013) proposed "the cell signaling disruption theory of aging", which postulates that free radicals can alter the cell's signaling network. They conclude: "If the cell can handle the stress caused by the action of free radicals, then adaptation takes place and damage does not occur. However, if the cell is overwhelmed by the action of radicals, subcellular damage and aging will take place" (Vina et al. 2013).

Harman's 1956 study suggested that mitochondria are key organelles involved in aging (Harman 1956). Miquel et al. (Miquel et al. 1980) proposed "the mitochondrial free radical theory of aging", pointing out that mitochondria are important sources of free radicals and mtDNA is a critical target to explain the age-associated damage in cells. Later on, Sastre et al. (Sastre et al. 1996) provided the first evidence that mitochondria are damaged inside cells due to an intrinsic damage which occurs in cells as they age.

Mitochondrial Function Declines with Age

Mitochondrial function declines with aging, concurrent with the appearance of mitochondrial morphological alterations (Shigenaga et al. 1994). Bratic and Larsson demonstrated that the number of mitochondria decreases with age in mice, rats and humans (Bratic and Larsson 2013). This is concomitant with a decrease in mtDNA abundance, as Herbert et al. (2010) demonstrated. Reduced mtDNA template availability could negatively impact the levels of mitochondrial gene transcripts, and as a consequence, reduce the levels of mitochondrial proteins. Additionally, Electron Transport Chain (ETC) function was found to be reduced up to 40% in rat liver and spleen mitochondria of older animals, although it was not found in the brain (Stocco et al. 1977). In humans, ETC decline has been demonstrated in liver, heart and skeletal muscle (Short et al. 2005, Ojaimi et al. 1999). In fact, the activity of specific ETC complexes and certain nuclear-encoded mitochondrial proteins decline as mammals' age (Benzi et al. 1992, Manczak et al. 2005). Despite of these studies related to the contribution of mitochondrial dysfunction to aging, it is important to take into account that mitochondrial biogenesis is controlled simultaneously at many different levels. Some studies demonstrated that caloric restriction and exercise increased the expression of genes involved in mitochondrial biogenesis (Civitarese et al. 2007, Barja 2004, Sanz et al. 2005). Reduction of available calories starves the ETC for electrons, reducing ROS and protecting the mitochondria and mtDNA (Sohal et al. 1994, Wolf et al. 2005).

mtDNA Mutations and Aging

There is sufficient evidence that mtDNA mutations accumulate with age (Hebert et al. 2010), including deletions, duplications and point mutations, although it remains unclear whether these alterations are a cause or consequence of the aging process. The first attempts to study whether mtDNA mutations are driving forces in aging were carried out by two independent groups (Trifunovic et al. 2004,

Kujoth et al. 2005). They created mtDNA mutator mice containing a point mutation in the proof-reading domain of the catalytic subunit of mtDNA polymerase-γ (PolgA^{mut}). This enzyme had a significant reduction of exonuclease activity with no decrease in DNA polymerase activity. The mtDNA mutator mice displayed three different forms of mtDNA mutations: random point mutations, linear molecules with large deletions, and, in certain tissues, molecules containing multimers of the control region. The mitochondrial mutation frequencies were increased by at least three to eleven-fold in multiple tissues, suggesting that the mutations occurred during embryonic and/or fetal development. The phenotype of these mtDNA mutator mice resembled many characteristics of the aging process (accelerated and premature aging): hair graying and loss, reduced bone density and increased incidence of kyphosis, reduced muscle mass, severe reduction in body fat, early loss of fertility, dilated cardiac hypertrophy, accelerated thymic atrophy, presbycusis, anemia, intestinal dysplasia. The life spans were reduced. The median life span was approximately 46–59 weeks with some subjects surviving to the age of 61–66 weeks. Due to mtDNA point mutations, ETC units were synthesized with amino acid substitutions that cause instability of the ETC complexes, reducing the respiratory function. However, these mice didn't show signs of increased oxidative damage to proteins, lipids and DNA. In addition, expression levels of antioxidant defense enzymes were not upregulated, indicating the absence of or only minor oxidative stress in tissues from mtDNA mutator mice. It seemed that increased levels of mtDNA mutations were not associated with increased markers of oxidative stress. In contrast, there was a strong correlation with the induction of apoptosis in target tissues, containing elevated levels of activated caspase-3, one of the effector caspases as early as three months of age. Also, several tissues of the mtDNA mutator mice exhibited increased DNA fragmentation. Thus, mtDNA mutation accumulation was associated with the activation of apoptosis, leading to cell death (Thompson 2006). These findings contradict the concept of a vicious cycle.

Although somatic mtDNA mutations can be compensated by heteroplasmy, these mutations accumulate in postmitotic tissues until a certain tissue-specific threshold in the level of mutant to normal mtDNA molecules is exceeded and cells become energetically compromised (Rossignol et al. 1999, Wallace 2010), inducing an impairment of energy metabolism, leading to aging effects on tissues that present high energetic demands, like the heart, skeletal muscle and brain (Kujoth et al. 2007). The accumulation of mtDNA mutations has been found in a variety of tissues during aging in human, monkeys and rodents (Wei and Lee 2002, Cortopassi and Arnheim 1990, Soong et al. 1992, Schwarze et al. 1995, Khaidakov et al. 2003).

Exposure of mtDNA to free radicals can result in a number of oxidative modifications to mtDNA, the most common, 8-oxo-2'-deoxyguanosine (oxo^{8}G), is probably due to a decline in the reparation mechanism with age (de Souza-Pinto et al. 2001, Hayakawa et al. 1991).

mtDNA deletions are most likely occurring during the repair of damaged mtDNA through exonuclease activity at double-strand breaks (Krishnan et al. 2008). An age-related accumulation of various deletions has been shown in multiple tissues of different species. Although some studies have shown that the overall percentage of deletions is relatively low (Gerhard et al. 2002, Zheng et al. 2012, Corral-Debrinski et al. 1992b), others demonstrated a high percentage of deleted mtDNA (Bua et al. 2006,

Bender et al. 2006). As described above, accumulation of mtDNA deletions was found to be tissue-specific and more prominent in tissues with greater energetic demands like muscle (Simonetti et al. 1992) and brain (Guo et al. 2010, Reeve et al. 2009, Reeve et al. 2008). A number of deletions can occur, however, the most commonly studied in humans is a 4977 bp deletion, also called "common deletion". This deletion removes all or part of the genes for NADH dehydrogenase subunits, cytochrome c oxidase subunit III and ATP synthase subunits 6 and 8 (Hebert et al. 2010). The accumulation of this deletion is tissue-dependent, being higher in tissues with high oxygen consumption of aged individuals (Corral-Debrinski et al. 1992a) than in all other tissues studied, like liver (Ross et al. 2002, Gerhard et al. 2002). An extensive study developed by Meissner et al. (Meissner et al. 2008) found an increase in this deletion with aging in the brain, heart and skeletal muscle of 92 individuals from 2 months to 102 years old. However, the levels were highly variable between individuals of the same decade and among different tissues within a single individual. This is in agreement with other studies (Zhang et al. 1997, Corral-Debrinski et al. 1992a, Cortopassi et al. 1992). In brain tissue, the highest amounts of this deletion were detected in cortex, putamen and Substancia Nigra (SN), corresponding to the brain areas characterized by a high dopamine metabolism (Bender et al. 2008).

A clear association was found between mtDNA point mutations accumulation and aging in an extensive variability of tissues, indicating the presence of a differential pattern of mtDNA point mutation accumulation between tissues. These point mutations were mainly found to occur in the control region of the mitochondrial genome. Different studies have found an accumulation of point mutations in this region including A189G, T408A and T414G with aging (Michikawa et al. 1999, Calloway et al. 2000, Del Bo et al. 2002, Del Bo et al. 2003, Theves et al. 2006, McInerny et al. 2009, Wang et al. 2001). Theves et al. (Theves et al. 2006) analyzed A189G mutation in the buccal cells as well as the skeletal muscle of unrelated individuals age 1–97. They found a higher percentage of mutations in the skeletal muscle of older individuals than in their buccal cells. Wang et al. (Wang et al. 2001) studied eight different tissues from 40 individuals and reported that A189G and T408G mutations have a tissue-specific frequency, presenting their highest values in muscle. However, the most frequent fibroblast-specific mutation, T414G, has been described in skin, but not in muscle. This tissue-specific accumulation of point mutations may be related to the metabolic characteristics of the tissue (McInerny et al. 2009).

Recent studies have investigated point mutations and deletions in single cells; only one particular point mutation or deletion was found in a single cell, and the percentage of mutant DNA molecules within each cell increases with age (Bodyak et al. 2001, Nekhaeva et al. 2002). Skeletal muscle develops a mosaic pattern of increasing COX deficient muscle fibers with increasing age, and multiple mtDNA alterations (point mutations, length variations, deletions) were found (Del Bo et al. 2003, Fayet et al. 2002). A study of deletions in COX negative muscle fibers from aged rats showed that > 90% of the mtDNA at the site of the electron transport system abnormality contained deletion mutations (Herbst et al. 2007).

Role of Haplogroups in Human Aging

As described above, human mtDNA is highly polymorphic. On the basis of certain SNPs present in mtDNA, which are well established and widely distributed, the human population can be divided into haplogroups (Malhi et al. 2003). Analysis of these polymorphisms from a wide range of human populations have revealed sets of ancestral mutations that define these haplogroups that have common ancestry and, because of uniparental inheritance, evolve independently from each other. Each of these haplogroups is defined by specific sets of associated mutations, thus allowing for a quick and precise classification of the mtDNA molecules within a certain population (De Benedictis et al. 1999). These associated mutations were suggested to affect coupling efficiency of the ETC enabling adaptation to life in different climatic conditions (Wallace 2005).

The relationship between mitochondrial haplogroups and aging in humans is based on analysis of centenarians in various populations, where certain haplogroups were found to be overrepresented in relation to controls. One of the hypotheses proposed argues that successful aging is due to uncoupling of the electron chain. Persons with more uncoupled haplogroups would produce more heat (an adaptation to colder climates), be more calorie restricted, and age better (Wallace 2005). This has not been confirmed directly, as comparison of uncoupling in cell lines with identical genetic backgrounds repopulated with mitochondria from haplogroups H and T did not show any differences (Amo et al. 2008). Although there are interesting suggestions of correlations between specific haplogroups and aging in some populations (Niemi et al. 2005, Niemi et al. 2003, Tanaka et al. 2004, Chen et al. 2012, Ross et al. 2001), further population studies are required.

Conclusions

While it is still not clear if mtDNA mutations are the cause or a consequence of aging, it is known that these mutations increase with age and play a key role in this process, leading to mitochondrial dysfunction. Further research is needed in this field to increase understanding of the role of mtDNA mutations in the aging process.

Keywords: Mitochondrial DNA mutations (mtDNA), free radical theory of aging, mitochondrial theory of aging, aging, heteroplasmy, point mutations, deletions, single cell, muscle, brain, heart, haplogroups

References

Amo, T., N. Yadava, R. Oh, D. G. Nicholls and M. D. Brand. 2008. Experimental assessment of bioenergetic differences caused by the common European mitochondrial DNA haplogroups H and T. Gene 411(1-2): 69–76. doi: 10.1016/j.gene.2008.01.007.

Barja, G. 2004. Aging in vertebrates, and the effect of caloric restriction: a mitochondrial free radical production-DNA damage mechanism? Biol Rev Camb Philos Soc 79(2): 235–51.

Bender, A., K. J. Krishnan, C. M. Morris, G. A. Taylor, A. K. Reeve, R. H. Perry et al. 2006. High levels of mitochondrial DNA deletions in substantia nigra neurons in aging and Parkinson disease. Nat Genet 38(5): 515–7. doi: 10.1038/ng1769.

Bender, A., R. M. Schwarzkopf, A. McMillan, K. J. Krishnan, G. Rieder, M. Neumann et al. 2008. Dopaminergic midbrain neurons are the prime target for mitochondrial DNA deletions. J Neurol 255(8): 1231–5. doi: 10.1007/s00415-008-0892-9.

Benzi, G., O. Pastoris, F. Marzatico, R. F. Villa, F. Dagani and D. Curti. 1992. The mitochondrial electron transfer alteration as a factor involved in the brain aging. Neurobiol Aging 13(3): 361–8.

Bodyak, N. D., E. Nekhaeva, J. Y. Wei and K. Khrapko. 2001. Quantification and sequencing of somatic deleted mtDNA in single cells: evidence for partially duplicated mtDNA in aged human tissues. Hum Mol Genet 10(1): 17–24.

Bratic, A. and N. G. Larsson. 2013. The role of mitochondria in aging. J Clin Invest 123(3): 951–7. doi: 10.1172/JCI64125.

Bua, E., J. Johnson, A. Herbst, B. Delong, D. McKenzie, S. Salamat et al. 2006. Mitochondrial DNA-deletion mutations accumulate intracellularly to detrimental levels in aged human skeletal muscle fibers. Am J Hum Genet 79(3): 469–80. doi: 10.1086/507132.

Calloway, C. D., R. L. Reynolds, G. L. Herrin, Jr. and W. W. Anderson. 2000. The frequency of heteroplasmy in the HVII region of mtDNA differs across tissue types and increases with age. Am J Hum Genet 66(4): 1384–97. doi: 10.1086/302844.

Chen, A., N. Raule, A. Chomyn and G. Attardi. 2012. Decreased reactive oxygen species production in cells with mitochondrial haplogroups associated with longevity. PLoS One 7(10): e46473. doi: 10.1371/journal.pone.0046473.

Civitarese, A. E., S. Carling, L. K. Heilbronn, M. H. Hulver, B. Ukropcova, W. A. Deutsch et al. 2007. Calorie restriction increases muscle mitochondrial biogenesis in healthy humans. PLoS Med 4(3): e76. doi: 10.1371/journal.pmed.0040076.

Corral-Debrinski, M., T. Horton, M. T. Lott, J. M. Shoffner, M. F. Beal and D. C. Wallace. 1992a. Mitochondrial DNA deletions in human brain: regional variability and increase with advanced age. Nat Genet 2(4): 324–9. doi: 10.1038/ng1292-324.

Corral-Debrinski, M., J. M. Shoffner, M. T. Lott and D. C. Wallace. 1992b. Association of mitochondrial DNA damage with aging and coronary atherosclerotic heart disease. Mutat Res 275(3-6): 169–80.

Cortopassi, G. A. and N. Arnheim. 1990. Detection of a specific mitochondrial DNA deletion in tissues of older humans. Nucleic Acids Res 18(23): 6927–33.

Cortopassi, G. A., D. Shibata, N. W. Soong and N. Arnheim. 1992. A pattern of accumulation of a somatic deletion of mitochondrial DNA in aging human tissues. Proc Natl Acad Sci U S A 89(16): 7370–4.

De Benedictis, G., G. Rose, G. Carrieri, M. De Luca, E. Falcone, G. Passarino et al. 1999. Mitochondrial DNA inherited variants are associated with successful aging and longevity in humans. FASEB J 13(12): 1532–6.

de Souza-Pinto, N. C., L. Eide, B. A. Hogue, T. Thybo, T. Stevnsner, E. Seeberg et al. 2001. Repair of 8-oxodeoxyguanosine lesions in mitochondrial DNA depends on the oxoguanine DNA glycosylase (OGG1) gene and 8-oxoguanine accumulates in the mitochondrial DNA of OGG1-defective mice. Cancer Res 61(14): 5378–81.

Del Bo, R., A. Bordoni, F. Martinelli Boneschi, M. Crimi, M. Sciacco, N. Bresolin et al. 2002. Evidence and age-related distribution of mtDNA D-loop point mutations in skeletal muscle from healthy subjects and mitochondrial patients. J Neurol Sci 202(1-2): 85–91.

Del Bo, R., M. Crimi, M. Sciacco, G. Malferrari, A. Bordoni, L. Napoli et al. 2003. High mutational burden in the mtDNA control region from aged muscles: a single-fiber study. Neurobiol Aging 24(6): 829–38.

Fayet, G., M. Jansson, D. Sternberg, A. R. Moslemi, P. Blondy, A. Lombes et al. 2002. Aging muscle: clonal expansions of mitochondrial DNA point mutations and deletions cause focal impairment of mitochondrial function. Neuromuscul Disord 12(5): 484–93.

Gerhard, G. S., F. A. Benko, R. G. Allen, M. Tresini, A. Kalbach, V. J. Cristofalo et al. 2002. Mitochondrial DNA mutation analysis in human skin fibroblasts from fetal, young, and old donors. Mech Ageing Dev 123(2-3): 155–66.

Guo, X., E. Kudryavtseva, N. Bodyak, A. Nicholas, I. Dombrovsky, D. Yang et al. 2010. Mitochondrial DNA deletions in mice in men: substantia nigra is much less affected in the mouse. Biochim Biophys Acta 1797(6-7): 1159–62. doi: 10.1016/j.bbabio.2010.04.005.

Harman, D. 1956. Aging: a theory based on free radical and radiation chemistry. J Gerontol 11(3): 298–300.

Hayakawa, M., K. Torii, S. Sugiyama, M. Tanaka and T. Ozawa. 1991. Age-associated accumulation of 8-hydroxydeoxyguanosine in mitochondrial DNA of human diaphragm. Biochem Biophys Res Commun 179(2): 1023–9.

Hebert, S. L., I. R. Lanza and K. S. Nair. 2010. Mitochondrial DNA alterations and reduced mitochondrial function in aging. Mech Ageing Dev 131(7-8): 451–62. doi: 10.1016/j.mad.2010.03.007.

Herbst, A., J. W. Pak, D. McKenzie, E. Bua, M. Bassiouni and J. M. Aiken. 2007. Accumulation of mitochondrial DNA deletion mutations in aged muscle fibers: evidence for a causal role in muscle fiber loss. J Gerontol A Biol Sci Med Sci 62(3): 235–45.

Khaidakov, M., R. H. Heflich, M. G. Manjanatha, M. B. Myers and A. Aidoo. 2003. Accumulation of point mutations in mitochondrial DNA of aging mice. Mutat Res 526(1-2): 1–7.

Krishnan, K. J., A. K. Reeve, D. C. Samuels, P. F. Chinnery, J. K. Blackwood, R. W. Taylor et al. 2008. What causes mitochondrial DNA deletions in human cells? Nat Genet 40(3): 275–9. doi: 10.1038/ng.f.94.

Kujoth, G. C., A. Hiona, T. D. Pugh, S. Someya, K. Panzer, S. E. Wohlgemuth et al. 2005. Mitochondrial DNA mutations, oxidative stress, and apoptosis in mammalian aging. Science 309(5733): 481–4. doi: 10.1126/science.1112125.

Kujoth, G. C., P. C. Bradshaw, S. Haroon and T. A. Prolla. 2007. The role of mitochondrial DNA mutations in mammalian aging. PLoS Genet 3(2): e24. doi: 10.1371/journal.pgen.0030024.

Malhi, R. S., H. M. Mortensen, J. A. Eshleman, B. M. Kemp, J. G. Lorenz, F. A. Kaestle et al. 2003. Native American mtDNA prehistory in the American Southwest. Am J Phys Anthropol 120(2): 108–24. doi: 10.1002/ajpa.10138.

Manczak, M., Y. Jung, B. S. Park, D. Partovi and P. H. Reddy. 2005. Time-course of mitochondrial gene expressions in mice brains: implications for mitochondrial dysfunction, oxidative damage, and cytochrome c in aging. J Neurochem 92(3): 494–504. doi: 10.1111/j.1471-4159.2004.02884.x.

McInerny, S. C., A. L. Brown and D. W. Smith. 2009. Region-specific changes in mitochondrial D-loop in aged rat CNS. Mech Ageing Dev 130(5): 343–9. doi: 10.1016/j.mad.2009.01.008.

Meissner, C., P. Bruse, S. A. Mohamed, A. Schulz, H. Warnk, T. Storm et al. 2008. The 4977 bp deletion of mitochondrial DNA in human skeletal muscle, heart and different areas of the brain: a useful biomarker or more? Exp Gerontol 43(7): 645–52. doi: 10.1016/j.exger.2008.03.004.

Michikawa, Y., K. Laderman, K. Richter and G. Attardi. 1999. Role of nuclear background and *in vivo* environment in variable segregation behavior of the aging-dependent T414G mutation at critical control site for human fibroblast mtDNA replication. Somat Cell Mol Genet 25(5-6): 333–42.

Miquel, J., A. C. Economos, J. Fleming and J. E. Johnson, Jr. 1980. Mitochondrial role in cell aging. Exp Gerontol 15(6): 575–91.

Nekhaeva, E., N. D. Bodyak, Y. Kraytsberg, S. B. McGrath, N. J. Van Orsouw, A. Pluzhnikov et al. 2002. Clonally expanded mtDNA point mutations are abundant in individual cells of human tissues. Proc Natl Acad Sci U S A 99(8): 5521–6. doi: 10.1073/pnas.072670199.

Niemi, A. K., A. Hervonen, M. Hurme, P. J. Karhunen, M. Jylha and K. Majamaa. 2003. Mitochondrial DNA polymorphisms associated with longevity in a Finnish population. Hum Genet 112(1): 29–33. doi: 10.1007/s00439-002-0843-y.

Niemi, A. K., J. S. Moilanen, M. Tanaka, A. Hervonen, M. Hurme, T. Lehtimaki et al. 2005. A combination of three common inherited mitochondrial DNA polymorphisms promotes longevity in Finnish and Japanese subjects. Eur J Hum Genet 13(2): 166–70. doi: 10.1038/sj.ejhg.5201308.

Ojaimi, J., C. L. Masters, K. Opeskin, P. McKelvie and E. Byrne. 1999. Mitochondrial respiratory chain activity in the human brain as a function of age. Mech Ageing Dev 111(1): 39–47.

Reeve, A. K., K. J. Krishnan, J. L. Elson, C. M. Morris, A. Bender, R. N. Lightowlers et al. 2008. Nature of mitochondrial DNA deletions in substantia nigra neurons. Am J Hum Genet 82(1): 228–35. doi: 10.1016/j.ajhg.2007.09.018.

Reeve, A. K., K. J. Krishnan, G. Taylor, J. L. Elson, A. Bender, R. W. Taylor et al. 2009. The low abundance of clonally expanded mitochondrial DNA point mutations in aged substantia nigra neurons. Aging Cell 8(4): 496–8. doi: 10.1111/j.1474-9726.2009.00492.x.

Ross, O. A., R. McCormack, M. D. Curran, R. A. Duguid, Y. A. Barnett, I. M. Rea et al. 2001. Mitochondrial DNA polymorphism: its role in longevity of the Irish population. Exp Gerontol 36(7): 1161–78.

Ross, O. A., P. Hyland, M. D. Curran, B. P. McIlhatton, A. Wikby, B. Johansson et al. 2002. Mitochondrial DNA damage in lymphocytes: a role in immunosenescence? Exp Gerontol 37(2-3): 329–40.

Rossignol, R., M. Malgat, J. P. Mazat and T. Letellier. 1999. Threshold effect and tissue specificity. Implication for mitochondrial cytopathies. J Biol Chem 274(47): 33426–32.

Sanz, A., R. Gredilla, R. Pamplona, M. Portero-Otin, E. Vara, J. A. Tresguerres et al. 2005. Effect of insulin and growth hormone on rat heart and liver oxidative stress in control and caloric restricted animals. Biogerontology 6(1): 15–26. doi: 10.1007/s10522-004-7380-0.

Sastre, J., F. V. Pallardo, R. Pla, A. Pellin, G. Juan, J. E. O'Connor et al. 1996. Aging of the liver: age-associated mitochondrial damage in intact hepatocytes. Hepatology 24(5): 1199–205. doi: 10.1002/hep.510240536.

Schwarze, S. R., C. M. Lee, S. S. Chung, E. B. Roecker, R. Weindruch and J. M. Aiken. 1995. High levels of mitochondrial DNA deletions in skeletal muscle of old rhesus monkeys. Mech Ageing Dev 83(2): 91–101.

Shigenaga, M. K., T. M. Hagen and B. N. Ames. 1994. Oxidative damage and mitochondrial decay in aging. Proc Natl Acad Sci U S A 91(23): 10771–8.

Short, K. R., M. L. Bigelow, J. Kahl, R. Singh, J. Coenen-Schimke, S. Raghavakaimal et al. 2005. Decline in skeletal muscle mitochondrial function with aging in humans. Proc Natl Acad Sci U S A 102(15): 5618–23. doi: 10.1073/pnas.0501559102.

Simonetti, S., X. Chen, S. DiMauro and E. A. Schon. 1992. Accumulation of deletions in human mitochondrial DNA during normal aging: analysis by quantitative PCR. Biochim Biophys Acta 1180(2): 113–22.

Sohal, R. S., S. Agarwal, M. Candas, M. J. Forster and H. Lal. 1994. Effect of age and caloric restriction on DNA oxidative damage in different tissues of C57BL/6 mice. Mech Ageing Dev 76(2-3): 215–24.

Sohal, R. S. and R. Weindruch. 1996. Oxidative stress, caloric restriction, and aging. Science 273(5271): 59–63.

Sohal, R. S. and W. C. Orr. 2012. The redox stress hypothesis of aging. Free Radic Biol Med 52(3): 539–55. doi: 10.1016/j.freeradbiomed.2011.10.445.

Soong, N. W., D. R. Hinton, G. Cortopassi and N. Arnheim. 1992. Mosaicism for a specific somatic mitochondrial DNA mutation in adult human brain. Nat Genet 2(4): 318–23. doi: 10.1038/ng1292-318.

Stocco, D. M., J. Cascarano and M. A. Wilson. 1977. Quantitation of mitochondrial DNA, RNA, and protein in starved and starved-refed rat liver. J Cell Physiol 90(2): 295–306. doi: 10.1002/jcp.1040900215.

Tanaka, M., V. M. Cabrera, A. M. Gonzalez, J. M. Larruga, T. Takeyasu, N. Fuku et al. 2004. Mitochondrial genome variation in eastern Asia and the peopling of Japan. Genome Res 14(10A): 1832–50. doi: 10.1101/gr.2286304.

Theves, C., C. Keyser-Tracqui, E. Crubezy, J. P. Salles, B. Ludes and N. Telmon. 2006. Detection and quantification of the age-related point mutation A189G in the human mitochondrial DNA. J Forensic Sci 51(4): 865–73. doi: 10.1111/j.1556-4029.2006.00163.x.

Thompson, L. V. 2006. Oxidative stress, mitochondria and mtDNA-mutator mice. Exp Gerontol 41(12): 1220–2. doi: 10.1016/j.exger.2006.10.018.

Trifunovic, A., A. Wredenberg, M. Falkenberg, J. N. Spelbrink, A. T. Rovio, C. E. Bruder et al. 2004. Premature ageing in mice expressing defective mitochondrial DNA polymerase. Nature 429(6990): 417–23. doi: 10.1038/nature02517.

Viña, J., C. Borras, K. M. Abdelaziz, R. Garcia-Valles and M. C. Gomez-Cabrera. 2013. The free radical theory of aging revisited: the cell signaling disruption theory of aging. Antioxid Redox Signal 19(8): 779–87. doi: 10.1089/ars.2012.5111.

Wallace, D. C. 2005. A mitochondrial paradigm of metabolic and degenerative diseases, aging, and cancer: a dawn for evolutionary medicine. Annu Rev Genet 39: 359–407. doi: 10.1146/annurev.genet.39.110304.095751.

Wallace, D. C. 2010. Mitochondrial DNA mutations in disease and aging. Environ Mol Mutagen 51(5): 440–50. doi: 10.1002/em.20586.

Wang, Y., Y. Michikawa, C. Mallidis, Y. Bai, L. Woodhouse, K. E. Yarasheski et al. 2001. Muscle-specific mutations accumulate with aging in critical human mtDNA control sites for replication. Proc Natl Acad Sci U S A 98(7): 4022–7. doi: 10.1073/pnas.061013598.

Wei, Y. H. and H. C. Lee. 2002. Oxidative stress, mitochondrial DNA mutation, and impairment of antioxidant enzymes in aging. Exp Biol Med (Maywood) 227(9): 671–82.

Wolf, F. I., S. Fasanella, B. Tedesco, G. Cavallini, A. Donati, E. Bergamini et al. 2005. Peripheral lymphocyte 8-OHdG levels correlate with age-associated increase of tissue oxidative DNA damage in Sprague-Dawley rats. Protective effects of caloric restriction. Exp Gerontol 40(3): 181–8. doi: 10.1016/j.exger.2004.11.002.

Zhang, C., M. Bills, A. Quigley, R. J. Maxwell, A. W. Linnane and P. Nagley. 1997. Varied prevalence of age-associated mitochondrial DNA deletions in different species and tissues: a comparison between human and rat. Biochem Biophys Res Commun 230(3): 630–5. doi: 10.1006/bbrc.1996.6020.

Zheng, Y., X. Luo, J. Zhu, X. Zhang, Y. Zhu, H. Cheng et al. 2012. Mitochondrial DNA 4977 bp deletion is a common phenomenon in hair and increases with age. Bosn J Basic Med Sci 12(3): 187–92.

19

Mitochondrial DNA Mutations and Mitochondrial Diseases

María Elena Gómez-Gómez

INTRODUCTION

Mitochondrial diseases are not uncommon as a whole, it is estimated that they affect at least one in every 5,000 individuals (Horvath et al. 2008). From a genetic standpoint, primary mitochondrial diseases can be classified into two major categories, depending on which genome, mitochondrial or nuclear, carries the responsible mutations.

mitochondrial DNA (mtDNA) mutations include point mutations, either homo- or heteroplasmic, and (invariably heteroplasmic) large-scale rearrangements. Heteroplasmic point mutations have been found in all mitochondrial genes and lead to different clinical phenotypes, including some canonical syndromes such as MELAS, MERRF, NARP and LS. The main disease entity associated with homoplasmic mtDNA mutations is LHON. Rearrangements (single deletions or duplications) of mtDNA are responsible for sporadic PEO, KSS and Pearson's sindrome (Viscomi et al. 2015).

Nuclear mutations have been found in a huge number of genes directly or indirectly related to the respiratory chain encoding, for instance, (i) proteins involved in mtDNA maintenance and/or replication machinery; (ii) structural subunits of the respiratory chain complexes; (iii) assembly factors of the respiratory complexes; (iv) components of the translation apparatus; and (v) proteins of the execution pathways, such as fission/fusion and apoptosis (Viscomi et al. 2015).

The clinical syndromes associated with mtDNA mutations are extremely variable and patients can present at any stage in life. On the whole, the age of onset reflects the level of mutation and the severity of the biochemical defect, but other factors (presumably nuclear genetic or environmental) also affect the expression of disease.

BR Salud Hospital Infanta Sofía-Laboratorio Central 28702 S.S. de los Reyes, Madrid, Spain.
Email: Sealen2000@hotmail.com

For the purposes of this review, we will concentrate solely on primary mtDNA diseases (Tuppen et al. 2010).

mtDNA Mutations and Diseases

Point mutations

mtDNA point mutations may occur within protein, tRNA or rRNA genes. However, more than half of disease-related point mutations reported are located within mt-tRNA genes. Point mutations in mitochondrial protein-coding genes specifically affect the function of the respiratory chain complex to which the corresponding protein belongs, whereas mt-tRNA mutations may impair overall mitochondrial translation by reducing the availability of functional mt-tRNAs (Tuppen et al. 2010).

Due to their heteroplasmic nature, point mutations display considerable clinical heterogeneity and are highly recessive. There is an increase in the number of pathogenic homoplasmic mutations, affecting just a single tissue and characterised by incomplete penetrance. Given the high mutational rate of the mitochondrial genome and the presence of numerous family- or population-specific polymorphisms, the distinction between neutral mtDNA variants and disease-causing mutation can often be difficult. Also, it has been frequently recognised that certain nucleotide changes that are not pathogenic per se may modulate the effects of deleterious mtDNA mutations (McFarland et al. 2002, McFarland et al. 2008, McFarland et al. 2007, McFarland et al. 2004, Taylor et al. 2003, Temperley et al. 2003, Yang et al. 2009, Swalwell et al. 2008, Cai et al. 2008).

MELAS syndrome (mitochondrial encephalopathy lactic acidosis, and stroke like episodes)

These patients often present the following symptoms: stroke-like episodes, causing subacute brain dysfunction and changes in brain structure, lactic acidosis and/or RRFS. Seizures are frequent in these patients associated with the episode or as an isolated phenomena. Other symptoms include intermittent episodes of encephalopathy, headaches, deafness, dementia and diabetes. Over 80% of patients with MELAS have the 3243A > G mutation in the *MT-TL1* gene. Other mutations in this mt-tRNA gene (e.g., 3271T > C), other mt-tRNA genes (e.g., 1642G > A, *MT-TV* gene) and protein-encoding genes (e.g., 9957T > C in the *MT-CO3* gene, several *MT-ND5* mutations (12770A > G, 13045A > C, 13513G > A and 13514A > G), and *MT-ND1* mutations) may also cause MELAS (Goto et al. 1991, 1990, Taylor et al. 1996, Manfredi et al. 1995, Santorelli et al. 1997, Shanske et al. 2008, Corona et al. 2001, Liolitsa et al. 2003, Kirby et al. 2004).

MERRF syndrome (myoclonus epilepsy and ragged-red fibers)

MERRF syndrome is a progressive, neurodegenerative disease which often presents in childhood or early adulthood following normal development. MERRF syndrome is characterized by myoclonic epilepsy, cerebellar ataxia and myopathy. In some cases it can be accompanied by dementia, deafness, optic atrophy, peripheral neuropathy, retinopathy and ophthalmoparesis. It is caused most commonly by a point mutation in the *MT-TK* gene, 8344A > G. Other mutations, such as 8356 T > C and 8363G > A are associated with this syndrome. The 8344A > G mutation always occurs in heteroplasmy and requires 95% of the mutated genomes to manifest the pathology. This mutation is also associated with other clinical phenotypes such as Leigh syndrome, myopathy, CPEO and MSL (Shoffner et al. 1990, Yoneda et al. 1990, Silvestri et al. 1993).

NARP (neurogenic muscle weakness, ataxia, and retinitis pigmentosa)

NARP is a multisystem disorder, characterized by neurogenic muscle weakness, developmental delay, sensory neuropathy, seizures, ataxia, dementia and pigmentary retinopathy. This combination of symptoms has been described in several families and is usually due to the *MT-ATP6* m.8993T > G mutation. It has been subsequently recognised that patients with a mutant load greater than 95% m.8993T > G have an onset in childhood with MILS (Tatuch et al. 1992, Tuppen et al. 2010).

LHON (leber hereditary optic neuropathy)

LHON is predominantly an organ-specific disease, targeting the retinal ganglion cells of the optic nerve. It is characterized by acute or subacute bilateral blindness due to a trophy of the optic nerve, which is manifested in the second or third decade of life and is more prevalent in men than in women, a fact that suggests the role of a nuclear modulated factor encoded in X chromosome (Vilkki et al. 1991). There are also cases with cardiac involvement, peripheral neuropathy and ataxia cerebellum.

There are three primary LHON mtDNA mutations (11778G > A, 3460G > A, and 14484T > C), which in total are present in at least 95% of LHON cases. LHON is usually due to a homoplasmic mtDNA mutation and all maternal offsprings will inherit the mutation; however, whilst 50% of males will be affected, only 10% of females will develop vision loss. The progression of the disease seems to depend upon the responsible mutation, with 71% of patients with the 14484T > C mutation showing some recovery, compared to 25% with 11778G > A. Whilst the majority of patients with 11778A > G who have symptoms develop LHON, there have been a few reports of other neurodegenerative phenotypes including early-onset dystonia (Man et al. 2002, Man et al. 2003, Harding et al. 1995).

Leigh syndrome maternally inherited

Leigh syndrome is a very heterogeneous disease characterized by multisystem degenerative disorders appearing in the first year of life. Leigh syndrome is characterized by devastating brain stem and basal ganglia dysfunction, demyelination, psychomotor regression, delayed development, seizures, ataxia and peripheral neuropathy. Leigh syndrome is due to severe failure of oxidative metabolism and can be due to a variety of different genetic defects affecting either the mitochondrial (e.g., 8993T > C/G, 10158T > C, 10191T > C) or nuclear genome (e.g., *SURF1* gene) (Finsterer 2008).

Rearrangement mutations

The majority of mtDNA rearrangement mutations are large-scale deletions, varying in size from 1.3 to 8 kb and span several genes. Single mtDNA deletions occur sporadically early in development and the identical deletion is present in all cells within affected tissues. These deletions may be due to inherited mutations in nuclear genes, whose products are involved in mtDNA maintenance and replication (e.g., POLG and PEO1 encoding Twinkle) and mitochondrial nucleotide metabolism (e.g., SLC25A4) (Schon et al. 1989, Chen et al. 1995, Hudson and Chinnery 2006, Kaukonen et al. 2000, Spelbrink et al. 2001). These deletions occur between the origins of replication of heavy (H) and light (L) strands, OH and OL, and are typically flanked by short direct repeats. Their mechanism of generation is also thought to be identical; although some researchers consider replication to be the most likely mechanism of deletion formation, others recently proposed that mtDNA deletions arise during the repair of damaged mtDNA (Bua et al. 2006, Samuels et al. 2004, Reeve et al. 2008, Shoffner et al. 1989, Krishnan et al. 2008).

The amount and tissue distribution of the deleted mtDNA are the most important factors in determining clinical symptoms (Moraes et al. 1995, Zeviani et al. 1988, Vielhaber et al. 2000).

The diseases induced by these mutations include PEO, KSS and PS.

PEO (Progressive external ophthalmoplegia)

PEO is most common in adults. It is characterized by ophthalmoplegia, ptosis of eyelids and myopathy. Ptosis is frequently the presenting symptom and may be asymmetrical. However, patients usually progress to bilateral disease. There may be other features depending on the underlying genetic defect but myopathy and fatigue are common in all patients (Moraes et al. 1989, Van Goethem et al. 2003). Muscle biopsy usually has RRFS and COX negative fibers (not exhibit activity of complex IV of OXPHOS system) (Sundaram et al. 2011). In general PEO is a benign disease with no family history and is typically caused by sporadic large-scale single deletions or multiple mtDNA deletions, although mtDNA point mutations are detected in some patients (e.g., 3243A > G, 12316G > A) (Moraes et al. 1993, Cardaioli et al. 2008).

KSS (Kearns-Sayre syndrome)

KSS is a multisystemic progressive disease occurring before the age of twenty and it is characterized by progressive external ophthalmoplegia and pigmentary retinopathy, usually detected as a "salt and pepper" retinopathy of the posterior fundus without the visual field defects. Patients often develop other neurological complications including cerebellar ataxia, cognitive impairment and deafness, as well as non-neurological features of cardiomyopathy, complete heart block, deafness, short stature, endocrinopathies and dysphagia (Maceluch and Niedziela 2006). Protein levels in cerebrovascular fluid are over 100 mg/dl and muscular biopsy presents RRFS and COX negative fibers. The disease is associated with large deletions in mtDNA of sporadic origin and patients rarely exceed 40 years of age (Filosto et al. 2007).

PS (Pearson syndrome)

Pearson Syndrome is a rare disorder of infancy that affects hematopoiesis and exocrine pancreatic function. The most common clinical features are usually sideroblastic anemia with vacuolization of bone marrow precursors, associated with exocrine pancreatic dysfunction. In addition, they tend to develop progressive liver failure, intestinal villus atrophy, diabetes mellitus and renal tubular dysfunction (Leonard and Schapira 2000). The clinical course in these children can be severe leading to early death. In those that survive, the blood disorder improves but they later develop the clinical features of Kearns-Sayre phenotype. In these children, there is a very high level of large-scale single mtDNA deletion present in all tissues, which are generally sporadic, but there are some cases of maternal inheritance (Rotig et al. 1990).

Nuclear DNA Defects and Diseases

Since the specific cause of more than 50% of patients diagnosed with mitochondrial disease is unknown, probably most of these diseases are due to mutations in nuclear genes, encoding proteins forming the 1500 mitochondrial proteome. Only half of these genes are known, which implies a great difficulty in properly diagnosing mitochondrial diseases of nuclear origin (Calvo et al. 2006).

Mutations in nuclear genes directly affecting the respiratory chain can be classified into three main groups:

- Mutations in proteins involved in mtDNA maintenance and/or replication machinery.
- Mutations in structural subunits of the respiratory chain complex.
- Mutations in genes encoding assembly factors of the respiratory complexes.

Apart from these three groups, there are those mutations that indirectly affect the respiratory chain.

Mutations in proteins involved in mtDNA maintenance and/or replication machinery

The mtDNA depends on numerous proteins encoded in nuclear DNA, for its successful replication, repair and stabilization. Mutations in these proteins can produce quantitative damage (depletion) or qualitative damage (deletions) in the mtDNA.

Among the diseases associated with a decrease in the amount of mtDNA, mitochondrial depletion syndrome stands out because of its severity. This syndrome is usually presented in two different clinical models: a hepatocerebral form with a neonatal debut and a degree of severe mtDNA depletion, and a myopathic form, with a later debut and a lower degree of depletion. The hepatocerebral form is associated in almost all cases with the gene encoding DGUOK mutations, while the myopathic form is associated with the gene encoding TK2 mutations (Saada et al. 2001).

Mutations in the gene encoding TP cause Mitochondrial Neurogastrointestinal Encephalopathy syndrome (MNGIE) normally associated with a less severe mtDNA depletion that is observed in the depletion syndrome (Nishino et al. 1999). These three genes are involved in nucleotide metabolism.

Other genes involved in mitochondrial depletion are MPV17 (Spinazzola et al. 2006), which encodes a protein located in the inner mitochondrial membrane of unknown function, SUCLA2 (Elpeleg et al. 2005), RRM2B that encodes ribonucleotide reductase controlled by p53 (p53R2) (Bourdon et al. 2007) and SUCLG1 encoding the alpha subunit of the succinate-coenzyme A ligase (SUCL) (Ostergaard et al. 2007).

The gene encoding POLG, deserves particular attention. About 50 pathogenic mutations have been described in this gene associated with a wide number of mitochondrial diseases and accompanied by both depletion as well as deletions of mtDNA (Longley et al. 2005).

The first described mutations were in families with external progressive ophthalmoplegia in its adPEO associated with multiple deletions in mtDNA (Van Goethem et al. 2001). The disease has also been associated with C10orf2 mutations in genes encoding for mitochondrial helicase (Spelbrink et al. 2001) and ANT1 (Kaukonen et al. 2000).

All mutations in the POLG gene associated with autosomal dominant progressive external ophthalmoplegia (adPEO) affect the domain of the polymerase protein. Biochemical characterization studies based on recombinant proteins carrying mutations in this domain, show a correlation between the severity of the phenotype observed in patients and polymerase decreasing activity and processivity caused by mutations (Graziewicz et al. 2004). Heterozygous composite mutations in patients with sporadic forms and autosomal recessive progressive external ophthalmoplegia (arPEO) have also been described. These mutations often affect the exonuclease domain or the spacer region of the protein.

Biochemical studies on recombinant proteins carrying mutations in the spacer region show a drastic decrease in catalytic effectiveness as well as the ability to interact with subunit accessory (Chan et al. 2005).

In recent years, the clinical spectrum caused by mutations in the POLG gene has increased considerably.

A high incidence of psychiatric disease, Parkinson syndrome and primary gonadal failure, causing early menopause in families with dominant mutations in the POLG gene has been recorded (Luoma et al. 2004, Pagnamenta et al. 2006).

Other conditions that have been associated with recessive mutations in the gene POLG are MIRAS (Winterthun et al. 2005) and Parkinson's disease associated with peripheral neuropathy (Davidzon et al. 2006). In all cases, the disease is accompanied by the presence of multiple deletions in mtDNA.

Additionally, recessive mutations in the gene POLG are a major cause of Alpers-Huttenlocher syndrome, a form of severe hepato encephalopathy in children, accompanied by mtDNA depletion (Naviaux and Nguyen 2004).

Thus, there is great interest in the study of this gene, given the increasingly greater number of mutations described and its association with a wide range of human pathologies.

Mutations in structural subunits of the respiratory chain complex

Although 72 of the 85 subunits that form the respiratory chain complexes are encoded in the nucleus, mutations in such genes are rare. This might be a reflection of the highly deleterious nature of these mutations, which would result in lethality during embryogenesis. To date, mutations have been described in subunits of complex I (NDUFS 1, 2, 3, 4, 7 and 8, NDUFV1 and 2, and NDUFA1 and 11), associated with Leigh syndrome, encephalomyopathy, leukodystrophy and the four subunits of complex II (SDHA, B, C and D) associated with Leigh syndrome and ataxia in the case of the A subunit, and more rarely paraganglioma and feocromocitoma in the case of subunits B, C, and D. Mutations have also been described in complex III subunits UQCRB and UQCRQ associated with hypoglycemia, lactic acidosis and severe psychomotor delay with extra pyramidal signs, as well as a mutation in the complex IV (COX6B1) associated with childhood encephalomyopathy (Zhu et al. 2009).

Mutations in genes encoding assembly factors of the respiratory complexes

There are diseases caused by defects in complex I, III, IV and V by mutations in nuclear genes encoding proteins involved in the correct assembly of these complexes.

A clear example is the COX deficiencies where mutations are described in numerous genes involved in the assembly and stability, as SCO1, SCO2, SURF1, COX10, COX15 and LRPPRC.

BSCL1 gene mutations have also been described, involved in assembly of complex III, and ATP12 involved in complex assembly V (Schapira 2006), as well as mutation in an assembly gene of complex I (B17.2L) associated with leukoencephalopathy (Ogilvie et al. 2005).

Mutations in genes that indirectly alter the function OXPHOS

Within this group are found mutations in a growing number of genes encoding proteins essential for different processes in mitochondrial physiology. For example, those affecting the composition of lipid membrane, such as mutations in the gene G4.5, encoding a group of proteins, tafazzines, altering the cardiolipin biosynthesis and cause Barth syndrome consisting of heart disease, myopathy and cyclic neutropenia. Mutations in genes involved in processes of mitochondrial fusion are also included, such as mutations in MFN2 causing CMT 2A subtype (Zuchner et al. 2004) or mutations in OPA1 causing the autosomal DOA and mutations in GDAP1 associated with CMT4A subtype (Niemann et al. 2005).

mtDNA Mutations and Cancer

Cancer cells produce ATP through glycolysis and lactic acid fermentation rather than oxidative phosphorylation. For that reason, it is suggested that mitochondria are involved in carcinogenesis through respiration alterations. Indeed, an increase in mtDNA mutations has been observed in a variety of cancer types: prostate, thyroid, oral cancer, vulvar cancer, hepatocellular carcinoma, colon, bladder, head and neck, lung and a number of blood cancers. Mutations are often found in primary tumors, but not in surrounding tissues. The main characteristics of mtDNA mutations, common to all tumor types, are: the majority of the mutations are base substitutions; mutations occur in all protein-coding mitochondrial genes; the D-loop region is the most frequent site of somatic mutations across tumor types; and the mutations are homoplasmic in nature (Taylor and Turnbull 2005).

Deletions, point mutations, insertions and duplications are reported in many kinds of cancer. The homoplasmic nature of mutated mtDNA suggests the possibility that some mutations are involved in tumorigenesis by affecting energy metabolism and/or ROS production. For example, some studies have observed decreased nuclear and mitochondrial hOGG1 expression in human lung cancer. Moreover, mitochondria play a key role in apoptosis. Another study showed that specific point mutations in mtDNA accelerate growth and reduce apoptosis in a variety of tumors, supporting the idea that some mtDNA mutations in tumors have functional advantages that promote tumor growth (Taylor and Turnbull 2005).

Breast cancer

Several studies have examined the presence of mtDNA mutations in breast cancer. Tan's studies (Tan et al. 2002) reported somatic mutations in 74% of the patients, with the majority of the mutations (81.5%) restricted to the control regions of the mtDNA. The remaining 18.5% of the mutations were detected in the 16S rRNA, ND2 and ATPase 6 genes. Parrella et al. (Parrella et al. 2001) found mtDNA mutations in 61% of the fine needle aspirates from primary breast tumors. In that study, it was reported that 42% of the mutations were present in the control region. The most common mtDNA mutations detected in breast cancer have been largely single base substitutions or insertions, a large deletion of 4977 bp has been detected in both the malignant and paired normal breast tissues of patients with breast cancer (Bianchi et al. 1995).

Colorectal cancer

Studies by Polyak et al. (Polyak et al. 1998) analysed 70% of mtDNA sequence in primary tumor samples. The mutations were found in regions encoding ND1, ND4L, ND5, Cytochrome b, COXI, COXII and COXIII as well as in the 12S and 16S rRNA genes. These mutations were also found in cell lines. Alonso et al. (Alonso et al. 1997) described somatic mutations in 23% of colorectal patients in the D-loop region, and mainly found deletions and insertions. Lately, Habano et al. (Habano et al. 1999, Habano et al. 1998) found mutations in the same region in the 44% of sporadic human colorectal carcinomas.

Ovarian cancer

According to the study of Liu et al. (Liu et al. 2001), 20% of samples from primary ovarian carcinomas presented single or multiple somatic mtDNA mutations. The four regions of the mitochondrial genome primarily affected by these mutations were the

D-loop, 12S rRNA, 16S rRNA and cytochrome b, suggesting that these regions may be mutational hotspots in ovarian cancer.

Gastric carcinoma

The study of Zhao et al. (Zhao et al. 2005) revealed 18 gene mutations in D-loop region of gastric carcinoma tissue. These results were confirmed by Wu et al. (Wu et al. 2005), finding somatic mutations in this region in 48% of samples. Also, mutations have been found in the 12S rRNA region in gastric tissues, with a higher frequency of these changes in the intestinal type than diffuse type of gastric carcinoma (Han et al. 2005). The analysis of gastric adenocarcinomas by Burgart et al. (1995) found a deletion in the D-loop region, present only in adenocarcinomas arising from the gastroesophageal junction, and absent in the distal tumors.

Hepatocellular cancer

The studies of Nomoto et al. (Nomoto et al. 2002) identified mutations in the D-loop region as a frequent event in this type of tumor. Besides, they found C-tract deletion/ insertion mutations in 42% of cases and five other missense and deletion/insertion mutations in 26% of cases. Another study confirmed a high frequency in mutations in the D-loop region in HCC patients, correlating this frequency with the degree of malignancy (Nishikawa et al. 2001). Other studies have found D-loop mutations in tumor and matched normal tissue from patients with HBV infection (Wheelhouse et al. 2005), with the higher frequency in subjects with HCC. Tamori et al. (Tamori et al. 2004) identified 12 specific mutation sites in the D-loop for HCC.

Pancreatic cancer

After the sequencing of pancreatic cancer cell lines and ductal adenocarcinoma xenografts, several homoplasmic mutations were found: rRNA genes, NADH dehydrogenase genes coding for complex I proteins (mtND1, mtND2, mtND3, mtND4, mtND4L and mtND5). Mutations were also found in complex III, mtCytB gene, complex IV mtCOX1, mtCOX2, and mtCOX3 genes and complex V mtATP6 and mtATP8 genes and the D-loop regulatory region (Jones et al. 2001).

Prostate cancer

Jessie et al. (Jessie et al. 2001) identified mtDNA deletions in prostate cancer. The same year another study (Jeronimo et al. 2001) found several mtDNA mutations in early stage prostate cancer. More recently, Petros et al. (Petros et al. 2005) showed that 11–12% of all prostate cancer patients harbored COX I mutations that altered conserved amino acids.

Lung cancer

One study (Sanchez-Cespedes et al. 2001) identified mutations in primary lung cancers, mainly transversions and transitions, although some tumors showed deletions or insertions in the C-tract. A later study (Suzuki et al. 2003) confirmed these base substitutions and D-loop changes in primary tumor cases of NSCLC.

Renal cell carcinoma

A 2003 study (Nagy et al. 2003) of end-stage renal disease found 94 mtDNA polymorphisms. In another study from the same group (Nagy et al. 2002) involving eight chromophobe RCCs found 28% of the sequence variants in the D-loop region. Previously deletions of mtDNA and the mRNA coding for NADH dehydrogenase subunit 3 were reported (Selvanayagam and Rajaraman 1996). The presence of deletions was confirmed by (Horton et al. 1996). Two studies (Simonnet et al. 2003, Simonnet et al. 2002) demonstrated a decrease in protein content and Electron Transport Chain enzyme activity of this type of carcinoma.

Thyroid cancer

Stefaneanu and Tasca (Stefaneanu and Tasca 1979) demonstrated that thyroid tumors contain abnormally high numbers of mitochondria. Another study (Abu-Amero et al. 2005) found mutations in complex I and severe defects in its activity in primary papillary thyroid carcinomas. Further studies (Maximo et al. 2002, Tong et al. 2003) demonstrated the presence of somatic mutations in these tumors. In patients exposed to Chernobyl radiation, high mtDNA content in these tumors and large-scale deletions were reported (Rogounovitch et al. 2002).

Brain tumors

An increase in the copy number of mtDNA in low-grade tumors and gliomas has been described (Liang and Hays 1996, Liang 1996). Although alterations in D-loop region were found in 36% of malignant gliomas, there was no correlation with aggressiveness (Montanini et al. 2005). The study by Wong et al. (Wong et al. 2003) detected mtDNA mutations in medulloblastoma.

Conclusions

As demonstrated above, mtDNA mutations are an important cause of genetic disease that must be considered. The clinical variability of these disorders and the heterogeneity of the mitochondrial genome makes the recognition of patients with mtDNA disease a real challenge.

The implementation of more high-throughput screening approaches should hopefully improve the situation, considering the continually increasing number of reported pathogenic mutations within both the mitochondrial and nuclear genomes. Several experimental approaches for mtDNA disease diagnosis and treatment are currently being investigated and may hopefully make their way into the clinic in the near future.

Furthermore, mitochondrial DNA mutations have been observed in many types of human cancer with clinical phenotypic variability, which it will require future studies and new approaches for early detection.

Keywords: Mitochondrial diseases, mtDNA mutations, MELAS syndrome, MERRF syndrome, neurogenic muscle weakness ataxia and retinitis pigmentosa, leber hereditary optic neuropathy, leigh syndrome maternally inherited, progressive external opthalmoplegia, Kearns-Sayre syndrome, Pearson syndrome, mtDNA mutations and cancer

References

Abu-Amero, K. K., A. S. Alzahrani, M. Zou and Y. Shi. 2005. High frequency of somatic mitochondrial DNA mutations in human thyroid carcinomas and complex I respiratory defect in thyroid cancer cell lines. Oncogene 24(8): 1455–60. doi: 10.1038/sj.onc.1208292.

Alonso, A., P. Martin, C. Albarran, B. Aquilera, O. Garcia, A. Guzman et al. 1997. Detection of somatic mutations in the mitochondrial DNA control region of colorectal and gastric tumors by heteroduplex and single-strand conformation analysis. Electrophoresis 18(5): 682–5. doi: 10.1002/elps.1150180504.

Bianchi, M. S., N. O. Bianchi and G. Bailliet. 1995. Mitochondrial DNA mutations in normal and tumor tissues from breast cancer patients. Cytogenet Cell Genet 71(1): 99–103.

Bourdon, A., L. Minai, V. Serre, J. P. Jais, E. Sarzi, S. Aubert et al. 2007. Mutation of RRM2B, encoding p53-controlled ribonucleotide reductase (p53R2), causes severe mitochondrial DNA depletion. Nat Genet 39(6): 776–80. doi: 10.1038/ng2040.

Bua, E., J. Johnson, A. Herbst, B. Delong, D. McKenzie, S. Salamat et al. 2006. Mitochondrial DNA-deletion mutations accumulate intracellularly to detrimental levels in aged human skeletal muscle fibers. Am J Hum Genet 79(3): 469–80. doi: 10.1086/507132.

Burgart, L. J., J. Zheng, Q. Shu, J. G. Strickler and D. Shibata. 1995. Somatic mitochondrial mutation in gastric cancer. Am J Pathol 147(4): 1105–11.

Cai, W., Q. Fu, X. Zhou, J. Qu, Y. Tong and M. X. Guan. 2008. Mitochondrial variants may influence the phenotypic manifestation of Leber's hereditary optic neuropathy-associated ND4 G11778A mutation. J Genet Genomics 35(11): 649–55. doi: 10.1016/S1673-8527(08)60086-7.

Calvo, S., M. Jain, X. Xie, S. A. Sheth, B. Chang, O. A. Goldberger et al. 2006. Systematic identification of human mitochondrial disease genes through integrative genomics. Nat Genet 38(5): 576–82. doi: 10.1038/ng1776.

Cardaioli, E., P. Da Pozzo, E. Malfatti, G. N. Gallus, A. Rubegni, A. Malandrini et al. 2008. Chronic progressive external ophthalmoplegia: a new heteroplasmic tRNA(Leu(CUN)) mutation of mitochondrial DNA. J Neurol Sci 272(1-2): 106–9. doi: 10.1016/j.jns.2008.05.005.

Chan, S. S., M. J. Longley and W. C. Copeland. 2005. The common A467T mutation in the human mitochondrial DNA polymerase (POLG) compromises catalytic efficiency and interaction with the accessory subunit. J Biol Chem 280(36): 31341–6. doi: 10.1074/jbc.M506762200.

Chen, X., R. Prosser, S. Simonetti, J. Sadlock, G. Jagiello and E. A. Schon. 1995. Rearranged mitochondrial genomes are present in human oocytes. Am J Hum Genet 57(2): 239–47.

Corona, P., C. Antozzi, F. Carrara, L. D'Incerti, E. Lamantea, V. Tiranti et al. 2001. A novel mtDNA mutation in the ND5 subunit of complex I in two MELAS patients. Ann Neurol 49(1): 106–10.

Davidzon, G., P. Greene, M. Mancuso, K. J. Klos, J. E. Ahlskog, M. Hirano et al. 2006. Early-onset familial parkinsonism due to POLG mutations. Ann Neurol 59(5): 859–62. doi: 10.1002/ana.20831.

Elpeleg, O., C. Miller, E. Hershkovitz, M. Bitner-Glindzicz, G. Bondi-Rubinstein, S. Rahman et al. 2005. Deficiency of the ADP-forming succinyl-CoA synthase activity is associated with encephalomyopathy and mitochondrial DNA depletion. Am J Hum Genet 76(6): 1081–6. doi: 10.1086/430843.

Filosto, M., G. Tomelleri, P. Tonin, M. Scarpelli, G. Vattemi, N. Rizzuto et al. 2007. Neuropathology of mitochondrial diseases. Biosci Rep 27(1-3): 23–30. doi: 10.1007/s10540-007-9034-3.

Finsterer, J. 2008. Leigh and Leigh-like syndrome in children and adults. Pediatr Neurol 39(4): 223–35. doi: 10.1016/j.pediatrneurol.2008.07.013.

Goto, Y., I. Nonaka and S. Horai. 1990. A mutation in the tRNA(Leu)(UUR) gene associated with the MELAS subgroup of mitochondrial encephalomyopathies. Nature 348(6302): 651–3. doi: 10.1038/348651a0.

Goto, Y., I. Nonaka and S. Horai. 1991. A new mtDNA mutation associated with mitochondrial myopathy, encephalopathy, lactic acidosis and stroke-like episodes (MELAS). Biochim Biophys Acta 1097(3): 238–40.

Graziewicz, M. A., M. J. Longley, R. J. Bienstock, M. Zeviani and W. C. Copeland. 2004. Structure-function defects of human mitochondrial DNA polymerase in autosomal dominant progressive external ophthalmoplegia. Nat Struct Mol Biol 11(8): 770–6. doi: 10.1038/nsmb805.

Habano, W., S. Nakamura and T. Sugai. 1998. Microsatellite instability in the mitochondrial DNA of colorectal carcinomas: evidence for mismatch repair systems in mitochondrial genome. Oncogene 17(15): 1931–7. doi: 10.1038/sj.onc.1202112.

Habano, W., T. Sugai, T. Yoshida and S. Nakamura. 1999. Mitochondrial gene mutation, but not large-scale deletion, is a feature of colorectal carcinomas with mitochondrial microsatellite instability. Int J Cancer 83(5): 625–9.

Han, C. B., J. M. Ma, Y. Xin, X. Y. Mao, Y. J. Zhao, D. Y. Wu et al. 2005. Mutations of mitochondrial 12S rRNA in gastric carcinoma and their significance. World J Gastroenterol 11(1): 31–5.

Harding, A. E., M. G. Sweeney, G. G. Govan and P. Riordan-Eva. 1995. Pedigree analysis in Leber hereditary optic neuropathy families with a pathogenic mtDNA mutation. Am J Hum Genet 57(1): 77–86.

Horton, T. M., J. A. Petros, A. Heddi, J. Shoffner, A. E. Kaufman, S. D. Graham, Jr. et al. 1996. Novel mitochondrial DNA deletion found in a renal cell carcinoma. Genes Chromosomes Cancer 15(2): 95–101. doi: 10.1002/(SICI)1098-2264(199602)15:2<95::AID-GCC3>3.0.CO;2-Z.

Horvath, R., G. Gorman and P. F. Chinnery. 2008. How can we treat mitochondrial encephalomyopathies? Approaches to therapy. Neurotherapeutics 5(4): 558–68. doi: 10.1016/j.nurt.2008.07.002.

Hudson, G. and P. F. Chinnery. 2006. Mitochondrial DNA polymerase-gamma and human disease. Hum Mol Genet 15 Spec No 2: R244-52. doi: 10.1093/hmg/ddl233.

Jeronimo, C., S. Nomoto, O. L. Caballero, H. Usadel, R. Henrique, G. Varzim et al. 2001. Mitochondrial mutations in early stage prostate cancer and bodily fluids. Oncogene 20(37): 5195–8. doi: 10.1038/sj.onc.1204646.

Jessie, B. C., C. Q. Sun, H. R. Irons, F. F. Marshall, D. C. Wallace and J. A. Petros. 2001. Accumulation of mitochondrial DNA deletions in the malignant prostate of patients of different ages. Exp Gerontol 37(1): 169–74.

Jones, J. B., J. J. Song, P. M. Hempen, G. Parmigiani, R. H. Hruban and S. E. Kern. 2001. Detection of mitochondrial DNA mutations in pancreatic cancer offers a "mass"-ive advantage over detection of nuclear DNA mutations. Cancer Res 61(4): 1299–304.

Kaukonen, J., J. K. Juselius, V. Tiranti, A. Kyttala, M. Zeviani, G. P. Comi et al. 2000. Role of adenine nucleotide translocator 1 in mtDNA maintenance. Science 289(5480): 782–5.

Kirby, D. M., R. McFarland, A. Ohtake, C. Dunning, M. T. Ryan, C. Wilson et al. 2004. Mutations of the mitochondrial ND1 gene as a cause of MELAS. J Med Genet 41(10): 784–9. doi: 10.1136/jmg.2004.020537.

Krishnan, K. J., A. K. Reeve, D. C. Samuels, P. F. Chinnery, J. K. Blackwood, R. W. Taylor et al. 2008. What causes mitochondrial DNA deletions in human cells? Nat Genet 40(3): 275–9. doi: 10.1038/ng.f.94.

Leonard, J. V. and A. H. Schapira. 2000. Mitochondrial respiratory chain disorders I: mitochondrial DNA defects. Lancet 355(9200): 299–304.

Liang, B. C. 1996. Evidence for association of mitochondrial DNA sequence amplification and nuclear localization in human low-grade gliomas. Mutat Res 354(1): 27–33.

Liang, B. C. and L. Hays. 1996. Mitochondrial DNA copy number changes in human gliomas. Cancer Lett 105(2): 167–73.

Liolitsa, D., S. Rahman, S. Benton, L. J. Carr and M. G. Hanna. 2003. Is the mitochondrial complex I ND5 gene a hot-spot for MELAS causing mutations? Ann Neurol 53(1): 128–32. doi: 10.1002/ana.10435.

Liu, V. W., H. H. Shi, A. N. Cheung, P. M. Chiu, T. W. Leung, P. Nagley et al. 2001. High incidence of somatic mitochondrial DNA mutations in human ovarian carcinomas. Cancer Res 61(16): 5998–6001.

Longley, M. J., M. A. Graziewicz, R. J. Bienstock and W. C. Copeland. 2005. Consequences of mutations in human DNA polymerase gamma. Gene 354: 125–31. doi: 10.1016/j.gene.2005.03.029.

Luoma, P., A. Melberg, J. O. Rinne, J. A. Kaukonen, N. N. Nupponen, R. M. Chalmers et al. 2004. Parkinsonism, premature menopause, and mitochondrial DNA polymerase gamma mutations: clinical and molecular genetic study. Lancet 364(9437): 875–82. doi: 10.1016/S0140-6736(04)16983-3.

Maceluch, J. A. and M. Niedziela. 2006. The clinical diagnosis and molecular genetics of kearns-sayre syndrome: a complex mitochondrial encephalomyopathy. Pediatr Endocrinol Rev 4(2): 117–37.

Man, P. Y., D. M. Turnbull and P. F. Chinnery. 2002. Leber hereditary optic neuropathy. J Med Genet 39(3): 162–9.

Man, P. Y., P. G. Griffiths, D. T. Brown, N. Howell, D. M. Turnbull and P. F. Chinnery. 2003. The epidemiology of Leber hereditary optic neuropathy in the North East of England. Am J Hum Genet 72(2): 333–9. doi: 10.1086/346066.

Manfredi, G., E. A. Schon, C. T. Moraes, E. Bonilla, G. T. Berry, J. T. Sladky et al. 1995. A new mutation associated with MELAS is located in a mitochondrial DNA polypeptide-coding gene. Neuromuscul Disord 5(5): 391–8.

Maximo, V., P. Soares, J. Lima, J. Cameselle-Teijeiro and M. Sobrinho-Simoes. 2002. Mitochondrial DNA somatic mutations (point mutations and large deletions) and mitochondrial DNA variants in human thyroid pathology: a study with emphasis on Hurthle cell tumors. Am J Pathol 160(5): 1857–65. doi: 10.1016/S0002-9440(10)61132-7.

McFarland, R., K. M. Clark, A. A. Morris, R. W. Taylor, S. Macphail, R. N. Lightowlers et al. 2002. Multiple neonatal deaths due to a homoplasmic mitochondrial DNA mutation. Nat Genet 30(2): 145–6. doi: 10.1038/ng819.

McFarland, R., A. M. Schaefer, J. L. Gardner, S. Lynn, C. M. Hayes, M. J. Barron et al. 2004. Familial myopathy: new insights into the T14709C mitochondrial tRNA mutation. Ann Neurol 55(4): 478–84. doi: 10.1002/ana.20004.

McFarland, R., P. F. Chinnery, E. L. Blakely, A. M. Schaefer, A. A. Morris, S. M. Foster et al. 2007. Homoplasmy, heteroplasmy, and mitochondrial dystonia. Neurology 69(9): 911–6. doi: 10.1212/01.wnl.0000267843.10977.4a.

McFarland, R., H. Swalwell, E. L. Blakely, L. He, E. J. Groen, D. M. Turnbull et al. 2008. The m.5650G > A mitochondrial tRNAAla mutation is pathogenic and causes a phenotype of pure myopathy. Neuromuscul Disord 18(1): 63–7. doi: 10.1016/j.nmd.2007.07.007.

Montanini, L., C. Regna-Gladin, M. Eoli, R. Albarosa, F. Carrara, M. Zeviani et al. 2005. Instability of mitochondrial DNA and MRI and clinical correlations in malignant gliomas. J Neurooncol 74(1): 87–9. doi: 10.1007/s11060-004-4036-5.

Moraes, C. T., S. DiMauro, M. Zeviani, A. Lombes, S. Shanske, A. F. Miranda et al. 1989. Mitochondrial DNA deletions in progressive external ophthalmoplegia and Kearns-Sayre syndrome. N Engl J Med 320(20): 1293–9. doi: 10.1056/NEJM198905183202001.

Moraes, C. T., F. Ciacci, G. Silvestri, S. Shanske, M. Sciacco, M. Hirano et al. 1993. A typical clinical presentations associated with the MELAS mutation at position 3243 of human mitochondrial DNA. Neuromuscul Disord 3(1): 43–50.

Moraes, C. T., M. Sciacco, E. Ricci, C. H. Tengan, H. Hao, E. Bonilla et al. 1995. Phenotype-genotype correlations in skeletal muscle of patients with mtDNA deletions. Muscle Nerve Suppl 3: S150–3.

Nagy, A., M. Wilhelm, F. Sukosd, B. Ljungberg and G. Kovacs. 2002. Somatic mitochondrial DNA mutations in human chromophobe renal cell carcinomas. Genes Chromosomes Cancer 35(3): 256–60. doi: 10.1002/gcc.10118.

Nagy, A., M. Wilhelm and G. Kovacs. 2003. Mutations of mtDNA in renal cell tumours arising in end-stage renal disease. J Pathol 199(2): 237–42. doi: 10.1002/path.1273.

Naviaux, R. K. and K. V. Nguyen. 2004. POLG mutations associated with Alpers' syndrome and mitochondrial DNA depletion. Ann Neurol 55(5): 706–12. doi: 10.1002/ana.20079.

Niemann, A., M. Ruegg, V. La Padula, A. Schenone and U. Suter. 2005. Ganglioside-induced differentiation associated protein 1 is a regulator of the mitochondrial network: new implications for Charcot-Marie-Tooth disease. J Cell Biol 170(7): 1067–78. doi: 10.1083/jcb.200507087.

Nishikawa, M., S. Nishiguchi, S. Shiomi, A. Tamori, N. Koh, T. Takeda et al. 2001. Somatic mutation of mitochondrial DNA in cancerous and noncancerous liver tissue in individuals with hepatocellular carcinoma. Cancer Res 61(5): 1843–5.

Nishino, I., A. Spinazzola and M. Hirano. 1999. Thymidine phosphorylase gene mutations in MNGIE, a human mitochondrial disorder. Science 283(5402): 689–92.

Nomoto, S., K. Yamashita, K. Koshikawa, A. Nakao and D. Sidransky. 2002. Mitochondrial D-loop mutations as clonal markers in multicentric hepatocellular carcinoma and plasma. Clin Cancer Res 8(2): 481–7.

Ogilvie, I., N. G. Kennaway and E. A. Shoubridge. 2005. A molecular chaperone for mitochondrial complex I assembly is mutated in a progressive encephalopathy. J Clin Invest 115(10): 2784–92. doi: 10.1172/JCI26020.

Ostergaard, E., E. Christensen, E. Kristensen, B. Mogensen, M. Duno, E. A. Shoubridge et al. 2007. Deficiency of the alpha subunit of succinate-coenzyme A ligase causes fatal infantile lactic acidosis with mitochondrial DNA depletion. Am J Hum Genet 81(2): 383–7. doi: 10.1086/519222.

Pagnamenta, A. T., J. W. Taanman, C. J. Wilson, N. E. Anderson, R. Marotta, A. J. Duncan et al. 2006. Dominant inheritance of premature ovarian failure associated with mutant mitochondrial DNA polymerase gamma. Hum Reprod 21(10): 2467–73. doi: 10.1093/humrep/del076.

Parrella, P., Y. Xiao, M. Fliss, M. Sanchez-Cespedes, P. Mazzarelli, M. Rinaldi et al. 2001. Detection of mitochondrial DNA mutations in primary breast cancer and fine-needle aspirates. Cancer Res 61(20): 7623–6.

Petros, J. A., A. K. Baumann, E. Ruiz-Pesini, M. B. Amin, C. Q. Sun, J. Hall et al. 2005. mtDNA mutations increase tumorigenicity in prostate cancer. Proc Natl Acad Sci U S A 102(3): 719–24. doi: 10.1073/pnas.0408894102.

Polyak, K., Y. Li, H. Zhu, C. Lengauer, J. K. Willson, S. D. Markowitz et al. 1998. Somatic mutations of the mitochondrial genome in human colorectal tumours. Nat Genet 20(3): 291–3. doi: 10.1038/3108.

Reeve, A. K., K. J. Krishnan, J. L. Elson, C. M. Morris, A. Bender, R. N. Lightowlers et al. 2008. Nature of mitochondrial DNA deletions in substantia nigra neurons. Am J Hum Genet 82(1): 228–35. doi: 10.1016/j.ajhg.2007.09.018.

Rogounovitch, T. I., V. A. Saenko, Y. Shimizu-Yoshida, A. Y. Abrosimov, E. F. Lushnikov, P. O. Roumiantsev et al. 2002. Large deletions in mitochondrial DNA in radiation-associated human thyroid tumors. Cancer Res 62(23): 7031–41.

Rotig, A., V. Cormier, S. Blanche, J. P. Bonnefont, F. Ledeist, N. Romero et al. 1990. Pearson's marrow-pancreas syndrome. A multisystem mitochondrial disorder in infancy. J Clin Invest 86(5): 1601–8. doi: 10.1172/JCI114881.

Saada, A., A. Shaag, H. Mandel, Y. Nevo, S. Eriksson and O. Elpeleg. 2001. Mutant mitochondrial thymidine kinase in mitochondrial DNA depletion myopathy. Nat Genet 29(3): 342–4. doi: 10.1038/ng751.

Samuels, D. C., E. A. Schon and P. F. Chinnery. 2004. Two direct repeats cause most human mtDNA deletions. Trends Genet 20(9): 393–8. doi: 10.1016/j.tig.2004.07.003.

Sanchez-Cespedes, M., P. Parrella, S. Nomoto, D. Cohen, Y. Xiao, M. Esteller et al. 2001. Identification of a mononucleotide repeat as a major target for mitochondrial DNA alterations in human tumors. Cancer Res 61(19): 7015–9.

Santorelli, F. M., K. Tanji, R. Kulikova, S. Shanske, L. Vilarinho, A. P. Hays et al. 1997. Identification of a novel mutation in the mtDNA ND5 gene associated with MELAS. Biochem Biophys Res Commun 238(2): 326–8. doi: 10.1006/bbrc.1997.7167.

Schapira, A. H. 2006. Mitochondrial disease. Lancet 368(9529): 70–82. doi: 10.1016/S0140-6736(06)68970-8.

Schon, E. A., R. Rizzuto, C. T. Moraes, H. Nakase, M. Zeviani and S. DiMauro. 1989. A direct repeat is a hotspot for large-scale deletion of human mitochondrial DNA. Science 244(4902): 346–9.

Selvanayagam, P. and S. Rajaraman. 1996. Detection of mitochondrial genome depletion by a novel cDNA in renal cell carcinoma. Lab Invest 74(3): 592–9.

Shanske, S., J. Coku, J. Lu, J. Ganesh, S. Krishna, K. Tanji et al. 2008. The G13513A mutation in the ND5 gene of mitochondrial DNA as a common cause of MELAS or Leigh syndrome: evidence from 12 cases. Arch Neurol 65(3): 368–72. doi: 10.1001/archneurol.2007.67.

Shoffner, J. M., M. T. Lott, A. S. Voljavec, S. A. Soueidan, D. A. Costigan and D. C. Wallace. 1989. Spontaneous Kearns-Sayre/chronic external ophthalmoplegia plus syndrome associated with a mitochondrial DNA deletion: a slip-replication model and metabolic therapy. Proc Natl Acad Sci U S A 86(20): 7952–6.

Shoffner, J. M., M. T. Lott, A. M. Lezza, P. Seibel, S. W. Ballinger and D. C. Wallace. 1990. Myoclonic epilepsy and ragged-red fiber disease (MERRF) is associated with a mitochondrial DNA tRNA(Lys) mutation. Cell 61(6): 931–7.

Silvestri, G., E. Ciafaloni, F. M. Santorelli, S. Shanske, S. Servidei, W. D. Graf et al. 1993. Clinical features associated with the A-->G transition at nucleotide 8344 of mtDNA ("MERRF mutation"). Neurology 43(6): 1200–6.

Simonnet, H., N. Alazard, K. Pfeiffer, C. Gallou, C. Beroud, J. Demont et al. 2002. Low mitochondrial respiratory chain content correlates with tumor aggressiveness in renal cell carcinoma. Carcinogenesis 23(5): 759–68.

Simonnet, H., J. Demont, K. Pfeiffer, L. Guenaneche, R. Bouvier, U. Brandt et al. 2003. Mitochondrial complex I is deficient in renal oncocytomas. Carcinogenesis 24(9): 1461–6. doi: 10.1093/carcin/bgg109.

Spelbrink, J. N., F. Y. Li, V. Tiranti, K. Nikali, Q. P. Yuan, M. Tariq et al. 2001. Human mitochondrial DNA deletions associated with mutations in the gene encoding Twinkle, a phage T7 gene 4-like protein localized in mitochondria. Nat Genet 28(3): 223–31. doi: 10.1038/90058.

Spinazzola, A., C. Viscomi, E. Fernandez-Vizarra, F. Carrara, P. D'Adamo, S. Calvo et al. 2006. MPV17 encodes an inner mitochondrial membrane protein and is mutated in infantile hepatic mitochondrial DNA depletion. Nat Genet 38(5): 570–5. doi: 10.1038/ng1765.

Stefaneanu, L. and C. Tasca. 1979. An electron-microscopic study of human thyroid cancer. Endocrinologie 17(4): 233–9.

Sundaram, C., A. K. Meena, M. S. Uppin, P. Govindaraj, A. Vanniarajan, K. Thangaraj et al. 2011. Contribution of muscle biopsy and genetics to the diagnosis of chronic progressive external opthalmoplegia of mitochondrial origin. J Clin Neurosci 18(4): 535–8. doi: 10.1016/j.jocn.2010.06.014.

Suzuki, M., S. Toyooka, K. Miyajima, T. Iizasa, T. Fujisawa, N. B. Bekele et al. 2003. Alterations in the mitochondrial displacement loop in lung cancers. Clin Cancer Res 9(15): 5636–41.

Swalwell, H., E. L. Blakely, R. Sutton, K. Tonska, M. Elstner, L. He et al. 2008. A homoplasmic mtDNA variant can influence the phenotype of the pathogenic m.7472Cins MTTS1 mutation: are two mutations better than one? Eur J Hum Genet 16(10): 1265–74. doi: 10.1038/ejhg.2008.65.

Tamori, A., S. Nishiguchi, M. Nishikawa, S. Kubo, N. Koh, K. Hirohashi et al. 2004. Correlation between clinical characteristics and mitochondrial D-loop DNA mutations in hepatocellular carcinoma. J Gastroenterol 39(11): 1063–8. doi: 10.1007/s00535-004-1445-3.

Tan, D. J., R. K. Bai and L. J. Wong. 2002. Comprehensive scanning of somatic mitochondrial DNA mutations in breast cancer. Cancer Res 62(4): 972–6.

Tatuch, Y., J. Christodoulou, A. Feigenbaum, J. T. Clarke, J. Wherret, C. Smith et al. 1992. Heteroplasmic mtDNA mutation (T----G) at 8993 can cause Leigh disease when the percentage of abnormal mtDNA is high. Am J Hum Genet 50(4): 852–8.

Taylor, R. W., P. F. Chinnery, F. Haldane, A. A. Morris, L. A. Bindoff, J. Wilson et al. 1996. MELAS associated with a mutation in the valine transfer RNA gene of mitochondrial DNA. Ann Neurol 40(3): 459–62. doi: 10.1002/ana.410400318.

Taylor, R. W., C. Giordano, M. M. Davidson, G. d'Amati, H. Bain, C. M. Hayes et al. 2003. A homoplasmic mitochondrial transfer ribonucleic acid mutation as a cause of maternally inherited hypertrophic cardiomyopathy. J Am Coll Cardiol 41(10): 1786–96.

Taylor, R. W. and D. M. Turnbull. 2005. Mitochondrial DNA mutations in human disease. Nat Rev Genet 6(5): 389–402. doi: 10.1038/nrg1606.

Temperley, R. J., S. H. Seneca, K. Tonska, E. Bartnik, L. A. Bindoff, R. N. Lightowlers et al. 2003. Investigation of a pathogenic mtDNA microdeletion reveals a translation-dependent deadenylation decay pathway in human mitochondria. Hum Mol Genet 12(18): 2341–8. doi: 10.1093/hmg/ddg238.

Tong, B. C., P. K. Ha, K. Dhir, M. Xing, W. H. Westra, D. Sidransky et al. 2003. Mitochondrial DNA alterations in thyroid cancer. J Surg Oncol 82(3): 170–3. doi: 10.1002/jso.10202.

Tuppen, H. A., E. L. Blakely, D. M. Turnbull and R. W. Taylor. 2010. Mitochondrial DNA mutations and human disease. Biochim Biophys Acta 1797(2): 113–28. doi: 10.1016/j.bbabio.2009.09.005.

Van Goethem, G., B. Dermaut, A. Lofgren, J. J. Martin and C. Van Broeckhoven. 2001. Mutation of POLG is associated with progressive external ophthalmoplegia characterized by mtDNA deletions. Nat Genet 28(3): 211–2. doi: 10.1038/90034.

Van Goethem, G., J. J. Martin and C. Van Broeckhoven. 2003. Progressive external ophthalmoplegia characterized by multiple deletions of mitochondrial DNA: unraveling the pathogenesis of human mitochondrial DNA instability and the initiation of a genetic classification. Neuromolecular Med 3(3): 129–46. doi: 10.1385/NMM:3:3:129.

Vielhaber, S., D. Kunz, K. Winkler, F. R. Wiedemann, E. Kirches, H. Feistner et al. 2000. Mitochondrial DNA abnormalities in skeletal muscle of patients with sporadic amyotrophic lateral sclerosis. Brain 123(Pt 7): 1339–48.

Vilkki, J., J. Ott, M. L. Savontaus, P. Aula and E. K. Nikoskelainen. 1991. Optic atrophy in Leber hereditary optic neuroretinopathy is probably determined by an X-chromosomal gene closely linked to DXS7. Am J Hum Genet 48(3): 486–91.

Viscomi, C., E. Bottani and M. Zeviani. 2015. Emerging concepts in the therapy of mitochondrial disease. Biochim Biophys Acta 1847(6-7): 544–57. doi: 10.1016/j.bbabio.2015.03.001.

Wheelhouse, N. M., P. B. Lai, S. J. Wigmore, J. A. Ross and D. J. Harrison. 2005. Mitochondrial D-loop mutations and deletion profiles of cancerous and noncancerous liver tissue in hepatitis B virus-infected liver. Br J Cancer 92(7): 1268–72. doi: 10.1038/sj.bjc.6602496.

Winterthun, S., G. Ferrari, L. He, R. W. Taylor, M. Zeviani, D. M. Turnbull et al. 2005. Autosomal recessive mitochondrial ataxic syndrome due to mitochondrial polymerase gamma mutations. Neurology 64(7): 1204–8. doi: 10.1212/01.WNL.0000156516.77696.5A.

Wong, L. J., M. Lueth, X. N. Li, C. C. Lau and H. Vogel. 2003. Detection of mitochondrial DNA mutations in the tumor and cerebrospinal fluid of medulloblastoma patients. Cancer Res 63(14): 3866–71.

Wu, C. W., P. H. Yin, W. Y. Hung, A. F. Li, S. H. Li, C. W. Chi et al. 2005. Mitochondrial DNA mutations and mitochondrial DNA depletion in gastric cancer. Genes Chromosomes Cancer 44(1): 19–28. doi: 10.1002/gcc.20213.

Yang, J., Y. Zhu, Y. Tong, L. Chen, L. Liu, Z. Zhang et al. 2009. Confirmation of the mitochondrial ND1 gene mutation G3635A as a primary LHON mutation. Biochem Biophys Res Commun 386(1): 50–4. doi: 10.1016/j.bbrc.2009.05.127.

Yoneda, M., Y. Tanno, S. Horai, T. Ozawa, T. Miyatake and S. Tsuji. 1990. A common mitochondrial DNA mutation in the t-RNA(Lys) of patients with myoclonus epilepsy associated with ragged-red fibers. Biochem Int 21(5): 789–96.

Zeviani, M., C. T. Moraes, S. DiMauro, H. Nakase, E. Bonilla, E. A. Schon et al. 1988. Deletions of mitochondrial DNA in Kearns-Sayre syndrome. Neurology 38(9): 1339–46.

Zhao, M. D., X. M. Hu, D. J. Sun, Q. Zhang, Y. H. Zhang and W. Meng. 2005. Expression of some tumor associated factors in human carcinogenesis and development of gastric carcinoma. World J Gastroenterol 11(21): 3217–21.

Zhu, X., X. Peng, M. X. Guan and Q. Yan. 2009. Pathogenic mutations of nuclear genes associated with mitochondrial disorders. Acta Biochim Biophys Sin (Shanghai) 41(3): 179–87.

Zuchner, S., I. V. Mersiyanova, M. Muglia, N. Bissar-Tadmouri, J. Rochelle, E. L. Dadali et al. 2004. Mutations in the mitochondrial GTPase mitofusin 2 cause Charcot-Marie-Tooth neuropathy type 2A. Nat Genet 36(5): 449–51. doi: 10.1038/ng1341.

20

Reproductive Approaches to Prevent the Transmission of Mitochondrial Diseases

María Jesús Sánchez-Calabuig,[a] *Noelia Fonseca Balvís,*[b]
Serafín Pérez-Cerezales[c] *and Pablo Bermejo-Álvarez**

INTRODUCTION

Mitochondria are small endosymbiotic organelles that function using genes encoded by their own circular genomes (mtDNA) as well as from nuclear DNA (nDNA). The dual origin of the DNA sequences (mtDNA or nDNA) required for proper functioning entails two possible causes of inheritable mitochondrial diseases: nDNA mutations or mtDNA mutations. The inheritable mitochondrial diseases, caused by mutations in nDNA, follow essentially the same Mendelian patterns of inheritance than any other inheritable disease caused by alterations in the chromosomic DNA sequence (Houstek et al. 2004). However, mitochondrial diseases caused by mtDNA mutations are subjected to two major differential features of mtDNA which contradicts Mendelian rules: (1) the exclusive maternal inheritance and (2) the presence of multiple copies of mtDNA per cell, which determines that most mitochondrial diseases occur in heteroplasmy (i.e., the coexistence of two populations of mutant and wild type mtDNA).

In this chapter, we will first discuss the biological mechanisms determining the maternal-exclusive inheritance and mtDNA selection during oogenesis (the so called "mitochondrial bottleneck"), essential for understanding the appearance and inheritance of mitochondrial diseases. Finally, we will discuss the novel reproductive therapies designed to prevent the transmission of mtDNA diseases.

INIA, Department of Animal Reproduction, Avda Puerta de Hierro 12 Local 10 Madrid, Madrid Spain, 28040.
[a] Email: mariasanchezcalabuig@gmail.com
[b] Email: btcnfb00@gmail.com
[c] Email: s.perez.cerezales@gmail.com
* Corresponding author: borrillobermejo@hotmail.com

Spermatozoa Mitochondria does not Contribute to the Offspring

In most animals, mitochondria and mtDNA are solely inherited from the oocyte, in contrast with other organism such as fungi and plants, where mtDNA inheritance is biparental (Barr et al. 2005). During mammalian fertilization, with some exceptions such as the Chinese hamster (Pickworth and Change 1969), spermatozoa mitochondria enter into the oocyte and thereby should be eliminated to achieve maternally-exclusive inheritance. The mechanism for elimination seems to be species-specific and limited to spermatozoa mitochondria, as inter-species but not intra-species crosses results in spermatozoa mtDNA transmission to the next generation and mitochondria derived from liver or spermatid derived mitochondria injected into the oocyte are able to replicate (Shitara et al. 1998). This mechanism was reported to be mediated by the ubiquitination of mitochondrial proteins of the spermatozoa (Sutovsky et al. 1999), perhaps prohibitin, occurring during spermatogenesis (Sutovsky et al. 2000). The ubiquitin-marked mitochondria are subsequently degraded by proteasomes (Sutovsky et al. 2003) and/or lysosomes (Sutovsky et al. 2000, Hiraoka and Hirao 1988) during early embryogenesis. Autophagosomes have been also associated with paternal mitochondria elimination in nematodes and autophagy components have been detected around the midpiece of the paternal spermatozoa in fertilized mouse eggs (Al Rawi et al. 2011).

The evolutionary reason for the maternally-specific inheritance remains unclear. One possible explanation may be that paternal mtDNA is removed because it could be damaged due to the high production of reactive oxygen species during the long and tortuous trip to the fertilization site. However, isogamous organisms also show monoparental inheritance (reviewed in (Breton et al. 2007)). Another hypothesis specific to mammals is that the highly active spermatozoa mitochondria may impair the mostly anaerobic and quiet oocyte's metabolism (Bermejo-Alvarez et al. 2010), whose mitochondria are rounded or oval-shaped, with few crest and some vacuoles, thereby with a limited oxidative capacity (Dumollard et al. 2004). Finally, the elimination of paternal mitochondria may have evolved as means to prevent heteroplasmy, together with the "mtDNA bottleneck".

mtDNA bottleneck

The offspring from heteroplasmic individuals contains different "doses" of mutant mtDNA and these different mutant:wild type proportions occur already in their oocytes, suggesting that a "mitochondrial bottleneck" occurs before oocytes are matured (Marchington et al. 1998). This mitochondrial bottleneck, leading to the unequal segregation of mtDNA variants in the offspring, is also confirmed by the rapid reversion of heteroplasmy into homoplasmy observed in the offspring of bovine (Olivo et al. 1983), mice (Meirelles and Smith 1997) and humans (Larsson et al. 1992), but it still remains controversial as to when and how it occurs.

Two major models have been proposed to explain the mitochondrial bottleneck: (1) the random limitation model, driven by a critical reduction in mtDNA number conducting to the selection of a mtDNA variant or (2) the selective replication model, driven by the positive selection of a mtDNA variant (Fig. 1). Early studies

(Bergstrom and Pritchard 1998, Nogawa et al. 1988) reported a very limited number of mitochondria in Primordial Germ Cells (PGCs), the precursors of gametogenesis, and proposed the random limitation model. Supporting this hypothesis, a drastic mtDNA reduction in PGC was observed, with ~200 copies in Embryonic Day 8.5 (E8.5) (Cree et al. 2008), in agreement with the ~200 segregation units previously calculated in mouse (Jenuth et al. 1996). Other authors reported a similar reduction of mtDNA in E8.5 PGC, but observed no shifts in mtDNA genotypic variance in germ cells during embryonic oogenesis, suggesting that the bottleneck was not due to mtDNA reduction but due to a selective replication of mtDNA variants during postnatal oogenesis (Wai et al. 2008). In partial contrast, another study observed no mtDNA reduction in PGCs, with ~3500 copies of mtDNA/cell (Cao et al. 2007). The reduction of mtDNA in PGC remains controversial probably due to technical issues such as different methods for PGC isolation (Cao et al. 2009) or the difficulties in quantifying mtDNA copy number with proper standard curves by qPCR (Bermejo-Alvarez et al. 2008). However, both studies (Cao et al. 2007, Wai et al. 2008) agree that the bottleneck is not caused by a drastic mtDNA reduction in PGCs, but by a selective replication of mtDNA variants.

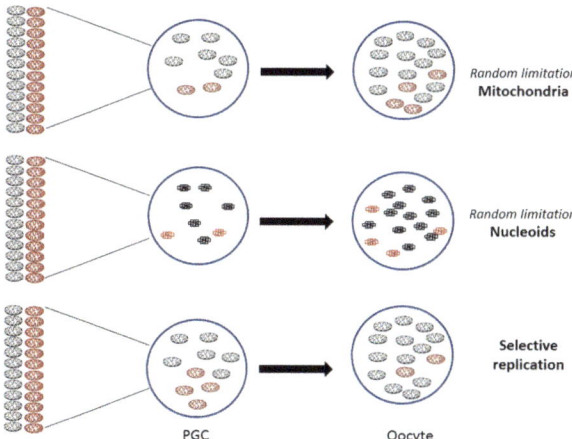

Figure 1. Two models have been proposed to explain the unequal segregation of mtDNA variants in the germ line: Random limitation (where the segregation unit can be the mitochondria or nucleoids) and selective replication of mtDNA variants.

The random limitation model can be explained by either a very limited number of mtDNA molecules in PGCs (Cree et al. 2008) or by the nucleoid hypothesis (Fig. 1). The nucleoid hypothesis proposes that the segregation unit is not the mtDNA molecule itself, but the nucleoid: a structure formed by the assembly of identical mtDNA molecules. These assembly units would drive selection through partitioning a reduced number of segregation units (Chen and Butow 2005, Jacobs et al. 2000). In this perspective, if nucleoid formation achieves a 20-fold reduction in the number of segregation units, it may reconcile the random limitation model with the high mtDNA copy number obtained by (Cao et al. 2007), as the ~3500 segregation units would be reduced to ~175. Nevertheless, the random limitation model is still incompatible with the observation that heteroplasmy skew does not occur at the stage with the lowest mtDNA amount (PGC) but later, after a substantial increase in mtDNA copy number has occurred (Wai et al. 2008). This observation, technically criticized by (Freyer et al. 2012), fits into the selective replication model. The selective replication model is also supported by several mouse-based studies that reported a selection against mutant mtDNA at the cellular (Fan et al. 2008) or organelle levels (Stewart et al. 2008). However, disease-linked mtDNA mutations in humans often appears in patients whose mothers are often healthy (Larsson et al. 1992), suggesting that the selection of mtDNA variants occurs randomly. A random selection by selective replication at the organelle level would still be possible, although it should be oocyte-specific, as no shifts in mutation loads have been observed before (Wai et al. 2008). The reduction of mtDNA copy numbers during oogenesis to virtually one mtDNA/mitochondria (Piko and Matsumoto 1976) may help in the selection.

Although the mechanism governing mitochondrial bottleneck is still disputed, it is widely accepted that heteroplasmy occurs at the oocyte level (Fan et al. 2008). The proportion of mutant mtDNA in the oocyte determines that of the offspring, but mitochondria segregation and mtDNA variant selection may occur also during early development (Freyer et al. 2012). This secondary mitochondrial segregation into somatic tissues may also lead to different proportions of mutant mtDNA in different tissues of the same individual, which in turn determines the clinical expression of the disease. For this reason, it is difficult to predict with confidence the clinical outcome of a mitochondrial disease based on prenatal diagnosis (Poulton and Marchington 2002). In any case, it is clear that reducing the amount of mutant mtDNA in oocytes lowers the odds of clinical appearance in the offspring.

Artificial Reproductive Techniques Aimed to Prevent the Appearance of Mitochondrial Diseases in the Offspring

Heteroplasmic oocytes may contain mtDNA mutations responsible for mitochondrial diseases. These mtDNA mutations may be already present in heteroplasmic women or generated by *de novo* mutations during oogenesis in homoplasmic women, a phenomenon that seems to increase with maternal age (Rebolledo-Jaramillo et al. 2014, Barritt et al. 2000). Different ARTs have been developed to reduce the dose of mutant mtDNA present in heteroplasmic oocytes or to select against those embryos harbouring a high dose. The proportion of mutant mtDNA that can be tolerated without developing clinical signs of a mitochondrial disease varies greatly (60–95% of mutant mtDNA) depending on the severity of the mutation (reviewed in (Wallace and Chalkia 2013)) and, as previously mentioned, it can be also influenced by the stochastic mitochondrial segregation in different somatic tissues and organs. The ARTs that can be used to prevent the transmission of mitochondrial diseases are (1) Preimplantation Genetic Diagnosis (2) Ooplasm transplantation, (3) Nuclear genome transfer and (4) Selective elimination of mutant mtDNA (Fig. 2).

Preimplantation Genetic Diagnosis (PGD) is aimed to select those embryos harbouring a low amount of mutant mtDNA, requiring a biopsy from a preimplantation embryo to estimate the proportions of mutant and wt DNA in the rest of the embryo. PGD is an established screening procedure used after IVF to select healthy embryos

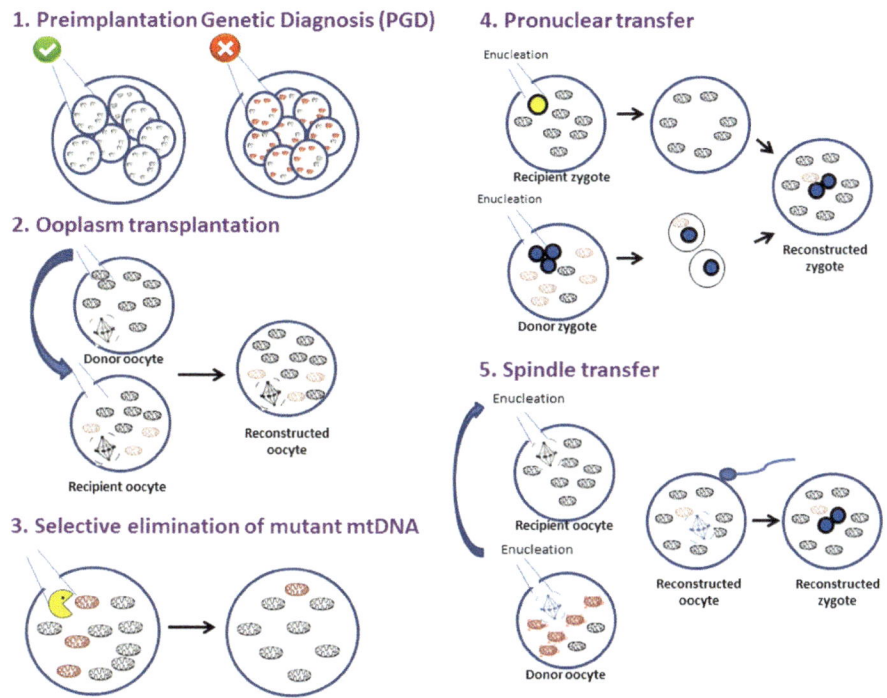

Figure 2. Artificial Reproductive Techniques aimed to prevent the appearance of mitochondrial diseases in the offspring. (1) Preimplantation Genetic Diagnosis consists of the selection of those embryos harbouring a lower dose of mutant mtDNA; (2) Ooplasm transplantation shifts heteroplasmy by introducing healthy mitochondria from a donor oocyte into a recipient oocyte; (3) Site-specific endonucleases have been used to selectively eliminate mutant mtDNA; and nuclear genetic material can be transferred from a donor zygote into an enucleated recipient zygote; (4) Pronuclear transfer, from a donor oocyte into a enucleated recipient oocyte; (5) Spindle transfer, that can be fertilized later by Intracytoplasmic Sperm Injection (ICSI).

before transferring them into the uterus. PGD is generally performed to exclude aneuploid embryos or embryos carrying a particular genetic mutations present in the parent. The use of PGD to prevent the transmission of mtDNA mutations, however, holds several limitations. Polar bodies, containing the genetic material discarded after meiosis and therefore not required for embryo development, are commonly used for PGD, but they contain no or very little mtDNA. Besides, the unequal segregation of mtDNA in different blastomeres reduces the predictive value of a biopsy of a few trophectoderm cells, i.e., the outer cells of the blastocyst destined exclusively to be part of the placenta, therefore less precious than those forming the inner cell mass. For these reasons, the biopsy of blastomeres at the 8-cell stage is preferred for PGD intended to select against mtDNA mutations (Mitalipov et al. 2014).

Ooplasmic transplantation consists of the transfer of ooplasm–oocyte's cytoplasm–from a fertile donor oocyte into a potentially compromised or heteroplasmic patient oocyte (Cohen et al. 1997). The technique was initially developed to treat reproductive infertility caused by a low number of mitochondria in the offspring: the healthy mitochondria present in the transferred ooplasm restore the developmental potential of incompetent oocytes (Van Blerkom et al. 1998, Cohen et al. 1998). Later, the transferred mitochondria were proved to survive and replicate through embryonic development (Brenner et al. 2000, Barritt et al. 2001), thus broadening the use of the technique to prevent the transmission of mtDNA-linked diseases. The technique is performed by direct injection of ooplasm into the oocyte as electrofusion yields poorer results (Cohen et al. 1998). The main disadvantage of this technique is that, given that host oocyte's mitochondria are not removed and that the amount of ooplasm injected is limited, oocyte transplantation can only vary the proportion of mutant mtDNA to a limited extend.

Nuclear genome transfer goes a step ahead than ooplasm transplantation by transferring the chromosomic content from a heteroplasmic oocyte or zygote into a healthy oocyte or zygote containing no mtDNA, harbouring the mutation responsible for the disease. By this technique, the only mtDNA from the donor oocyte or zygote left in the resulting reconstructed oocyte or zygote comes from the few mitochondria present in the biopsy performed to isolate the nuclear genome from the donor oocyte or zygote, which has been estimated to be below 2% (Craven et al. 2010, Paull et al. 2013, Tachibana et al. 2013). There are two ways to perform nuclear genome transfer: before or after fertilization. Nuclear genome transfer was initially performed in humans after fertilization by transferring the chromosomes enclosed in the parental pronuclei from the donor zygote into an enucleated healthy zygote (pronuclear transfer (Craven et al. 2010)). This technique involves an ethical controversy, as it implies the destruction of a healthy embryo to "cure" another. Nuclear genome transfer before fertilization avoids this issue by transferring the maternal chromosomes forming the meiotic spindle from a donor oocyte into an enucleated healthy recipient oocyte (maternal spindle transfer (Paull et al. 2013, Tachibana et al. 2013)). The reconstructed oocyte is later fertilized by intracytoplasmic sperm injection (ICSI). Another similar technique consists of transferring the first or second polar bodies from a donor oocyte into a healthy enucleated oocyte. This technique, reported in the mouse model (Wang et al. 2014), takes advantage of the reduced mtDNA present in the polar bodies to achieve an even lower mtDNA transmission rate from the donor oocyte.

Both oocyte transplantation and nuclear genome transfer results in an embryo carrying genetic material from three different donors, which raises biological, medical and ethical concerns (Hayden 2013, Reinhardt et al. 2013). In contrast to these techniques, selective elimination of mutant mtDNA does not introduce any exogenous genetic material: it only reduces the load of mutant mtDNA. The reduction is achieved by using site-specific endonucleases: Zinc Finger Nucleases (ZFN), Transcription Activator-like Effector Nucleases (TALEN) or Clustered Regularly Interspaced Short Palindromic Repeats (CRISPR). These three genome engineering systems are able to find specific loci across the genome and induce a Double-strand breaks (DSB) close to their target site. DSB in chromosomes is repaired by NHEJ or, less frequently, by HR. As NHEJ, the most frequent repair mechanism, is an error-prone mechanism, it often results in point mutations, which may alter the open reading frame of a particular gene. These systems have revolutionized the transgenesis field by allowing the efficient and easy-to-tailor mean (especially CRISPR) to generate knock-outs (Shen et al. 2013, Wang et al. 2013). However, in contrast to chromosomic DNA, DSB in mtDNA are usually not repaired by NHEJ or HR and, therefore, DSB often results in mtDNA degradation (Shokolenko et al. 2013). Therefore, elimination of mutant mtDNA molecules can be achieved by directing site-specific endonucleases against the mutant sequence.

Selective elimination of mtDNA has been shown to effectively shift heteroplasmy in cultured cells (Srivastava and Moraes 2001, Alexeyev et al. 2008), *in vivo* tissues (Bacman et al. 2010) and mouse zygotes (Reddy et al. 2015). Both mitochondrial directed ZFN (Minczuk et al. 2010, Minczuk et al. 2008) and TALEN (Reddy et al. 2015) have been used and it is expected that a CRISPR system will be soon developed. A recent pioneer study observed a ~5 fold reduction in the proportion of the targeted mtDNA variant in both embryos and offspring after injecting mito-TALEN in mouse zygotes (Reddy et al. 2015). This reduction can prevent the appearance of clinical symptoms without introducing any exogenous genetic material, but the shift in heteroplasmy occurs at the expense of a proportional reduction in the oocyte's mtDNA content. The mitochondrial loss associated with this technique may impair subsequent embryo development, as low mtDNA copy number has been observed to adversely affect oocyte fertilizability (Reynier et al. 2001).

Concluding Remarks

mtDNA constitutes an exception to Mendelian rules. The selective degradation of spermatozoa mitochondria occurring right after fertilization determines the exclusive maternal inheritance, whereas the presence of more than 10,000 copies per oocyte (Reynier et al. 2001), subjected to a previous unequal segregation (the mitochondrial bottleneck), leads to an unequal and unpredictable inheritance of disease-carrier mtDNA mutations. Fortunately, several ARTs have been developed to reduce the percentage of heteroplasmy by embryo selection after PGD, ooplasm transplant from a healthy oocyte, pronuclear transfer into a healthy enucleated zygote, spindle or polar body transfer into a health oocyte, and selective mtDNA degradation by site-specific endonucleases. Some of these techniques are currently approved in different countries or are in the process to be approved and they will constitute an effective way to prevent transmission of mitochondrial diseases.

Acknowledgements

NFB is supported by a FPI Grant, SPC by a Juan de La Cierva Grant and PBA by Grants RYC-2012-10193 and AGL2014-58739-R from the Spanish Ministry of Economy and Competitiveness.

Keywords: Mitochondrial bottleneck, heteroplasmy, spermatozoa, mtDNA replication model, artificial reproductive techniques, mitochondrial replacement, preimplantation genetic diagnosis, ooplasm transplantation, nuclear transplantation, CRISPR

References

Al Rawi, S., S. Louvet-Vallee, A. Djeddi, M. Sachse, E. Culetto, C. Hajjar et al. 2011. Postfertilization autophagy of sperm organelles prevents paternal mitochondrial DNA transmission. Science 334(6059): 1144–7.
Alexeyev, M. F., N. Venediktova, V. Pastukh, I. Shokolenko, G. Bonilla and G. L. Wilson. 2008. Selective elimination of mutant mitochondrial genomes as therapeutic strategy for the treatment of NARP and MILS syndromes. Gene Therapy 15(7): 516–23.

Bacman, S. R., S. L. Williams, S. Garcia and C. T. Moraes. 2010. Organ-specific shifts in mtDNA heteroplasmy following systemic delivery of a mitochondria-targeted restriction endonuclease. Gene Therapy 17(6): 713–20.

Barr, C. M., M. Neiman and D. R. Taylor. 2005. Inheritance and recombination of mitochondrial genomes in plants, fungi and animals. New Phytol 168(1): 39–50.

Barritt, J. A., J. Cohen and C. A. Brenner. 2000. Mitochondrial DNA point mutation in human oocytes is associated with maternal age. Reprod Biomed Online 1(3): 96–100.

Barritt, J. A., C. A. Brenner, H. E. Malter and J. Cohen. 2001. Mitochondria in human offspring derived from ooplasmic transplantation. Hum Reprod 16(3): 513–6.

Bergstrom, C. T. and J. Pritchard. 1998. Germline bottlenecks and the evolutionary maintenance of mitochondrial genomes. Genetics 149(4): 2135–46.

Bermejo-Alvarez, P., D. Rizos, D. Rath, P. Lonergan and A. Gutierrez-Adan. 2008. Epigenetic differences between male and female bovine blastocysts produced *in vitro*. Physiol Genomics 32(2): 264–72.

Bermejo-Alvarez, P., P. Lonergan, D. Rizos and A. Gutierrez-Adan. 2010. Low oxygen tension during IVM improves bovine oocyte competence and enhances anaerobic glycolysis. Reprod Biomed Online 20(3): 341–349.

Brenner, C. A., J. A. Barritt, S. Willadsen and J. Cohen. 2000. Mitochondrial DNA heteroplasmy after human ooplasmic transplantation. Fertility and Sterility 74(3): 573–8.

Breton, S., H. D. Beaupre, D. T. Stewart, W. R. Hoeh and P. U. Blier. 2007. The unusual system of doubly uniparental inheritance of mtDNA: isn't one enough? Trends Genet 23(9): 465–74.

Cao, L., H. Shitara, T. Horii, Y. Nagao, H. Imai, K. Abe et al. 2007. The mitochondrial bottleneck occurs without reduction of mtDNA content in female mouse germ cells. Nature Genetics 39(3): 386–90.

Cao, L., H. Shitara, M. Sugimoto, J. Hayashi, K. Abe and H. Yonekawa. 2009. New evidence confirms that the mitochondrial bottleneck is generated without reduction of mitochondrial DNA content in early primordial germ cells of mice. PLoS Genet 5(12): e1000756.

Cohen, J., R. Scott, T. Schimmel, J. Levron and S. Willadsen. 1997. Birth of infant after transfer of anucleate donor oocyte cytoplasm into recipient eggs. Lancet 350(9072): 186–7.

Cohen, J., R. Scott, M. Alikani, T. Schimmel, S. Munne, J. Levron et al. 1998. Ooplasmic transfer in mature human oocytes. Mol Hum Reprod 4(3): 269–80.

Craven, L., H. A. Tuppen, G. D. Greggains, S. J. Harbottle, J. L. Murphy, L. M. Cree et al. 2010. Pronuclear transfer in human embryos to prevent transmission of mitochondrial DNA disease. Nature 465(7294): 82–5.

Cree, L. M., D. C. Samuels, S. C. de Sousa Lopes, H. K. Rajasimha, P. Wonnapinij, J. R. Mann et al. 2008. A reduction of mitochondrial DNA molecules during embryogenesis explains the rapid segregation of genotypes. Nature Genetics 40(2): 249–54.

Chen, X. J. and R. A. Butow. 2005. The organization and inheritance of the mitochondrial genome. Nat Rev Genet 6(11): 815–25.

Dumollard, R., P. Marangos, G. Fitzharris, K. Swann, M. Duchen and J. Carroll. 2004. Sperm-triggered [Ca2+] oscillations and Ca2+ homeostasis in the mouse egg have an absolute requirement for mitochondrial ATP production. Development 131(13): 3057–67.

Fan, W., K. G. Waymire, N. Narula, P. Li, C. Rocher, P. E. Coskun et al. 2008. A mouse model of mitochondrial disease reveals germline selection against severe mtDNA mutations. Science 319 (5865): 958–62.

Freyer, C., L. M. Cree, A. Mourier, J. B. Stewart, C. Koolmeister, D. Milenkovic et al. 2012. Variation in germline mtDNA heteroplasmy is determined prenatally but modified during subsequent transmission. Nature Genetics 44(11): 1282–5.

Hayden, E. C. 2013. Regulators weigh benefits of 'three-parent' fertilization. Nature 502(7471): 284–5.

Hiraoka, J. and Y. Hirao. 1988. Fate of sperm tail components after incorporation into the hamster egg. Gamete Res 19(4): 369–80.

Houstek, J., T. Mracek, A. Vojtiskova and J. Zeman. 2004. Mitochondrial diseases and ATPase defects of nuclear origin. Biochimica et Biophysica Acta 1658(1-2): 115–21.

Jacobs, H. T., S. K. Lehtinen and J. N. Spelbrink. 2000. No sex please, we're mitochondria: a hypothesis on the somatic unit of inheritance of mammalian mtDNA. BioEssays 22(6): 564–72.

Jenuth, J. P., A. C. Peterson, K. Fu and E. A. Shoubridge. 1996. Random genetic drift in the female germline explains the rapid segregation of mammalian mitochondrial DNA. Nat Genet 14(2): 146–51.

Larsson, N. G., M. H. Tulinius, E. Holme, A. Oldfors, O. Andersen, J. Wahlstrom et al. 1992. Segregation and manifestations of the mtDNA tRNA(Lys) A-->G(8344) mutation of myoclonus epilepsy and ragged-red fibers (MERRF) syndrome. American Journal of Human Genetics 51(6): 1201–12.

Marchington, D. R., V. Macaulay, G. M. Hartshorne, D. Barlow and J. Poulton. 1998. Evidence from human oocytes for a genetic bottleneck in an mtDNA disease. American Journal of Human Genetics 63(3): 769–75.

Meirelles, F. V. and L. C. Smith. 1997. Mitochondrial genotype segregation in a mouse heteroplasmic lineage produced by embryonic karyoplast transplantation. Genetics 145(2): 445–51.

Minczuk, M., M. A. Papworth, J. C. Miller, M. P. Murphy and A. Klug. 2008. Development of a single-chain, quasi-dimeric zinc-finger nuclease for the selective degradation of mutated human mitochondrial DNA. Nucleic Acids Research 36(12): 3926–38.

Minczuk, M., P. Kolasinska-Zwierz, M. P. Murphy and M. A. Papworth. 2010. Construction and testing of engineered zinc-finger proteins for sequence-specific modification of mtDNA. Nat Protoc 5(2): 342–56.

Mitalipov, S., P. Amato, S. Parry and M. J. Falk. 2014. Limitations of preimplantation genetic diagnosis for mitochondrial DNA diseases. Cell Rep 7(4): 935–7.

Nogawa, T., W. K. Sung, G. M. Jagiello and W. Bowne. 1988. A quantitative analysis of mitochondria during fetal mouse oogenesis. Journal of Morphology 195(2): 225–34.

Olivo, P. D., M. J. Van de Walle, P. J. Laipis and W. W. Hauswirth. 1983. Nucleotide sequence evidence for rapid genotypic shifts in the bovine mitochondrial DNA D-loop. Nature 306(5941): 400–2.

Paull, D., V. Emmanuele, K. A. Weiss, N. Treff, L. Stewart, H. Hua et al. 2013. Nuclear genome transfer in human oocytes eliminates mitochondrial DNA variants. Nature 493(7434): 632–7.

Pickworth, S. and M. C. Change. 1969. Fertilization of Chinese hamster eggs *in vitro*. Journal of Reproduction and Fertility 19(2): 371–4.

Piko, L. and L. Matsumoto. 1976. Number of mitochondria and some properties of mitochondrial DNA in the mouse egg. Dev Biol 49(1): 1–10.

Poulton, J. and D. R. Marchington. 2002. Segregation of mitochondrial DNA (mtDNA) in human oocytes and in animal models of mtDNA disease: clinical implications. Reproduction 123(6): 751–5.

Rebolledo-Jaramillo, B., M. S. Su, N. Stoler, J. A. McElhoe, B. Dickins, D. Blankenberg et al. 2014. Maternal age effect and severe germ-line bottleneck in the inheritance of human mitochondrial DNA. Proc Natl Acad Sci U S A 111(43): 15474–9.

Reddy, P., A. Ocampo, K. Suzuki, J. Luo, S. R. Bacman, S. L. Williams et al. 2015. Selective elimination of mitochondrial mutations in the germline by genome editing. Cell 161(3): 459–69.

Reinhardt, K., D. K. Dowling and E. H. Morrow. 2013. Medicine. Mitochondrial replacement, evolution, and the clinic. Science 341(6152): 1345–6.

Reynier, P., P. May-Panloup, M. F. Chretien, C. J. Morgan, M. Jean, F. Savagner et al. 2001. Mitochondrial DNA content affects the fertilizability of human oocytes. Mol Hum Reprod 7(5): 425–9.

Shen, B., J. Zhang, H. Wu, J. Wang, K. Ma, Z. Li et al. 2013. Generation of gene-modified mice via Cas9/RNA-mediated gene targeting. Cell Res 23(5): 720–3.

Shitara, H., J. I. Hayashi, S. Takahama, H. Kaneda and H. Yonekawa. 1998. Maternal inheritance of mouse mtDNA in interspecific hybrids: segregation of the leaked paternal mtDNA followed by the prevention of subsequent paternal leakage. Genetics 148(2): 851–7.

Shokolenko, I. N., G. L. Wilson and M. F. Alexeyev. 2013. Persistent damage induces mitochondrial DNA degradation. DNA Repair (Amst) 12(7): 488–99.

Srivastava, S. and C. T. Moraes. 2001. Manipulating mitochondrial DNA heteroplasmy by a mitochondrially targeted restriction endonuclease. Human Molecular Genetics 10(26): 3093–9.

Stewart, J. B., C. Freyer, J. L. Elson and N. G. Larsson. 2008. Purifying selection of mtDNA and its implications for understanding evolution and mitochondrial disease. Nat Rev Genet 9(9): 657–62.

Sutovsky, P., R. D. Moreno, J. Ramalho-Santos, T. Dominko, C. Simerly and G. Schatten. 1999. Ubiquitin tag for sperm mitochondria. Nature 402(6760): 371–2.

Sutovsky, P., R. D. Moreno, J. Ramalho-Santos, T. Dominko, C. Simerly and G. Schatten. 2000. Ubiquitinated sperm mitochondria, selective proteolysis, and the regulation of mitochondrial inheritance in mammalian embryos. Biology of Reproduction 63(2): 582–90.

Sutovsky, P., T. C. McCauley, M. Sutovsky and B. N. Day. 2003. Early degradation of paternal mitochondria in domestic pig (Sus scrofa) is prevented by selective proteasomal inhibitors lactacystin and MG132. Biology of Reproduction 68(5): 1793–800.

Tachibana, M., P. Amato, M. Sparman, J. Woodward, D. M. Sanchis, H. Ma et al. 2013. Towards germline gene therapy of inherited mitochondrial diseases. Nature 493(7434): 627–31.

Van Blerkom, J., J. Sinclair and P. Davis. 1998. Mitochondrial transfer between oocytes: potential applications of mitochondrial donation and the issue of heteroplasmy. Hum Reprod 13(1O): 2857–68.

Wai, T., D. Teoli and E. A. Shoubridge. 2008. The mitochondrial DNA genetic bottleneck results from replication of a subpopulation of genomes. Nature Genetics 40(12): 1484–8.

Wallace, D. C. and D. Chalkia. 2013. Mitochondrial DNA genetics and the heteroplasmy conundrum in evolution and disease. Cold Spring Harb Perspect Biol 5(11): a021220.

Wang, H., H. Yang, C. S. Shivalila, M. M. Dawlaty, A. W. Cheng, F. Zhang et al. 2013. One-step generation of mice carrying mutations in multiple genes by CRISPR/Cas-mediated genome engineering. Cell 153(4): 910–8.

Wang, T., H. Sha, D. Ji, H. L. Zhang, D. Chen, Y. Cao et al. 2014. Polar body genome transfer for preventing the transmission of inherited mitochondrial diseases. Cell 157(7): 1591–604.

21

Mitochondrial DNA Mutations and Neurodegenerative Diseases

Paula Núñez

INTRODUCTION

Mitochondrial diseases caused by mtDNA mutations are in general multi-symptomatic diseases with dysfunctions affecting different systems and tissues. The most affected tissues are the post-mitotic ones, such as myocytes and neurons. The causes for this specific susceptibility lie in the non-proliferative nature of these cells, making it difficult to remove cells with high levels of damaged or mutated mtDNA. Furthermore, myocytes and neurons are also cells with a high-energy demand. This distinction makes them more susceptible to the ATP depletion that can derive from the pathogenic mutations in the mtDNA and affect proteins involved in the respiratory chain (Schon and Manfredi 2003, Pinto and Moraes 2015).

Mutations in mtDNA often result in severe neurological impairment due to neurodegeneration in the Central Nervous System (CNS). The grade and pattern of neurodegeneration within patients with mitochondrial disease can vary depending on the kind of mutation and mutation load segregation throughout brain regions as a whole, as well as neurons individually. mtDNA changes have been hypothesized to have a role also in age-related neurodegenerative diseases like Alzheimer's, Parkinson's, amyotrophic lateral sclerosis and Huntington's disease.

University of Oviedo, School of Medicine, Department of Functional Biology (Physiology) Julian Claveria, 6 Oviedo, Asturias Spain, 33006.
Email: nunezpaula@uniovi.es

Alzheimer's Disease (AD)

The prevalence of neurodegenerative diseases such as Alzheimer's disease increases markedly with age. This prevalence increases with life expectancy and it is predicted that the prevalence will approximately reach over 100 million world-wide by the middle of this century. Population based surveys estimate that AD affects 7–10% of individuals > 65 years of age and possibly 50–60% of people over 85 years of age (Fukae and Mizuno 2007). Genetic and environmental factors contribute to the development of this disease, which is progressive and irreversible and results in cognitive and/or motor disorders. AD is characterized by neuronal and synaptic impairments in the cerebral cortex, as well as in certain subcortical regions. Here there are accumulations of senile plaques, composed mostly by beta-amyloid peptide and intraneuronal tau deposition as neurofibrillary tangles. The most studied pathogenetic model for this disease is the "beta-amyloid cascade", where an imbalance in the cleavage sequence of Amyloid Precursor Protein (APP) leads to an accumulation of toxic Aβ (Amyloid Beta) fragment, cytotoxic plaques, and consequent neurodegeneration (Pinto and Moraes 2015).

Mitochondrial accumulation of Aβ is one of the key mechanisms to cause mitochondrial dysfunction and leads to pathological process in AD (Humphries and Szweda 1998, Querfurth and LaFerla 2010). Cytoplasmic hybrid ('cybrid') cells, made from mitochondrial DNA of nonfamilial AD subjects, show antioxidant-reversible lowering of mitochondrial membrane potential. Expression of AD mitochondrial genes in cybrid cells depresses cytochrome c oxidase activity and increases oxidative stress, which in turn, lowers mitochondrial membrane potential. Under stress, cells with AD mitochondrial genes are more likely to activate cell death pathways, which drive caspase 3-mediated Aβ peptide secretion and may account for increased Aβ deposition in the AD brain. Aβ fragments negatively affect mitochondrial function, suggesting that mitochondrial dysfunction is a consequence of the Aβ toxicity (Chen and Yan 2010, Roses et al. 2013).

Moreover, AD is associated with aberrant processing of the APP by γ-secretase. The γ-secretase activity and the APP cleavage predominantly occur in the MAM (mitochondria-associated ER membranes), a compartment of the endoplasmic reticulum physically and biochemically connected to mitochondria. MAM function and ER–mitochondrial connectivity are increased in AD, so that AD is also proposed to be an ER–mitochondrial communication and MAM dysfunction disease (Area-Gomez et al. 2012, Muresan and Ladescu Muresan 2015). This hypothesis, rather than the β-amyloid cascade one, would explain also other biochemical changes occurring in the disease like mitochondrial dysfunction, elevated levels of cholesterol, altered metabolism of fatty acids and phospholipids and aberrant calcium homeostasis (Schon and Area-Gomez 2013).

Mitochondrial genomic dysfunction is also reported in AD pathology. Damaged DNA lesions by oxidative stress are much higher in mtDNA of AD post-mortem tissues. Degraded mtDNA and related proteins were also found in the mitochondria of AD brains (Reddy 2009) and base excision repair (BER) pathway is defective in AD post-mortem brain whole tissue lysates (Canugovi et al. 2014). Analysis of the mtDNA revealed that COX deficiency is caused by high levels of mtDNA deletions which accumulate with age. Krishnan et al. (2012) described a higher percentage of cytochrome c oxidase deficient neurons, which show a higher level of mtDNA mutations in the AD brains compared with age-matched controls. Another study showed that COX activity has been persistently found to be reduced in platelets from patients affected both with AD or Mild Cognitive Impairment (MCI), a syndrome associated with a high risk for the development of AD (Valla et al. 2006). Their results reveal that mtDNA mutation has a critical role in the AD-related mitochondrial dysfunction.

It has been suggested that inherited haplogroups may influence AD risk but to date, no clear result has been found (Mancuso 2008). Levels of the common 4977 nucleotide pair (np) mitochondrial DNA (mtDNA) deletion (mtDNA4977) were quantified in the cortex, putamen, and cerebellum of patients with AD and compared to age-matched controls. The levels of mtDNA deletions in control brains started low, but rose markedly after age 75, while those of AD patients started high and declined to low levels by age 80. Choosing age 75 to arbitrarily delineate between younger and older subjects, younger patients had 15 times more mtDNA deletions than younger controls, while older patients had one-fifth the deletion level of older controls. Younger AD patients also had fourfold more deletions than older AD patients. These results support the hypothesis that OXPHOS defects resulting from somatic mtDNA mutations may play a role in AD pathophysiology (Corral-Debrinski et al. 1994).

Wang et al. (2005) reported that mitochondrial DNA has higher levels of oxidized bases than nuclear DNA, and multiple oxidized bases are significantly higher in an AD brain compared to controls. The sequence of the mtDNA Control Region (CR) from AD brains was investigated to determine whether mtDNA mutations contribute more generally to the etiology of AD. Sixty-five percent of the AD brains harbored the T414G mutation, whereas this mutation was absent from all controls. Moreover, cloning and sequencing of the mtDNA CR from patient and control brains revealed that all AD brains had an average 63% increase in heteroplasmic mtDNA CR mutations and that AD brains from patients 80 years and older had a 130% increase in heteroplasmic CR mutations. In addition, these mutations preferentially altered known mtDNA regulatory elements. AD patient's brains also had an average 50% reduction in the mtDNA L-strand ND6 transcript and in the mtDNA/nuclear DNA ratio. Because reduced ND6 mRNA and mtDNA copy numbers would decrease brain oxidative phosphorylation, these CR mutations could account for some of the mitochondrial defects observed in AD (Coskun et al. 2004).

However, one report from Japanese patients failed to find any causal role of mtDNA mutations in the etiology of Alzheimer's disease among 153 AD patients and 129 normal control subjects (Tanaka et al. 2010). Furthermore, Elson et al. (2006) reported that pathogenic inherited mtDNA mutations do not constitute a major etiological factor in sporadic AD. It is apparent that at least a small proportion of AD patients carry a pathogenic mtDNA mutation and a small proportion of cognitively normal aged individuals carry a mtDNA mutation that reduces the risk of AD (Simoncini et al. 2015).

Parkinson's Disease (PD)

Parkinson's disease is associated with the selective loss of Dopamine (DA) neurons in the Substancia Nigra pars compacta (SNpc) and DA levels in the corpus striatum of the nigrostriatal DA pathway in the brain. This loss of DA causes a deregulation in the basal ganglia circuitries that leads to the appearance of motor symptoms such as bradykinesia, resting tremor, rigidity and postural instability as well as non-motor symptoms such as sleep disturbances, depression, and cognitive deficits (Rodriguez-Oroz et al. 2009). An estimated seven to ten million people worldwide are living with Parkinson's disease. Incidence of Parkinson increases with age, but an estimated four percent of people with PD are diagnosed before the age of 50.

Recent genetic studies have shown that genes, such as PINK1 and parkin, that act to maintain mitochondrial structural integrity, and DJ-1, that acts as a potential ROS scavenger, are important in familiar, early-onset forms of PD. Parkin, an ubiquitin ligase, may protect neuron mitochondria (Schapira 2011). Studies in mouse models and Drosophila have shown that either parkin deficiency or mutations in this gene lead to increased oxidative stress and have the potential to impair mitochondrial functions. Silvestri et al. (2005) describe that PINK1 mutations confer different autophosphorylation activity, which is regulated by the C-terminal portion of the protein. They also demonstrate the mitochondrial localization of both wild-type and mutant PINK1 proteins unequivocally, determining that a short N-terminal part of PINK1 is sufficient for its mitochondrial targeting. Mutations in PINK1 enzymatic activity may cause PD via impaired phosphorylation of PINK1's substrate, evidently at the mitochondrial level. However, the mitochondrial sub-localization of PINK1 is still debatable. Some reports indicate its localization in the inner mitochondrial membrane (Pridgeon et al. 2007) whereas some provide evidence of its presence in the mitochondrial membrane space (Plun-Favreau et al. 2007), outer mitochondrial membrane or cytoplasm (Zhou et al. 2008).

The mitochondrial chaperone mortalin has been linked to neurodegeneration in PD based on reduced protein levels in affected brain regions of PD patients and its interaction with the PD-associated protein DJ-1. Recently, two amino acid exchanges in the ATPase domain (R126W) and the substrate-binding domain (P509S) of mortalin were identified in PD patients. In neuronal and non-neuronal human cell lines, the disease-associated variants caused a mitochondrial phenotype of increased reactive oxygen species and reduced mitochondrial membrane potential, which were exacerbated upon proteolytic stress. These functional impairments correspond with characteristic alterations of the mitochondrial network in cells overexpressing mutant mortalin compared with wild-type. In line with a loss-of-function hypothesis, knockdown of mortalin in human cells caused impaired mitochondrial function that was rescued by wt mortalin, but not by the variants (Burbulla et al. 2011).

Whether mtDNA changes are sufficient to confer PD symptoms is still controversial. Studies on animal models (Song 2012) and the fact that patients with mutations in nDNA genes involved in mtDNA replication machinery also show Parkinsonism similar to cases of juvenile PD (Palin et al. 2013), suggest that they play a role in the pathogenesis. Moreover, in a 2-stage association study of mtDNA variants and PD, risk of PD with super-haplogroup JT has been reduced. The reduced risk of PD with haplogroups J, K, and T was mirrored by an increased risk of PD in super-haplogroup HV, which increases survival after sepsis. Antagonistic pleiotropy between mtDNA haplogroups may thus be determining the genetic site in humans, leading to an augmented risk of PD in later life (Hudson 2013).

Ikebe et al. (1990) for the first time reported the increased levels of common mtDNA deletion (4977 bp deletion) in PD brains. However, the study is lacking data from age-matched controls and some studies did not find an increase in mtDNA deletions in PD patients relative to age-matched controls. Bender (2006) confirmed that PD patients had a greater proportion of COX deficient neurons in the substantia nigra than controls. They showed that mtDNA deletion levels were higher in the COX-deficient neurons than those with normal COX activity underlying their respiratory chain defect. They detected discrete break points in mtDNA, implying intracellular clonal expansion as the underlying mechanisms of deletion formation. Furthermore, in the hippocampus, patients also had significantly higher deletion levels than controls. Taken together, these data imply that mtDNA deletions are higher in PD patients, possibly throughout the brain, somatically generated, and clonally expanded in some cases to levels able to cause COX deficiency.

Other studies have shown that mtDNA mutations, independent of mitochondrial respiratory chain function, correlate with PD. Lin and Beal (2006) indicated that mean somatic mtDNA point mutations in neurons were 250 mutations/106 bp higher in early PD compared to both controls and late stage disease. They found no difference between point mutations in established PD and controls. This suggests that mtDNA point mutations may predispose these neurons to early cell death, enabling the subsequent survival of neurons without mtDNA mutations.

Amyotrophic Lateral Sclerosis (ALS)

Amyotrophic lateral sclerosis is a motor neuron disease, characterized by rapidly progressive weakness, muscle atrophy and fasciculation, muscle spasticity, dysarthria, dysphagia, dyspnea caused by the degeneration of the upper and lower motor neurons. Most persons die within 2–5 years of receiving a diagnosis of ALS. ALS is more prevalent in men than in women, with a typical age of diagnosis at age 55–75 years. There is no known definitive cause of ALS, but a hereditary form of the disease, familial ALS, occurs in 5%–10% of cases (Mancuso and Navarro 2015). Increased mtDNA deletions have been found in the muscle and brain of ALS patients, although the levels are still relatively low and of unknown consequence (Keeney and Bennett 2010). Moreover, in some cases of patients with mtDNA mutations, an ALS phenotype has been diagnosed (Borthwick et al. 2006). Also in this case, the association of haplogroup with increased risk factor is still controversial (Ingram 2012).

Electron microscopy examinations of muscle, liver, spinal motor neurons and motor cortex revealed structural defects in the mitochondrial DNA (Menzies et al. 2002). A mutation in cytochrome c oxidase subunit I was found in a patient with a motor neuron disease phenotype (Comi et al. 1998). Another patient with motor neuron disease had a mutation in a mitochondrial tRNA gene. Other study examined the role of somatic mtDNA mutations in human aging by quantifying the accumulation of the common 4,977 nucleotide pair (np) deletion (mtDNA4977) in the cortex, putamen and cerebellum. They found a significant increase in the mtDNA4977 deletion in elderly individuals. Similar changes were observed with a different 7436 np deletion. These changes suggest that somatic mtDNA deletions might contribute to the neurological impairment often associated with aging (Corral-Debrinski et al.

1992). However, no significant accumulation of the 5 kb (4977) common deletion in mtDNA has been found by single-cell analysis of motor neurons from sporadic ALS cases (Mawrin et al. 2004).

There are also reports of increased levels of common mtDNA deletions as well as multiple deletions in tissues such as brain and skeletal muscle from patients affected with this disease (Ro et al. 2003). In isolated cases with ALS, COX negative fibers are observed in skeletal muscle, even if no precise correlation exists between severity of COX defect, age and duration of the disease. Oxidative metabolism defects in muscle fibers support the hypothesis that mitochondrial dysfunction must play a role in the pathogenesis of the disease (Crugnola et al. 2010). Mutations in the cytosolic SOD1 gene are the most common form of inherited ALS. SOD1 A4V is the most common mutation in North America, accounting for up to 50% of cases (Saeed et al. 2009). The precise mechanism by which cytosolic SOD1 causes mitochondrial dysfunction remains elusive. However, multiple disease-causing mutants, but not wild-type SOD1, are now demonstrated to be recruited to mitochondria, but only in affected tissues. This is independent of the copper chaperone for SOD1 and dismutase activity. Highly preferential association with spinal cord mitochondria is seen in human ALS for a mutant SOD1 that accumulates only to trace cytoplasmic levels. Despite variable proportions that are successfully imported, nearly constant amounts of SOD1 mutants and covalently damaged adducts of them accumulate as apparent import intermediates and/or are tightly aggregated or cross-linked onto integral membrane components on the cytoplasmic face of those mitochondria. These findings implicate damage from action of spinal cord-specific factors that recruit mutant SOD1 to spinal mitochondria as the basis for their selective toxicity in ALS (Liu et al. 2004).

Huntington's Disease (HD)

Huntington's disease is a genetic autosomal dominant neurodegenerative disorder characterized by progressive chorea, psychiatric disturbances and dementia and for which no disease modifying therapy currently exists. Moreover these abnormal symptoms worsen as the disease progresses. The mutation underlying the disease is due to expansion of a CAG trinucleotide repeat in the HTT gene resulting in the accumulation of mutant huntingtin protein in the brain. Symptoms usually appear between the ages of 30 and 45, although they may appear earlier or later (Ha et al. 2012).

Earlier biochemical studies conducted on brains from HD patients have shown impaired activity of respiratory chain complexes including enzymes from the TCA cycle. One study analyzed mitochondrial respiratory chain function in the caudate nucleus and platelets from patients with HD. In the caudate nucleus, severe defects of complexes II and III and a deficiency of complex IV activity were demonstrated. In HD neurotoxin models, a mitochondrial defect in HD caudate parallels was identified, supporting the role of abnormal energy metabolism in HD. Brain regions such as caudate and putamen from advanced disease patients also showed reduced activity of complexes II–IV (Beal 1997).

Lymphocytes, leucocytes and cortical tissues in HD patients have been found to have higher mtDNA deletions compared to controls. Deletions based in four areas of mitochondrial DNA have been investigated in a group of 60 Iranian patients clinically diagnosed with HD and 70 healthy controls. A total of 41 patients out of 60 had CAG expansion (group A). About 19 patients did not show expansion but had the clinical symptoms of HD (group B). One of the four-mtDNA deletions was found in at least 90% of samples. Multiple deletions have also been observed in 63% of HD patients. These results showed that HD patients had higher frequencies of mtDNA deletions in lymphocytes in comparison to the controls. It is thus proposed that CAG repeat instability and mutant HTT are causative factors in mtDNA damage (Banoei et al. 2007).

Conclusion

Mitochondrial DNA mutations can result in neurological diseases due to neurodegeneration. This is because cells of a non-proliferative nature, such as neurons, form the tissues exhibiting greatest amount of damage. The extent of neurodegeneration within patients with mitochondrial disease can vary depending on the mutation load segregation throughout brain regions. Mitochondrial DNA changes have been hypothesized to have a role in age-related neurodegenerative diseases such as Alzheimer's, Parkinson's, Amyotrophic Lateral Sclerosis and Huntington's disease.

Keywords: mtDNA, mitochondrial respiratory chain function, neurodegenerative disease, Alzheimer's disease, Parkinson's disease, amyotrophic lateral sclerosis, Huntington's disease

References

Area-Gomez, E., M. Del Carmen Lara Castillo, M. D. Tambini, C. Guardia-Laguarta, A. J. de Groof, Madra, et al. 2012. Upregulated function of mitochondria-associated ER membranes in Alzheimer disease. EMBO J 31: 4106–4123.

Banoei, M. M., M. Houshmand, M. S. Panahi, P. Shariati, M. Rostami, M. D. Manshadi and T. Majidizadeh. 2007. Huntington's disease and mitochondrial DNA deletions: event or regular mechanism for mutant huntingtin protein and CAG repeats expansion? Cell Mol Neurobiol 2007 Nov; 27(7): 867–75. Epub 2007 Oct 20.

Beal, M. F. 1997. New techniques for investigating mitochondrial DNA in neurodegenerative diseases. Neurology 1997 Oct; 49(4): 907–8.

Bender, A. 2006. High levels of mitochondrial DNA deletions in substantia nigra neurons in aging and Parkinson disease. Nat Genet 38: 515–517.

Borthwick, G. M., R. W. Taylor, T. J. Walls, K. Tonska, G. A. Taylor, P. J. Shaw et al. 2006. Motor neuron disease in a patient with a mitochondrial tRNAIle mutation. Ann Neurol 59: 570–574.

Burbulla, L. F., C. Schelling, H. Kato, D. Rapaport, D. Woitalla, C. Schiesling et al. 2011. Dissecting the role of the mitochondrial chaperone mortalin in Parkinson's disease: functional impact of disease-related variants on mitochondrial homeostasis. Hum Mol Genet 19: 4437–4452.

Canugovi, C., R. A. Shamanna, D. L. Croteau and V. A. Bohr. 2014. Base excision DNA repair levels in mitochondrial lysates of Alzheimer's disease. Neurobiol Aging 35: 1–8.

Chen, J. X. and S. S. Yan. 2010. Role of mitochondrial amyloid-beta in Alzheimer's disease. J Alzheimers Dis 20: 569–578.

Comi, G. P., A. Bordoni, S. Salani, L. Franeschina, M. Sciacco, A. Prelle et al. 1998. Cytochrome c oxidase subunit I microdeletion in a patient with motor neuron disease. Ann Neurol 43: 110–116.

Corral-Debrinski, M., T. Horton, M. T. Lott, J. M. Shoffner, M. F. Beal and D. C. Wallace. 1992. Mitochondrial DNA deletions in human brain: regional variability and increase with advanced age. Nat Genet 2: 324–329.

Corral-Debrinski, M., T. Horton, M. T. Lott, J. M. Shoffner, A. C. McKee, M. F. Beal et al. 1994. Marked changes in mitochondrial DNA deletion levels in Alzheimer brains. Genomics 23: 471–476.

Coskun, P. E., M. F. Beal and D. C. Wallace. 2004. Alzheimer's brains harbor somatic mtDNA control-region mutations that suppress mitochondrial transcription and replication. Proc Natl Acad Sci USA 101: 10726–10731.

Crugnola, V., C. Lamperti, V. Lucchini, D. Ronchi, L. Peverelli, A. Prelle et al. 2010. Mitochondrial respiratory chain dysfunction in muscle from patients with amyotrophic lateral sclerosis. Arch Neurol 67: 849–854.

Elson, J. L., C. Herrnstadt, G. Preston, L. Thal, C. M. Morris, J. A. Edwardson et al. 2006. Does the mitochondrial genome play a role in the etiology of Alzheimer's disease? Hum Genet 119: 241–254.

Fukae, J., Y. Mizuno and N. Hattori. 2007. Mitochondrial dysfunction in Parkinson's disease. Mitochondrion 7: 58–62.

Ha, A. D., C. A. Beck and J. Jankovic. 2012. Intermediate CAG repeats in Huntington's disease: analysis of COHORT. Tremor Other Hyperkinet Mov 2: tre-02-64-287-4.

Hudson, G. 2013. Two-stage association study and meta-analysis of mitochondrial DNA variants in Parkinson disease. Neurology 80: 2042–2048.

Humphries, K. M. and L. I. Szweda. 1998. Selective inactivation of alpha-ketoglutarate dehydrogenase and pyruvate dehydrogenase: reaction of lipoic acid with 4-hydroxy-2-nonenal. Biochemistry 37: 15835–15841.

Ikebe, S., M. Tanaka, K. Ohno, W. Sato, K. Hattori, T. Kondo et al. 1990. Increase of deleted mitochondrial DNA in the striatum in Parkinson's disease and senescence. Biochem Biophys Res Commun 170: 1044–1048.

Ingram, C. J. 2012. Analysis of European case-control studies suggests that common inherited variation in mitochondrial DNA is not involved in susceptibility to amyotrophic lateral sclerosis. Amyotroph Lateral Scler 13: 341–346.

Keeney, P. M. and J. P. Bennett Jr. 2010. ALS spinal neurons show varied and reduced mtDNA gene copy numbers and increased mtDNA gene deletions. Mol Neurodegener 5: 21.

Krishnan, K. J., T. E. Ratnaike, H. L. M. De Gruyter, E. Jaros and D. M. Turnbull. 2012. Mitochondrial DNA deletions cause the biochemical defect observed in Alzheimer's disease. Neurobiol Aging 33: 2210–2214.

Lin, M. T. and M. F. Beal. 2006. Mitochondrial dysfunction and oxidative stress in neurodegenerative diseases. Nature 443: 787–795.

Liu, J., C. Lillo, P. A. Jonsson, C. Vande Velde, C. M. Ward and T. M. Miller. 2004. Toxicity of familial ALS-linked SOD1 mutants from selective recruitment to spiral mitochondria. Neuron 43: 5–17.

Mancuso, M. 2008. Mitochondrial DNA sequence variation and neurodegeneration. Hum Genomics 3: 71–78.

Mancuso, R. and X. Navarro. 2015. Amyotrophic lateral sclerosis: Current perspectives from basic research to the clinic. Prog Neurobiol. In press.

Mawrin, C., E. Kirches, G. Krause, F. R. Wiedemann, C. K. Vorwerk, B. Bogerts et al. 2004. Single-cell analysis of mtDNA levels in sporadic amyotrophic lateral sclerosis. Neuroreport 15: 939–943.

Menzies, F. M., P. G. Ince and J. Shaw. 2002. Mitochondrial involvement in amyotrophic lateral sclerosis. Neurochem Int 40: 543–551.

Muresan, V. and Z. Ladescu Muresan. 2015. Amyloid-β precursor protein: Multiple fragments, numerous transport routes and mechanisms. Exp Cell Res 334: 45–53.

Palin, E. J., A. Paetau and A. Suomalainen. 2013. Mesencephalic complex I deficiency does not correlate with Parkinsonism in mitochondrial DNA maintenance disorders. Brain 136: 2379–2392.

Pinto, M. and C. T. Moraes. 2015. Mechanisms linking mtDNA damage and aging. Free Radic Biol Med 85: 250–258.

Plun-Favreau, H., K. Klupsch, N. Moisoi, S. Gandhi, S. Kjaer, D. Frith et al. 2007. The mitochondrial protease HtrA2 is regulated by Parkinson's disease-associated ase PINK1. Nat Cell Biol 9: 1243–1252.

Pridgeon, J. W., J. A. Olzmann, L. S. Chin and L. Li. 2007. PINK1 protects against oxidative stress by phosphorylating mitochondrial chaperone TRAP1. PLoS Biol 5: e172.

Querfurth, H. W. and F. M. LaFerla. 2010. Alzheimer's disease. N Engl J Med 362: 329–344.

Reddy, P. H. 2009. Amyloid beta, mitochondrial structural and functional dynamics in Alzheimer's disease. Exp Neurol 218: 286–292.

Ro, L. S., S. L. Lai, C. M. Chen and S. T. Chen. 2003. Deleted 4977-bp mitochondrial DNA mutation is associated with sporadic amyotrophic lateral sclerosis: a hospital-based case-control study. Muscle Nerve 28: 737–743.

Rodriguez-Oroz, M. C., M. Jahanshahi, P. Krack, I. Litvan, R. Macias, E. Bezard et al. 2009. Initial clinical manifestations of Parkinson's disease: features and pathophysiological mechanisms. Lancet Neurol 8: 1128–1139.

Roses, A. D., M. W. Lutz, D. G. Crenshaw, I. Grossman, A. M. Saunders and W. K. Gottschalk. 2013. TOMM40 and APOE. Requirements for replication studies of association with age of disease onset and enrichment of a clinical trial. Alzheimers Demen 9: 132–136.

Saeed, M., Y. Yang, H. X. Deng, W. Y. Hung, N. Siddique and L. Dollefave. 2009. Age and founder effect of SOD1 A4V mutation causing ALS. Neurology 72: 1634–1639.

Schapira, A. H. V. 2011. Mitochondrial pathology in Parkinson's disease, Mt. Sinai J Med 78: 872–881.

Schon, E. A. and G. Manfredi. 2003. Neuronal degeneration and mitochondrial dysfunction. J Clin Invest 111: 303–312.

Schon, E. A. and E. Area-Gomez. 2013. Mitochondria-associated ER membranes in Alzheimer disease. Mol Cell Neurosci 5: 26–36.

Silvestri, L., V. Caputo, E. Bellacchio, L. Atorino, B. Dallapiccola, E. M. Valente et al. 2005. Mitochondrial import and enzymatic activity of PINK1 mutants associated to recessive Parkinsonism. Hum Mol Genet 14: 3477–3492.

Simoncini, C., D. Orsucci, E. Caldarazzo Ienco, G. Siciliano, U. Bonuccelli and M. Mancuso. 2015. Alzheimer's pathogenesis and its link to the mitochondrion. Oxid Med Cell Longev 2015: 803942.

Song, L. 2012. Mutant twinkle increases dopaminergic neurodegeneration, mtDNA deletions and modulates Parkin expression. Hum Mol Genet 21: 5147–5158.

Tanaka, N., Y. I. Goto, J. Akanuma, M. Kato, T. Kinoshita, F. Yamashita et al. 2010. Mitochondrial DNA variants in a Japanese population of patients with Alzheimer's disease. Mitochondrion 10: 32–37.

Valla, J., L. Schneider, T. Niedzielko, K. D. Coon, R. Caselli, M. N. Sabbagh et al. 2006. Impaired platelet mitochondrial activity in Alzheimer's disease and mild cognitive impairment. Mitochondrion 6: 323–330.

Wang, J., S. Xiong, C. Xie, W. R. Markesbery and M. A. Lovell. 2005. Increased oxidative damage in nuclear and mitochondrial DNA in Alzheimer's disease. J Neurochem 93: 953–962.

Zhou, C., Y. Huang, Y. Shao, J. May, D. Prou, C. Perier et al. 2008. The kinase domain of mitochondrial PINK1 faces the cytoplasm. Proc Natl Acad Sci USA 105: 12022–12027.

22

An Overview of mtDNA Analysis for Age-at-Death Estimation in Forensic Sciences

Silvia Zoppis

INTRODUCTION

Age-at-death estimation is one of the main challenges in the forensic sciences since it contributes to the personal identification of human remains. There are many anthropological techniques to estimate the age-at-death in children and adults. However, in adults this methodology is less accurate and requires population specific references. Thus, new methodologies have been developed based on the natural process of aging, which leads to alterations of tissues and organs on different biochemical levels (Zapico and Ubelaker 2013). One of these methodologies is aspartic acid racemization which, in forensic casework, seems to be the most accurate. However, it is important to take into account other techniques due to the fact that the specific forensic context and the remains available will determine the possibility of application of one methodology or another.

One of these methodologies is the analysis of mtDNA mutations based on the "free radical theory" of aging, which implies that aging and associated degenerative diseases can be attributed to deleterious effects of reactive oxygen species. A variant of this theory is the "mitochondrial theory," which predicts that a "vicious cycle" within the mitochondria contributes to the aging process (Zapico and Ubelaker 2015). The oxidative stress increase could be the origin of cellular molecule damage and, in particular, the abundance of free radicals could cause heteroplasmic mutations in mtDNA. An augmentation of oxidative damage occurs with the decay in mitochondrial

University of Rome "Sapienza" Section of Legal Medicine – Laboratory of Forensic Genetics Viale Regina Elena, 336, 00161 Rome, Italy.
Email: silvia.zoppis@uniroma1.it

respiratory function and induces an accumulation of non-repaired lesions on mtDNA in various postmitotic tissues (Beckman and Ames 1998, Harman 1972, Wei and Lee 2002). The mitochondrial genome is especially at risk of deletion because it contains no introns (Anderson et al. 1981) and suffers frequent replication without the advantage of adequate repair mechanisms. However, this theory is controversial and in 2014, Shokolenko et al. presented doubts about this mitochondrial theory of aging. The article states that this theory has been increasingly losing ground and is undergoing extensive revision due to its inability to explain a growing body of emerging data, and the notion of the central role of mtDNA in the aging process is being met with increased skepticism. According to the authors, new evidence suggests that both experimentally induced oxidative stress and radiation therapy result in very low levels of mtDNA mutagenesis and recent advances provide evidence against the existence of the "vicious" cycle of mtDNA damage and ROS production. Finally, they suggest that increased ROS production in aging may be the result of adaptive signaling rather than a detrimental byproduct of normal respiration that drives aging.

Independently of how mtDNA mutations originate, it is clear that these somatic rearrangements in mtDNA, including point mutations, large scale deletions and tandem duplications, have been observed in various tissues like skin fibroblasts, brain, heart, testis and in the skeletal muscle (Calloway et al. 2000, Bodyak et al. 2001, Meissner et al. 2006, Wei et al. 2009, Yao et al. 2015). These mutations seem to accumulate with age, so they could help to improve the estimation of age at death.

mtDNA Mutations for Forensic Age Estimation

At least 20 different mtDNA deletions have been found in various tissues and the most common one is the 4977 bp deletion, which has been observed in biopsy and autopsy material from individuals aged 20 years and older (Simonetti et al. 1992). Another heteroplasmic mutation, which has been detected at significant levels in skeletal muscles from aged individuals (Del Bo et al. 2002, Wang et al. 2001), is an adenine to guanine transition at position 189 (A189G). The main analysis technique, PCR, is still one of the routine techniques in forensic laboratories and can be used also in tissue that has undergone putrefaction.

The studies of Meissner et al. in 1997 and 1999 in skeletal muscle showed a correlation between the common 4977-bp deletion and age. Their calculation allowed a rough estimation of the age-at-death and can discriminate between young and elderly persons in the identification process of human remains based solely on skeletal muscle.

Other authors looked for disease mutations like the transition mutation in MERRF syndrome. Munscher et al. in 1993, using PCR methodology, found an association of this mutation with age in postmortem specimens of extra ocular and skeletal muscle from healthy people. Other authors looked for general deletions in mitochondrial DNA. Papiha et al. in 1998 analyzed blood and bone samples from patients who had undergone orthopedic surgery; they only found mtDNA deletion in bones from patients up to 70 years old with osteoporosis/rheumatoid arthritis, but not in the blood. In wisdom teeth from healthy subjects, Mornstad et al. in 1999 used the PCR amplification of HV2 of the mitochondrial D-loop to demonstrate a decrease in the amount of mtDNA in dentin with age; this decrease is highest in the oldest age groups.

In the same year (2001), in an interesting longitudinal retrospective study by Lagerstrom-Fermer et al., the authors monitored the level of heteroplasmy at nt 309 and nt 16189 of the control region of human mtDNA. As a unique source of DNA, they analyzed multiple cervical-cell samples collected, during one or two decades, from four women with heteroplasmy at either nt 309 or nt 16189. According to accurate, quantitative analysis by solid-phase minisequencing, the level of heteroplasmy remained stable in the cervical-cell samples from all four women during the time studied. They also analyzed autopsy samples from several different tissues, all containing nt 309 in heteroplasmic form, of one of the women, who was deceased. They concluded their work stating that on the basis of the results, heteroplasmy in the control region of mtDNA seems to be inherited and is not the result of somatic age-related accumulation. On the contrary, in 2002, Wei and Lee reviewed work done to support their view that oxidative stress and oxidative damage are a result of concurrent accumulation of mtDNA mutations and defective antioxidant enzymes in human aging.

In 2002, Von Wurmb-Schwark et al. using a real-time PCR (qPCR) aimed to evaluate the accuracy with which age can be predicted, knowing only the frequency of the common 4977 bp deletion, and derived a statistical formula which describes the confidence with which the 4977 bp frequency predicts age. Their results indicated that the mutation frequency could be used to distinguish between tissue from young and old individuals. However in this data set, while there was considerable agreement of 4977 bp frequency among replicates from the same individual sample, there was substantial diversity of mean mutation frequency between individuals of the same or similar ages, thus the simplest interpretation of these results is that there are biological modifiers of 4977 bp frequency that are age-independent, which may limit the usefulness of this deletion frequency alone as a "molecular forensic clock."

Other authors whose results were not in agreement with these previous studies are Mohamed et al., who in 2004 aimed to study the cell types in human blood carrying the 4977 bp deleted mtDNA and their accumulation with regard to donor age. Their 10 donors displayed differences in the pattern of the accumulation with regard to the different cell types, but no age-dependence was observed, thus the authors concluded that differences of the accumulation pattern might be due to actual individual living behavior or environmental factors.

Later on, the studies of Theves et al. in 2006 focused on the analysis of the A189G mutations, using three different approaches: automated DNA sequencing, Southern blot hybridization using a digoxigenin-labeled oligonucleotide probe, and peptide nucleic acid real-time PCR, due to the fact that this method is more sensitive than DNA sequencing. The analysis of heteroplasmic A189G mutation by the PNA/qPCR method provided a sensitive detection as well as reliable and reproducible quantification of mutant level. This technique allowed the authors to demonstrate the absence of the A189G transition in buccal cells in young individuals and its presence in older individuals from the same maternal lineage, concluding that it is a somatic mutation. Moreover, they demonstrated for the first time the accumulation with age of the A189G mutation in mitotic buccal cells at levels lower than 13%. In postmitotic muscle tissues, mutation accumulation was age-related and reached very high levels in individuals 60 years of age or older. These results could have many implications in heteroplasmy interpretation in forensic caseworks but also in anthropological studies.

Lacan et al. in 2009 and 2011 demonstrated with the same methodology that this somatic point mutation occurs in bone tissue and is related to age. The detection of age-related mutations in bone tissue could help to estimate age-at-death within the context of forensic or/and anthropological identification procedures, when traditional osteological markers studied are absent or inefficient. The authors found that the number of duplications increases from 38 years and that at least one duplicated fragment is present in 50% of cases after 70 years in bone tissue. These results confirm that several age-related mutations can be detected in the D-loop of mtDNA. If the number of duplications does not correlate totally with age in DNA from bone samples, the presence of duplicated fragments can give a first indication of the age of the individual. Indeed in bone, one duplication is systematically detected in over 38-year-old individuals. This result is of course not sufficient to determine the age-at-death but the authors made the hypothesis that the appearance of various other heteroplasmic rearrangements could be found linked with age in the same region of mtDNA extracted from bone tissue.

In a recent article of Zapico and Ubelaker (2015), the authors aim to evaluate the mutations in mtDNA from dentin and pulp and their relation with age. This article analyzes tooth tissue, which has been largely ignored in studies examining the relationship between age and mtDNA mutations, reporting the analysis of age-dependent mtDNA mutations in tooth samples from two different Spanish populations and observing distinct differences between these populations, thus suggesting a possible ancestry effect on patterns of mtDNA mutations. The mtDNA damage was estimated using real-time PCR by the analysis of HV2 of the mitochondrial D-loop. This work studied mtDNA damage, not specific mutations, based on previous approaches developed by various authors using quantitative PCR taking into account that this damage reduces, but does not block, template amplification (Sikorky et al. 2004, Ayala-Torres et al. 2000, Dominguez-Garrido et al. 2009, Salazar and Van Houten 1997, Sawyer and Van Houten 1999). Based on these techniques, the authors developed a methodology that quantifies HV2 amplification in the mitochondria, through first normalizing the amount of mtDNA with respect to the amount of nuclear DNA, as this will be less affected by oxidative damage. Through the application of linear regression analysis, a strong negative correlation between HV2 amplification and age was found in dentin from both populations studied. These results are in agreement with previous correlations found in bone tissue (Lacan et al. 2009) and other soft tissues (Meissner et al. 2008), where they looked for a particular mutation. mtDNA damage in pulp, measured as amplification of HV2, was variable between different ages. With respect to the ancestry component, although there are several studies that correlate certain mitochondrial haplogroups with longevity in different populations (De Benedictis et al. 1999, Niemi et al. 2005), very few studies specifically correlate these haplogroups with the aging process (Dominguez-Garrido et al. 2009, Chen et al. 2012, Katzman et al. 2014), finding that specific polymorphism could be the cause of the decrease in oxidative damage. So, the role of the whole haplogroup in the aging process is not clear. Finally, the outcomes of this study are significant both for enriching the current studies related to aging and mitochondrial damage, leading also to potential ancestry studies, and providing a new and innovative tool for age-at-death estimation in a forensic context.

As described above, some studies have pointed out to the relationship between mtDNA mutations and age in different tissues, which are potentially interesting in aging research and in forensic identification because they could help to improve the estimation of age-at-death. Although the analysis of mtDNA mutations does not reach as high a correlation coefficient as aspartic acid racemization, in a methodological approach it results in an advantage as it is based on PCR technique on different tissues.

Conclusions

Some studies were developed toward aging research and some were conducted to improve the techniques for forensic age estimation. Throughout these two directions, these lines of inquiry added evidence of the accumulation of these mutations with the aging process, leading to an increase of knowledge in this field.

In a forensic scenario, bones and teeth are often preserved long after all other tissues have disappeared. In adults there are several techniques to estimate the age based on a physiological degeneration of skeletal and dental structures (Wittwer-Backofen et al. 2014). However, many endogenous and exogenous factors, pathological conditions and fragmentary remains influence this relationship. For this reason forensic scientists are currently searching for alternative and quantifiable methodologies for age estimation based on the natural process of aging (Speller et al. 2012, Zapico and Ubelaker 2013), for example, tooth-cementum annulations. In this context, the increasing research and knowledge concerning the mutations of mtDNA in different tissues—especially bones and teeth—and their demonstrated relation with age may have an important role for age-at-death estimation in forensic investigations, especially whenever the specific context and the remains available do not permit the application of other traditional identification methodologies.

Keywords: mtDNA, mutations, aging, age-at-death, estimation, forensic, identification

References

Anderson, S., A. T. Bankier, B. G. Barrell, M. H. de Bruijn, A. R. Coulson, J. Drouin et al. 1981. Sequence and organization of the human mitochondrial genome. Nature 290: 457–465.

Ayala-Torres, S., Y. Chen, T. Svoboda, J. Rosenblatt and B. Van Houten. 2000. Analysis of gene-specific DNA damage and repair using quantitative polymerase chain reaction. Methods 22: 135–147.

Beckman, K. B. and B. N. Ames. 1998. The free radical theory of aging matures. Physiol Rev 78: 547–581.

Bodyak, N. D., E. Nekhaeva, J. Y. Wei and K. Khrapko. 2001. Quantification and sequencing of somatic deleted mtDNA in single cells: evidence for partially duplicated mtDNA in aged human tissues. Hum Mol Gen 10(1): 17–24.

Calloway, C. D., R. L. Reynolds, Jr., G.L. Herrin and W. W. Anderson. 2000. The frequency of heteroplasmy in the HVII region of mtDNA differs across tissue types and increases with age. Am J Hum Genet 66: 1384–1397.

Chen, A., N. Raule, A. Chomyn and G. Attardi. 2012. Decreased reactive oxygen species production in cells with mitochondrial haplogroups associated with longevity. PLoS One 7: e46473. doi:10.1371/journal.pone.0046473.

De Benedictis, G., G. Rose, G. Carrieri, M. De Luca, E. Falcone, G. Passarino et al. 1999. Mitochondrial DNA inherited variants are associated with successful aging and longevity in humans. FASEB J 13(12): 1532–1536.

Del Bo, R., A. Bordoni, F. Martinelli Boneschi, M. Crimi, M. Sciacco, N. Bresolin et al. 2002. Evidence and age-related distribution of mtDNA d-loop point mutations in skeletal muscle from healthy subjects and mitochondrial patients. J Neurol Sci 202: 85–91.

Domínguez-Garrido, E., D. Martínez-Redondo, C. Martín-Ruiz, A. Gómez-Durán, E. Ruiz-Pesini, P. Madero et al. 2009. Association of mitochondrial haplogroup J and mtDNA oxidative damage in two different North Spain elderly populations. Biogerontology 10: 435–442.

Harman, D. 1972. The biologic clock: the mitochondria? J Am Geriatr Soc 20: 145–147.

Katzman, S. M., E. S. Strotmeyer, M. A. Nalls, Y. Zhao, S. Mooney, N. Schork et al. 2014. Mitochondrial DNA sequence variation associated with peripheral nerve function in the elderly. J Gerontol (A Biol Sci Med Sci) 2014. doi:10.1093/gerona/glu175.

Lacan, M., C. Theves, S. Amory, C. Keyser, E. Crubezy and J. P. Salles. 2009. Detection of the A189G mtDNA heteroplasmic mutation in relation to age in modern and ancient bones. Int J Legal Med 123: 161–167.

Lacan, M., C. Theves, C. Keyser, A. Farrugia, J. P. Baraybar, E. Crubézy et al. 2011. Detection of age-related duplications in mtDNA from human muscles and bones. Int J Legal Med 125: 293–300.

Lagerstrom-Fermer, M., C. Olsson, L. Forsgren and A. C. Syvanen. 2001. Heteroplasmy of the human mtDNA control region remains constant during life. Am J Hum Genet 68: 1299–1301.

Meissner, C., N. von Wurmb and M. Oehmichen. 1997. Detection of the age-dependent 4977 bp deletion of mitochondrial DNA. A pilot study. Int J Legal Med 110: 288–291.

Meissner, C., N. von Wurmb, B. Schimansky and M. Oehmichen. 1999. Estimation of age at death based on quantitation of the 4977-bp deletion of human mitochondrial DNA in skeletal muscle. Forensic Sci Int 105: 115–124.

Meissner, C., P. Bruse and M. Oehmichen. 2006. Tissue-specific deletion patterns of the mitochondrial genome with advancing age. Exp Geront 41(5): 518–24.

Meissner, C., P. Bruse, S. A. Mohamed, A. Schulz, H. Warnk, T. Storm et al. 2008. The 4977 bp deletion of mitochondrial DNA in human skeletal muscle, heart and different areas of the brain: a useful biomarker or more? Exp Geront 43(7): 645–652.

Mohamed, S. A., D. Wesch, A. Blumenthal, P. Bruse, K. Windler and M. Ernst. 2004. Detection of the 4977 bp deletion of mitochondrial DNA in different human blood cells. Exp Geront 39(2): 181–8.

Mornstad, H., H. Pfeiffer, C. Yoon and A. Teivens. 1999. Demonstration and semi-quantification of mtDNA from human dentine and its relation to age. Int J Legal Med 112: 98–100.

Munscher, C., T. Rieger, J. Muller-Hocker and B. Kadenbach. 1993. The point mutation of mitochondrial DNA characteristic for MERRF disease is found also in healthy people of different ages. FEBS Lett 317: 27–30.

Niemi, A. K., J. S. Moilanen, M. Tanaka, A. Hervonen, M. Hurme, T. Lehtimäki et al. 2005. A combination of three common inherited mitochondrial DNA polymorphisms promotes longevity in Finnish and Japanese subjects. Eur J Human Genet 13(2): 166–170.

Papiha, S. S., H. Rathod, I. Briceno, J. Pooley and H. K. Datta. 1998. Age related somatic mitochondrial DNA deletions in bone. J Clin Pathol 51: 117–120.

Salazar, J. J. and B. Van Houten. 1997. Preferential mitochondrial DNA injury caused by glucose oxidase as a steady generator of hydrogen peroxide in human fibroblasts. Mutat Res 385(2): 139–149.

Sawyer, D. E. and B. Van Houten. 1999. Repair of DNA damage in mitochondria. Mutat Res 434(3): 161–176.

Shokolenko, I. N., G. L. Wilson and M. F. Alexeyev. 2014. Aging: a mitochondrial DNA perspective, critical analysis and an update. World J Exp Med 4(4): 46–57.

Sikorsky, J. A., D. A. Primerano, T. W. Fenger and J. Denvir. 2004. Effect of DNA damage on PCR amplification efficiency with the relative threshold cycle method. Biochem Biophys Res Commun 323(3): 823–830.

Simonetti, S., X. Chen, S. DiMauro and E. A. Schon. 1992. Accumulation of deletions in human mitochondrial DNA during normal aging: analysis by quantitative PCR. Biochimica et Biophysica Acta 1180: 113–122.

Speller, C. F., K. L. Spalding, B. A. Buchholz, D. Hildebrand, J. Moore, R. Mathewes et al. 2012. Personal identification of cold case remains through combined contribution from anthropological, mtDNA, and bomb-pulse dating analyses. J Forensic Sci 57(5): 1354–1360.

Theves, C., C. Keyser-Tracqui, E. Crubezy, J.P. Salles, B. Ludes and N. Telmon. 2006. Detection and quantification of the age-related point mutation A189G in the human mitochondrial DNA. J Forensic Sci 51: 865–873.

von Wurmb-Schwark, N., R. Higuchi, A. P. Fenech, C. Elfstroem, C. Meissner and M. Oehmichen et al. 2002. Quantification of human mitochondrial DNA in a real time PCR. Forensic Sci Int 126(1): 34–9.

Wang, Y., Y. Michikawa, C. Mallidis, Y. Bai, L. Woodhouse, K. E. Yarasheski et al. 2001. Muscle-specific mutations accumulate with aging in critical human mtDNA control sites for replication. Proc National Academy of Sciences of the United States of America 98: 4022–4027.

Wei, Y. H. and H. C. Lee. 2002. Oxidative stress, mitochondrial DNA mutation, and impairment of antioxidant enzymes in aging. Exp Biol Med 227: 671–682.

Wei, Y. H., S. B. Wu, Y. S. Ma and H. C. Lee. 2009. Respiratory function decline and DNA mutation in mitochondria, oxidative stress and altered gene expression during aging. Chang Gung Medical Journal 32: 113–132.

Wittwer-Backofen, U., M. Kastner, D. Moller, M. Vohberger, S. Lutz-Bonengel and D. Speck. 2014. Ambiguous provenance? Experience with provenance analysis of human remains from Namibia in the Alexander Ecker collection. Anthropol Anz 71(1-2): 65–86.

Yao, Y. G., S. Kajigava and N. S. Young. 2015. Mitochondrial DNA mutations in single human blood cells. Mutat Res 779: 68–77.

Zapico, S. C. and D. H. Ubelaker. 2013. Applications of physiological bases of ageing to forensic sciences. Estimation of age-at-death. Ageing Res Rev 12(2): 605–617.

Zapico, S. C. and D. H. Ubelaker. 2015. Relationship between mitochondrial DNA mutations and aging. Estimation of Age-at-death. J Gerontol (A Biol Sci Med Sci) 1–6.

Section V
Epigenetics

23

Epigenetics
Its Role in Aging, Diseases and Biological Age Estimation

Christian Thomas[a] and *Sara C. Zapico**

INTRODUCTION

Epigenetics, a new emerging field in biomedical science, is defined as the study of heritable changes in gene function that do not change the DNA sequence, also describing the mechanisms that enable cells to respond quickly to environmental changes and providing a link between genes and the environment (Egger et al. 2004, Turan et al. 2010). DNA is wrapped around clusters (octamers) of globular histone proteins to form nucleosomes, which are organized into chromatin. Changes to the structure of chromatin influence gene expression because when chromatin is condensed, genes are inactivated, while they are expressed when chromatin is opened (Peterson and Laniel 2004, Rountree et al. 2001). These changes are controlled by reversible epigenetic patterns of DNA methylation and histone modifications (Feinberg and Tycko 2004). Alterations in these patterns can dysregulate gene expression, leading to diseases and accelerated aging.

The present chapter will give an overview of the implications of epigenetics in aging, diseases and its application to biological age estimation.

Smithsonian Institution, National Museum of Natural History, Anthropology Department 10th and Constitution Ave, NW, PO Box 37012 Washington, DC 20560 USA.
[a] Email: crf.thomas@gmail.com
* Corresponding author: saiczapico@gmail.com

Epigenetic Modifications

DNA methylation

DNA methylation is a covalent modification of cytosine residues in cytosine/guanine-rich regions, called CpGs islands. These islands are placed in gene regulatory elements like promoters and other genomic sites, such as intergenic regions and repetitive elements (Mehler 2008). Proteins with the Methyl-CpG-binding domain bind specifically to these islands, inducing, generally, transcriptional repression, although activation has also been reported. Enzymes involved in this process are DNA methyltransferases (DNMTs), catalyzing this reaction by transferring methyl groups from S-adenosylmethionine to cytosine residues (Qureshi and Mehler 2011) and Ten-eleven translocator enzymes (TETs), involved in methyl cytosine hydroxylation, which ultimately remove the mark (Bermejo-Alvarez et al. 2015). Demethylation can be achieved passively during DNA replication (Kagiwada et al. 2013), or actively mainly by TET enzymes (Tahiliani et al. 2009).

Chromatin conformation and changes in histone proteins

Changes in chromatin conformations modulate the accessibility of regulatory and functional genomic regions to other nuclear factors, including those that mediate transcription and DNA replication and repair (Qureshi and Mehler 2011). Moreover, chromatin can be modified and/or rearranged at the level of the nucleosome by post-translational modifications of histones. These modifications are carried out in a reversible manner by Histone Acetyltransferases (HAT), Histone Deacetylases (HDACs) and histone methyltransferases/demethylases. These enzymes are recruited to ensure that a receptive DNA region is either accessible for transcription or targeted for silencing (Espino et al. 2005, Elgin and Grewal 2003, Hendrich et al. 2001, Jenuwein and Allis 2001). Transcriptional and epigenetic factors with specific protein domains bind to these histone modifications (Ruthenburg et al. 2007).

Non-coding RNAs (ncRNA)

ncRNA regulation describes the action of RNA molecules derived from the genome but not translated into proteins (Mehler 2008), like transfer RNAs and ribosomal RNAs. ncRNAs are classified as "short" or "long" (> 200 nucleotides). Short ncRNAs include microRNAs (miRNAs), short-interfering RNAs (siRNAs), PIWI-interacting RNAs, 3'untranslated region-derived RNAs and many others (Mattick et al. 2010). Long ncRNAs (lncRNAs) include long intergenic ncRNAs and enhancer-like RNAs as well as those that are encoded in the genome in antisense, intronic and overlapping configurations relative to protein-coding genes (Qureshi et al. 2010). These ncRNAs have a broad range of roles like mediation of DNA methylation, histone modification and higher-order chromatin remodeling, transcription and RNA post-transcriptional processing, transport and translation (Mattick et al. 2009).

Epigenetics-based Technologies

The number of techniques to analyze the epigenetic changes in the genome is constantly growing. For DNA methylation, bisulfite DNA treatment followed by clonal sequencing has been traditionally applied. However, the increased availability of genome-wide approaches such as microarray or deep sequencing have allowed for high throughput analyses. Based on bisulfite conversion, BeadArray (Illumina) platform (with bisulfite conversion) analyzes a moderate number of CpGs across the genome in a large number of samples (Fouse et al. 2010). Another strategy has been the use of microarrays with methylation-sensitive and—insensitive restriction endonuclease digestion or affinity-based enrichment of methylated DNA by Chromatin Immunoprecipitation (ChIP), also known as Methylated DNA Immunoprecipitation (MeDIP) (Ben-Avraham et al. 2012, Qureshi and Mehler 2011). Next-Generation Sequencing (NGS) based methods, mostly linked to ChIP (MeDIP-seq) but also linked to Bisulfite Treatment (BS-seq), have been established for detection of DNA methylation across the entire genome, developing a variety of platforms to increase the number of CpGs for which methylation can be assessed (Ben-Avraham et al. 2012).

Histone modifications can be analyzed by ChIP in a similar way to the method used to analyze DNA methylation, i.e., coupled to qPCR, microarrays (ChIP-chip) or sequencing (ChIP-Sep). Other complementary techniques can be used to determine chromatin accessibility and nucleosome dynamics like sonication followed by sequencing (Sono-Seq) or FAIRE-Seq (formaldehyde-assisted isolation of regulatory elements followed by sequencing) (Qureshi and Mehler 2011). Likewise, Micrococcal nuclease digestion and sequencing (MNase-Seq) can be used to define maps of nucleosome locations throughout the genome (Schones et al. 2008, Kaplan et al. 2009).

The study of ncRNAs has been driven by ultra high throughput RNA-Seq. New NGS techniques will improve this analysis even in a single cell with very high resolution (Qureshi and Mehler 2011). Other approaches, like the use of real-time imaging of transcription in living cells, can capture the act of transcription, including the kinetics of RNA polymerase movement, the association of transcription factors and the progression of the polymerase of the gene. Likewise, the use of nanoparticles or antisense oligonucleotides, labeled with radionuclide, can be coupled with positron emission tomography or magnetic resonance imaging to noninvasively image cerebral RNAs in live animals (Lendvai et al. 2009, Chen et al. 2011, Liu et al. 2007, Hwang do et al. 2010).

Epigenetics and Aging

The "epigenetic theory of aging" postulates that non-adaptive epigenetic alterations are fundamental to aging. Epimutations accumulate with age, leading to activation of genes, normally epigenetically downregulated (Bennett-Baker et al. 2003, Salpea et al. 2012). Several studies in humans assessed this hypothesis, ratifying that epigenetic changes are a function of age in many tissues and can be used as a marker of chronological age (Ben-Avraham et al. 2012).

Bocker et al. (Bocker et al. 2011) assessed the methylation profile of CD34[+] hematopoietic progenitor cells and demonstrated that epigenetic changes occur with aging, with increased methylation of Polycomb chromatin genes. In CD4$^+$ blood cells, 360 CpG sites have been identified that were either hypo- or hyper-methylated with age. This aging-associated differentially-methylated regions signature has been demonstrated in buccal cells (Rakyan et al. 2010). This research has been extended to human brains, confirming the importance of methylation in the mechanism of aging (Hernandez et al. 2011).

Bocklandt et al. (Bocklandt et al. 2011) showed that 88 methylation sites (nearly 80 genes) in homozygote twins (21–55 years old) demonstrated significant changes with age, predicting individual's age with an accuracy of 5.2 years. Although these previous studies confirmed the correlation of epigenetic changes with age, these changes depend on the tissue. The study of Christensen et al. (Christensen et al. 2009) verified this hypothesis, illustrating that the methylation pattern with age can predict the tissue of origin. Gentilini et al. (Gentilini et al. 2015) analyzed the stochastic epigenetic mutations in the HUMARA locus, finding an increase with age and mediation of the skewing of X-Chromosome inactivation.

Regarding ncRNA, miRNAs affect gene expression during the aging process in mice and modulate senescence in human cell lines (Smith-Vikos and Slack 2012). For instance, miR-34a has been designated as an aging marker in several tissues and systems. Boon et al. (Boon et al. 2013) demonstrated that miR-34a is upregulated in the aging heart and its inhibition reduces cell death and fibrosis following acute myocardial infarction. miRNA-339 and miRNA-556 bind three untranslated regions of Klotho mRNA, an anti-aging protein whose expression decreases in normal aging of mice. Experimental results pointed out that these miRNAs can directly decrease this protein, playing a role in its regulation (Mehi et al. 2014). Apart from these intracellular miRNAs, several circulatory miRNAs seem to be common for the major age-related diseases since they have two opposite roles, activating and inhibiting inflammatory pathways (Olivieri et al. 2013).

Chang et al. (Chang et al. 2013) demonstrated the role of lncRNAs in aging and rejuvenated human skin, identifying 1293 rejuvenated genes, including these lncRNAs. Abdelmohsen et al. (Abdelmohsen et al. 2013) identified senescence-associated lncRNAs (SAL-RNAs) in human fibroblasts.

Epigenetics and Diseases

Common human age-related diseases are accompanied by a loss of genomic DNA methylation (Ben-Avraham et al. 2012).

Epigenetics play a potential role in the pathophysiology of neurodegenerative diseases (Qureshi and Mehler 2011). Alzheimer's Disease (AD) genes are subject to regulation by DNA methylation: tissues from AD patients show differential profiles of DNA methylation specifically for AD-related genes, such as Amyloid Precursor Protein (APP) and others (West et al. 1995, Mastroeni et al. 2009, Mastroeni et al. 2010). Abnormalities in histone and chromatin regulation have been also reported in AD brains (Chuang et al. 2009, Lee and Ryu 2010). Likewise, various miRNAs and lncRNAs have been demonstrated to play a role in regulating AD-related genes

(Guo et al. 2006, Cogswell et al. 2008, Faghihi et al. 2008). In the same line of research, epigenetic dysregulation is implicated in the pathophysiology of Huntington's Disease (Buckley et al. 2010) through DNA methylation (Seong et al. 2010) and miRNAs (Johnson et al. 2008, Packer et al. 2008, Marti et al. 2010). Genes associated with Parkinson's Disease (PD) are also epigenetically controlled: a decrease in DNA methylation of SNCA (alpha-synuclein) gene has been reported in idiopathic PD patients. This decrease induced an increase in SNCA expression, promoting PD pathogenesis (Jowaed et al. 2010, Matsumoto et al. 2010). Also, the expression of this protein has been regulated by miRNAs and lncRNAs (Junn et al. 2009, Guo et al. 2006). Amyothropic Lateral Sclerosis (ALS) is another disease associated with epigenetic modifications. Variants of the Elongator Protein 3 (ELP), a gene implicated in DNA demethylation, with HAT enzyme activity and that regulates nucleosome positioning, are associated with sporadic ALS (Simpson et al. 2009, Okada et al. 2010, Winkler et al. 2002, Matsumoto et al. 2011). The dysregulation of various miRNAs has also been implicated with ALS (Buratti et al. 2010, Williams et al. 2009).

Epigenetic changes also play a role in immunity and related disorders. Shifts in both acetylation and methylation are required to coordinate DNA accessibility and permit recombination, allowing cells to mount an immune response against a specific antigen (Wilson et al. 2005). Abnormal DNA methylation has been observed in patients with lupus and arthritis, which raises the possibility that altered DNA methylation patterns may contribute to display idiopathic autoimmunity (Oelke and Richardson 2004, Richardson 2003, Watts et al. 2005).

A number of risk factors for Cardiovascular Disease (CVD) such as nutrition, smoking, pollution, stress and the circadian rhythm, have been suggested to act through epigenetic modifications. Prospective investigations are needed to determine whether individuals who are exposed to various environmental challenges and develop risk factors accumulate epigenetic alterations over time whether these alterations increase the incidence of CVD and to determine whether these environmentally induced epigenetic changes associated with CVD risk are heritable (reviewed in (Ordovas and Smith 2010)).

Feinberg and Vogelstein were the first to associate differences in DNA methylation status to cancer (Feinberg and Vogelstein 1983). Since then, there has been an explosion of research regarding the implication of epigenetics in cancer. Altered DNA methylation patterns change the expression of cancer-associated genes (Rodenhiser and Mann 2006). DNA hypomethylation activates oncogenes and initiates chromosome instability (Kazazian 2004, Narayan et al. 1998, Gaudet et al. 2003). In contrast, DNA hypermethylation induces silencing of tumor suppressor genes. The incidence of hypermethylation, mainly in sporadic cancers, varies depending on the gene and the tumor type; for example, BRCA1 hypermethylation is primarily associated with 10–20% of sporadic breast and ovarian cancers (Esteller et al. 2001). These epigenetic changes can be used in the molecular diagnosis of a variety of cancers (Cottrell 2004). Likewise, several epigenetic therapies are currently being studied in clinical trials or have been approved for specific cancer types (Egger et al. 2004, Balch et al. 2005, Laird 2005).

Epigenetics for Age-at-Death Estimation

As described previously, epigenetic drift is associated with aging. For that reason, over the past few years, researchers have been trying to identify DNA methylation markers that are significantly correlated with age, in order to be used for age-at-death estimation in the forensic sciences. Bocklandt et al. (Bocklandt et al. 2011) found that methylation in the promoter of the EDARADD gene showed a linear correlation with age in saliva samples over a 5-decade period. More recently, EDARADD methylation in blood samples has been reported to be able to predict chronological age with an error of only 3.34 years. Other authors used ELOVL2, finding a good correlation with age in blood samples, although the difference between chronological and predicted age was approximately five years (Garagnani et al. 2013, Zbiec-Piekarska et al. 2015a). Another age prediction model (Weidner et al. 2014) used ASPA, ITG2B and EDARADD genes, finding an accurate age prediction of 5.4 years. A meta-analysis on the association between DNA methylation and age in blood resulted in 44 genes whose methylated levels were highly correlated with age, including EDARADD and ELOVL2 (Bacalini et al. 2015). Zbiec-Piekarska et al. have recently improved their age estimation prediction using ELOVL2, Clorf132, TRIM59, KLF14 and FHL2 to obtain a MAD of 3.9 years (Zbiec-Piekarska et al. 2015b). More recently, Bekaert et al. (Bekaert et al. 2015) evaluated the accuracy of using four age-associated genes ASPA, PDE4C, ELOVL2 and EDARADD for age prediction in blood samples, finding that ELOVL2 showed the highest accuracy with a MAD of 3.75 years. Moreover, the study was extended to samples of teeth, the first of its kind to use this type of sample, and obtained a MAD of 4.86 years. In spite of these previous works, further research is needed in other tissues and samples to improve age-at-death estimation using this valuable and innovative tool.

Conclusions

Epigenetics is a new and growing field in biomedical research. Although different studies pointed out its implication in aging and diseases and its potential application for age-at-death estimation in forensic science, further studies are needed to improve the knowledge in this field, leading to the development of new therapeutic approaches and novel applications for forensic laboratories that could be used routinely.

Keywords: Epigenetics, DNA methylation, histone, nucleosome, chromatin, ncRNA, lncRNA, miRNA, aging, neurodegenerative diseases, cardiovascular diseases, immune system, cancer, age-at-death estimation, ELOVL2

References

Abdelmohsen, K., A. Panda, M. J. Kang, J. Xu, R. Selimyan, J. H. Yoon et al. 2013. Senescence-associated lncRNAs: senescence-associated long noncoding RNAs. Aging Cell 12(5): 890–900. doi: 10.1111/acel.12115.
Bacalini, M. G., A. Boattini, D. Gentilini, E. Giampieri, C. Pirazzini, C. Giuliani et al. 2015. A meta-analysis on age-associated changes in blood DNA methylation: results from an original analysis pipeline for Infinium 450k data. Aging (Albany NY) 7(2): 97–109.

Balch, C., J. S. Montgomery, H. I. Paik, S. Kim, S. Kim, T. H. Huang et al. 2005. New anti-cancer strategies: epigenetic therapies and biomarkers. Front Biosci 10: 1897–931.

Bekaert, B., A. Kamalandua, S. C. Zapico, W. Van de Voorde and R. Decorte. 2015. Improved age determination of blood and teeth samples using a selected set of DNA methylation markers. Epigenetics 1–9. doi: 10.1080/15592294.2015.1080413.

Ben-Avraham, D., R. H. Muzumdar and G. Atzmon. 2012. Epigenetic genome-wide association methylation in aging and longevity. Epigenomics 4(5): 503–9. doi: 10.2217/epi.12.41.

Bennett-Baker, P. E., J. Wilkowski and D. T. Burke. 2003. Age-associated activation of epigenetically repressed genes in the mouse. Genetics 165(4): 2055–62.

Bermejo-Alvarez, P., P. Ramos-Ibeas, K. E. Park, A. P. Powell, L. Vansandt, B. Derek et al. 2015. Tet-mediated imprinting erasure in H19 locus following reprogramming of spermatogonial stem cells to induced pluripotent stem cells. Sci Rep 5: 13691. doi: 10.1038/srep13691.

Bocker, M. T., I. Hellwig, A. Breiling, V. Eckstein, A. D. Ho and F. Lyko. 2011. Genome-wide promoter DNA methylation dynamics of human hematopoietic progenitor cells during differentiation and aging. Blood 117(19): e182–9. doi: 10.1182/blood-2011-01-331926.

Bocklandt, S., W. Lin, M. E. Sehl, F. J. Sanchez, J. S. Sinsheimer, S. Horvath et al. 2011. Epigenetic predictor of age. PLoS One 6(6): e14821. doi: 10.1371/journal.pone.0014821.

Boon, R. A., K. Iekushi, S. Lechner, T. Seeger, A. Fischer, S. Heydt et al. 2013. MicroRNA-34a regulates cardiac ageing and function. Nature 495(7439): 107–10. doi: 10.1038/nature11919.

Buckley, N. J., R. Johnson, C. Zuccato, A. Bithell and E. Cattaneo. 2010. The role of REST in transcriptional and epigenetic dysregulation in Huntington's disease. Neurobiol Dis 39(1): 28–39. doi: 10.1016/j.nbd.2010.02.003.

Buratti, E., L. De Conti, C. Stuani, M. Romano, M. Baralle and F. Baralle. 2010. Nuclear factor TDP-43 can affect selected microRNA levels. FEBS J 277(10): 2268–81. doi: 10.1111/j.1742-4658.2010.07643.x.

Chang, A. L., P. H. Bitter, Jr., K. Qu, M. Lin, N. A. Rapicavoli and H. Y. Chang. 2013. Rejuvenation of gene expression pattern of aged human skin by broadband light treatment: a pilot study. J Invest Dermatol 133(2): 394–402. doi: 10.1038/jid.2012.287.

Chen, A. K., W. J. Rhee, G. Bao and A. Tsourkas. 2011. Delivery of molecular beacons for live-cell imaging and analysis of RNA. Methods Mol Biol 714: 159–74. doi: 10.1007/978-1-61779-005-8_10.

Christensen, B. C., E. A. Houseman, C. J. Marsit, S. Zheng, M. R. Wrensch, J. L. Wiemels et al. 2009. Aging and environmental exposures alter tissue-specific DNA methylation dependent upon CpG island context. PLoS Genet 5(8): e1000602. doi: 10.1371/journal.pgen.1000602.

Chuang, D. M., Y. Leng, Z. Marinova, H. J. Kim and C. T. Chiu. 2009. Multiple roles of HDAC inhibition in neurodegenerative conditions. Trends Neurosci 32(11): 591–601. doi: 10.1016/j.tins.2009.06.002.

Cogswell, J. P., J. Ward, I. A. Taylor, M. Waters, Y. Shi, B. Cannon et al. 2008. Identification of miRNA changes in Alzheimer's disease brain and CSF yields putative biomarkers and insights into disease pathways. J Alzheimers Dis 14(1): 27–41.

Cottrell, S. E. 2004. Molecular diagnostic applications of DNA methylation technology. Clin Biochem 37(7): 595–604. doi: 10.1016/j.clinbiochem.2004.05.010.

Egger, G., G. Liang, A. Aparicio and P. A. Jones. 2004. Epigenetics in human disease and prospects for epigenetic therapy. Nature 429(6990): 457–63. doi: 10.1038/nature02625.

Elgin, S. C. and S. I. Grewal. 2003. Heterochromatin: silence is golden. Curr Biol 13(23): R895–8.

Espino, P. S., B. Drobic, K. L. Dunn and J. R. Davie. 2005. Histone modifications as a platform for cancer therapy. J Cell Biochem 94(6): 1088–102. doi: 10.1002/jcb.20387.

Esteller, M., P. G. Corn, S. B. Baylin and J. G. Herman. 2001. A gene hypermethylation profile of human cancer. Cancer Res 61(8): 3225–9.

Faghihi, M. A., F. Modarresi, A. M. Khalil, D. E. Wood, B. G. Sahagan, T. E. Morgan et al. 2008. Expression of a noncoding RNA is elevated in Alzheimer's disease and drives rapid feed-forward regulation of beta-secretase. Nat Med 14(7): 723–30. doi: 10.1038/nm1784.

Feinberg, A. P. and B. Vogelstein. 1983. Hypomethylation distinguishes genes of some human cancers from their normal counterparts. Nature 301(5895): 89–92.

Feinberg, A. P. and B. Tycko. 2004. The history of cancer epigenetics. Nat Rev Cancer 4(2): 143–53. doi: 10.1038/nrc1279.

Fouse, S. D., R. O. Nagarajan and J. F. Costello. 2010. Genome-scale DNA methylation analysis. Epigenomics 2(1): 105–17. doi: 10.2217/epi.09.35.

Garagnani, P., C. Giuliani, C. Pirazzini, F. Olivieri, M. G. Bacalini, R. Ostan et al. 2013. Centenarians as super-controls to assess the biological relevance of genetic risk factors for common age-related diseases: a proof of principle on type 2 diabetes. Aging (Albany NY) 5(5): 373–85.

Gaudet, F., J. G. Hodgson, A. Eden, L. Jackson-Grusby, J. Dausman, J. W. Gray et al. 2003. Induction of tumors in mice by genomic hypomethylation. Science 300(5618): 489–92. doi: 10.1126/science.1083558.

Gentilini, D., P. Garagnani, S. Pisoni, M. G. Bacalini, L. Calzari, D. Mari et al. 2015. Stochastic epigenetic mutations (DNA methylation) increase exponentially in human aging and correlate with X chromosome inactivation skewing in females. Aging (Albany NY) 7(8): 568–78.

Guo, J. H., H. P. Cheng, L. Yu and S. Zhao. 2006. Natural antisense transcripts of Alzheimer's disease associated genes. DNA Seq 17(2): 170–3.

Hendrich, B., J. Guy, B. Ramsahoye, V. A. Wilson and A. Bird. 2001. Closely related proteins MBD2 and MBD3 play distinctive but interacting roles in mouse development. Genes Dev 15(6): 710–23. doi: 10.1101/gad.194101.

Hernandez, D. G., M. A. Nalls, J. R. Gibbs, S. Arepalli, M. van der Brug, S. Chong et al. 2011. Distinct DNA methylation changes highly correlated with chronological age in the human brain. Hum Mol Genet 20(6): 1164–72. doi: 10.1093/hmg/ddq561.

Hwang do, W., I. C. Song, D. S. Lee and S. Kim. 2010. Smart magnetic fluorescent nanoparticle imaging probes to monitor microRNAs. Small 6(1): 81–8. doi: 10.1002/smll.200901262.

Jenuwein, T. and C. D. Allis. 2001. Translating the histone code. Science 293(5532): 1074–80. doi: 10.1126/science.1063127.

Johnson, R., C. Zuccato, N. D. Belyaev, D. J. Guest, E. Cattaneo and N. J. Buckley. 2008. A microRNA-based gene dysregulation pathway in Huntington's disease. Neurobiol Dis 29(3): 438–45. doi: 10.1016/j.nbd.2007.11.001.

Jowaed, A., I. Schmitt, O. Kaut and U. Wullner. 2010. Methylation regulates alpha-synuclein expression and is decreased in Parkinson's disease patients' brains. J Neurosci 30(18): 6355–9. doi: 10.1523/JNEUROSCI.6119-09.2010.

Junn, E., K. W. Lee, B. S. Jeong, T. W. Chan, J. Y. Im and M. M. Mouradian. 2009. Repression of alpha-synuclein expression and toxicity by microRNA-7. Proc Natl Acad Sci U S A 106(31): 13052–7. doi: 10.1073/pnas.0906277106.

Kagiwada, S., K. Kurimoto, T. Hirota, M. Yamaji and M. Saitou. 2013. Replication-coupled passive DNA demethylation for the erasure of genome imprints in mice. EMBO J 32(3): 340–53. doi: 10.1038/emboj.2012.331.

Kaplan, N., I. K. Moore, Y. Fondufe-Mittendorf, A. J. Gossett, D. Tillo, Y. Field et al. 2009. The DNA-encoded nucleosome organization of a eukaryotic genome. Nature 458(7236): 362–6. doi: 10.1038/nature07667.

Kazazian, H. H., Jr. 2004. Mobile elements: drivers of genome evolution. Science 303(5664): 1626–32. doi: 10.1126/science.1089670.

Laird, P. W. 2005. Cancer epigenetics. Hum Mol Genet 14 Spec No 1: R65–76. doi: 10.1093/hmg/ddi113.

Lee, J. and H. Ryu. 2010. Epigenetic modification is linked to Alzheimer's disease: is it a maker or a marker? BMB Rep 43(10): 649–55. doi: 10.5483/BMBRep.2010.43.10.649.

Lendvai, G., S. Estrada and M. Bergstrom. 2009. Radiolabelled oligonucleotides for imaging of gene expression with PET. Curr Med Chem 16(33): 4445–61.

Liu, C. H., Y. R. Kim, J. Q. Ren, F. Eichler, B. R. Rosen and P. K. Liu. 2007. Imaging cerebral gene transcripts in live animals. J Neurosci 27(3): 713–22. doi: 10.1523/JNEUROSCI.4660-06.2007.

Marti, E., L. Pantano, M. Banez-Coronel, F. Llorens, E. Minones-Moyano, S. Porta et al. 2010. A myriad of miRNA variants in control and Huntington's disease brain regions detected by massively parallel sequencing. Nucleic Acids Res 38(20): 7219–35. doi: 10.1093/nar/gkq575.

Mastroeni, D., A. McKee, A. Grover, J. Rogers and P. D. Coleman. 2009. Epigenetic differences in cortical neurons from a pair of monozygotic twins discordant for Alzheimer's disease. PLoS One 4(8): e6617. doi: 10.1371/journal.pone.0006617.

Mastroeni, D., A. Grover, E. Delvaux, C. Whiteside, P. D. Coleman and J. Rogers. 2010. Epigenetic changes in Alzheimer's disease: decrements in DNA methylation. Neurobiol Aging 31(12): 2025–37. doi: 10.1016/j.neurobiolaging.2008.12.005.

Matsumoto, L., H. Takuma, A. Tamaoka, H. Kurisaki, H. Date, S. Tsuji et al. 2010. CpG demethylation enhances alpha-synuclein expression and affects the pathogenesis of Parkinson's disease. PLoS One 5(11): e15522. doi: 10.1371/journal.pone.0015522.

Matsumoto, T., C. S. Yun, H. Yoshikawa and H. Nishida. 2011. Comparative studies of genome-wide maps of nucleosomes between deletion mutants of elp3 and hos2 genes of Saccharomyces cerevisiae. PLoS One 6(1): e16372. doi: 10.1371/journal.pone.0016372.

Mattick, J. S., P. P. Amaral, M. E. Dinger, T. R. Mercer and M. F. Mehler. 2009. RNA regulation of epigenetic processes. Bioessays 31(1): 51–9. doi: 10.1002/bies.080099.

Mattick, J. S., R. J. Taft and G. J. Faulkner. 2010. A global view of genomic information--moving beyond the gene and the master regulator. Trends Genet 26(1): 21–8. doi: 10.1016/j.tig.2009.11.002.

Mehi, S. J., A. Maltare, C. R. Abraham and G. D. King. 2014. MicroRNA-339 and microRNA-556 regulate Klotho expression *in vitro*. Age (Dordr) 36(1): 141–9. doi: 10.1007/s11357-013-9555-6.

Mehler, M. F. 2008. Epigenetic principles and mechanisms underlying nervous system functions in health and disease. Prog Neurobiol 86(4): 305–41. doi: 10.1016/j.pneurobio.2008.10.001.

Narayan, A., W. Ji, X. Y. Zhang, A. Marrogi, J. R. Graff, S. B. Baylin et al. 1998. Hypomethylation of pericentromeric DNA in breast adenocarcinomas. Int J Cancer 77(6): 833–8.

Oelke, K. and B. Richardson. 2004. Decreased T cell ERK pathway signaling may contribute to the development of lupus through effects on DNA methylation and gene expression. Int Rev Immunol 23(3-4): 315–31. doi: 10.1080/08830180490452567.

Okada, Y., K. Yamagata, K. Hong, T. Wakayama and Y. Zhang. 2010. A role for the elongator complex in zygotic paternal genome demethylation. Nature 463(7280): 554–8. doi: 10.1038/nature08732.

Olivieri, F., M. R. Rippo, A. D. Procopio and F. Fazioli. 2013. Circulating inflamma-miRs in aging and age-related diseases. Front Genet 4: 121. doi: 10.3389/fgene.2013.00121.

Ordovas, J. M. and C. E. Smith. 2010. Epigenetics and cardiovascular disease. Nat Rev Cardiol 7(9): 510–9. doi: 10.1038/nrcardio.2010.104.

Packer, A. N., Y. Xing, S. Q. Harper, L. Jones and B. L. Davidson. 2008. The bifunctional microRNA miR-9/miR-9* regulates REST and CoREST and is downregulated in Huntington's disease. J Neurosci 28(53): 14341–6. doi: 10.1523/JNEUROSCI.2390-08.2008.

Peterson, C. L. and M. A. Laniel. 2004. Histones and histone modifications. Curr Biol 14(14): R546–51. doi: 10.1016/j.cub.2004.07.007.

Qureshi, I. A., J. S. Mattick and M. F. Mehler. 2010. Long non-coding RNAs in nervous system function and disease. Brain Res 1338: 20–35. doi: 10.1016/j.brainres.2010.03.110.

Qureshi, I. A. and M. F. Mehler. 2011. Advances in epigenetics and epigenomics for neurodegenerative diseases. Curr Neurol Neurosci Rep 11(5): 464–73. doi: 10.1007/s11910-011-0210-2.

Rakyan, V. K., T. A. Down, S. Maslau, T. Andrew, T. P. Yang, H. Beyan et al. 2010. Human aging-associated DNA hypermethylation occurs preferentially at bivalent chromatin domains. Genome Res 20(4): 434–9. doi: 10.1101/gr.103101.109.

Richardson, B. 2003. DNA methylation and autoimmune disease. Clin Immunol 109(1): 72–9.

Rodenhiser, D. and M. Mann. 2006. Epigenetics and human disease: translating basic biology into clinical applications. CMAJ 174(3): 341–8. doi: 10.1503/cmaj.050774.

Rountree, M. R., K. E. Bachman, J. G. Herman and S. B. Baylin. 2001. DNA methylation, chromatin inheritance, and cancer. Oncogene 20(24): 3156–65. doi: 10.1038/sj.onc.1204339.

Ruthenburg, A. J., H. Li, D. J. Patel and C. D. Allis. 2007. Multivalent engagement of chromatin modifications by linked binding modules. Nat Rev Mol Cell Biol 8(12): 983–94. doi: 10.1038/nrm2298.

Salpea, P., V. R. Russanova, T. H. Hirai, T. G. Sourlingas, K. E. Sekeri-Pataryas, R. Romero et al. 2012. Postnatal development- and age-related changes in DNA-methylation patterns in the human genome. Nucleic Acids Res 40(14): 6477–94. doi: 10.1093/nar/gks312.

Schones, D. E., K. Cui, S. Cuddapah, T. Y. Roh, A. Barski, Z. Wang et al. 2008. Dynamic regulation of nucleosome positioning in the human genome. Cell 132(5): 887–98. doi: 10.1016/j.cell.2008.02.022.

Seong, I. S., J. M. Woda, J. J. Song, A. Lloret, P. D. Abeyrathne, C. J. Woo et al. 2010. Huntingtin facilitates polycomb repressive complex 2. Hum Mol Genet 19(4): 573–83. doi: 10.1093/hmg/ddp524.

Simpson, C. L., R. Lemmens, K. Miskiewicz, W. J. Broom, V. K. Hansen, P. W. van Vught et al. 2009. Variants of the elongator protein 3 (ELP3) gene are associated with motor neuron degeneration. Hum Mol Genet 18(3): 472–81. doi: 10.1093/hmg/ddn375.

Smith-Vikos, T. and F. J. Slack. 2012. MicroRNAs and their roles in aging. J Cell Sci 125(Pt 1): 7–17. doi: 10.1242/jcs.099200.

Tahiliani, M., K. P. Koh, Y. Shen, W. A. Pastor, H. Bandukwala, Y. Brudno et al. 2009. Conversion of 5-methylcytosine to 5-hydroxymethylcytosine in mammalian DNA by MLL partner TET1. Science 324(5929): 930–5. doi: 10.1126/science.1170116.

Turan, N., S. Katari, C. Coutifaris and C. Sapienza. 2010. Explaining inter-individual variability in phenotype: is epigenetics up to the challenge? Epigenetics 5(1): 16–9.

Watts, G. M., F. J. Beurskens, I. Martin-Padura, C. M. Ballantyne, L. B. Klickstein, M. B. Brenner et al. 2005. Manifestations of inflammatory arthritis are critically dependent on LFA-1. J Immunol 174(6): 3668–75.

Weidner, C. I., Q. Lin, C. M. Koch, L. Eisele, F. Beier, P. Ziegler et al. 2014. Aging of blood can be tracked by DNA methylation changes at just three CpG sites. Genome Biol 15(2): R24. doi: 10.1186/gb-2014-15-2-r24.

West, R. L., J. M. Lee and L. E. Maroun. 1995. Hypomethylation of the amyloid precursor protein gene in the brain of an Alzheimer's disease patient. J Mol Neurosci 6(2): 141–6. doi: 10.1007/BF02736773.

Williams, A. H., G. Valdez, V. Moresi, X. Qi, J. McAnally, J. L. Elliott et al. 2009. MicroRNA-206 delays ALS progression and promotes regeneration of neuromuscular synapses in mice. Science 326(5959): 1549–54. doi: 10.1126/science.1181046.

Wilson, C. B., K. W. Makar, M. Shnyreva and D. R. Fitzpatrick. 2005. DNA methylation and the expanding epigenetics of T cell lineage commitment. Semin Immunol 17(2): 105–19. doi: 10.1016/j.smim.2005.01.005.

Winkler, G. S., A. Kristjuhan, H. Erdjument-Bromage, P. Tempst and J. Q. Svejstrup. 2002. Elongator is a histone H3 and H4 acetyltransferase important for normal histone acetylation levels *in vivo*. Proc Natl Acad Sci U S A 99(6): 3517–22. doi: 10.1073/pnas.022042899.

Zbiec-Piekarska, R., M. Spolnicka, T. Kupiec, Z. Makowska, A. Spas, A. Parys-Proszek et al. 2015a. Examination of DNA methylation status of the ELOVL2 marker may be useful for human age prediction in forensic science. Forensic Sci Int Genet 14: 161–7. doi: 10.1016/j.fsigen.2014.10.002.

Zbiec-Piekarska, R., M. Spolnicka, T. Kupiec, A. Parys-Proszek, Z. Makowska, A. Paleczka et al. 2015b. Development of a forensically useful age prediction method based on DNA methylation analysis. Forensic Sci Int Genet 17: 173–9. doi: 10.1016/j.fsigen.2015.05.001.

Index

About the Editor

Sara C. Zapico is a Research Collaborator in the Department of Anthropology at the National Museum of Natural History, Smithsonian Institution in Washington, DC. She received her PhD in molecular biology and biochemistry, with a focus on biomedical sciences, from the University Institute of Oncology of Principado de Asturias (IUOPA) at the University of Oviedo, Spain and her Master in Forensic Anthropology and Genetics from the University of Granada, Spain. This dual background defines her current research interests and projects, exploring the application of biochemical approaches to forensic science issues, like age-at-death estimation and the determination of post-mortem interval, with implications on aging and biomedical research. She is also involved in international multidisciplinary projects in the fields of forensic and biomedical sciences.